Climate-Resilient Structures and Infrastructures

The safety and serviceability of in-service structures and infrastructure systems are often threatened by natural hazards. Owners and decision-makers are especially concerned about the resilience of an asset (structure or infrastructure), namely its ability to be in readiness for, absorb, recover from, and adapt to disruptive events. Climate change is increasingly producing nonstationary natural hazards such as wind hazards, and concurrent multiple hazards may result in especially severe performance reduction.

This book focuses on hazard modeling and resilience methods and their application in engineering cases. It presents the state of the art of the theory, methodology, and applications related to infrastructure resilience and offers practical approaches to enhancing resilience for engineers, policymakers, and researchers.

Cao Wang is Senior Lecturer and ARC DECRA Fellow at the University of Technology Sydney, Australia. He is a member of the Committee on Probability and Statistics in the Physical Sciences ($C(PS)^2$) and the Bernoulli Society for Mathematical Statistics and Probability.

Bilal M. Ayyub is Professor and Director of the Center for Technology and Systems Management at the University of Maryland, USA. He is a distinguished member of ASCE, an honorary member of ASME, and a co-chair of the ASCE-NOAA Task Force on Climate Resilience in Engineering Practice, and he has chaired the ASCE Committee on Adaptation to a Changing Climate.

Michael Beer is Professor and Head of the Institute for Risk and Reliability, Leibniz Universität Hannover, Germany. He is also part-time Professor at the University of Liverpool and guest Professor at Tongji University and Tsinghua University, China. He is the Chairman of the European Safety and Reliability Association (ESRA) and serves on the Executive Board of the International Safety and Reliability Association (IASSAR), on the Executive Board of the European Association of Structural Dynamics (EASD), and on the Board of Directors of the International Association for Probabilistic Safety Assessment and Management (IAPSAM). He is a fellow of the Alexander von Humboldt-Foundation.

Climate-Resilient Structures and Infrastructures

Edited by

Cao Wang, Bilal M. Ayyub, and Michael Beer

CRC Press
Taylor & Francis Group
Boca Raton London New York

CRC Press is an imprint of the
Taylor & Francis Group, an **informa** business

Designed cover image: Wenwen Tang

First edition published 2026
by CRC Press
2385 NW Executive Center Drive, Suite 320, Boca Raton FL 33431

and by CRC Press
4 Park Square, Milton Park, Abingdon, Oxon, OX14 4RN

CRC Press is an imprint of Taylor & Francis Group, LLC

ISBN: 9781032936017 (hbk)
ISBN: 9781032937182 (pbk)
ISBN: 9781003567332 (ebk)

DOI: 10.1201/9781003567332

Typeset in Times
by codeMantra

Cao Wang thanks his mentors, colleagues, and students for their collaboration and support.

Bilal M. Ayyub dedicates his contributions in this book to his teachers and students.

Michael Beer appreciates the invaluable support of all his collaborators over the past decades.

Contents

Preface

Infrastructure systems such as power grid systems, transportation systems, water distribution systems, and others fundamentally support the functionalities of modern societies. These systems/networks are expected to be resilient during their service lives. There have been different definitions of the term "resilience," but they commonly emphasize on the robustness, redundancy, rapidity, and resourcefulness of the infrastructure system under consideration. Furthermore, in a changing environment, the deterioration process of the resistances of infrastructure components may be accelerated, while the occurrence rates and/or magnitudes of future hazardous events may increase due to the potential impacts of climate change. Hence, it is important to understand and quantify the resilience of infrastructure systems subjected to climate change, which is the basis of risk-informed decision-making for the infrastructures.

This book is themed climate-resilient structures and infrastructures, with a focus on hazard modeling, resilience methods, and applications in engineering cases. This book includes the state of the art of the theory, methodology, and application related to infrastructure resilience. This book will hopefully deepen our understanding of the potential impacts of climate change on infrastructure resilience, as well as practical approaches to enhancing resilience. The target audience of this book includes civil engineers, policymakers, and high-degree researchers.

The editors warmly thank the chapter authors for their valuable contributions to this book, including Ansumana L. M. Swarray, Bilal M. Ayyub, and Tareq A. Al-Ansari (Chapter 2); Julian Salomon, Niklas R. Winnewisser, and Michael Beer (Chapter 3); Francesco Simone and Riccardo Patriarca (Chapter 4); Henry J. Liu (Chapter 5); Yating Zhang and Zachary M. Labe (Chapter 6); Si Han Li and Jayantha T. Obeysekera (Chapter 7); Nathan Chan and Bilal M. Ayyub (Chapter 8); Wanting (Lisa) Wang and John W. van de Lindt (Chapter 9); De-Cheng Feng, Jia-Yi Ding, and Carmine Galasso (Chapter 10); Yiwen Cui, Nelson Lam, Shuangmin Shi, Emad Gad, and Lihai Zhang (Chapter 11); Shuangmin Shi, Nelson Lam, Yiwen Cui, Emad Gad, and Lihai Zhang (Chapter 12); Yuhao Cao, Xinjian Wang, and Zaili Yang (Chapter 13); Bawantha Rathnayaka, Dilan Robert, Varuna Adikariwattage, Chandana Siriwardana, and Sujeeva Setunge (Chapter 14); Yaodan Cui and Kairui Feng (Chapter 15); and Innocent Chirisa, Tinashe Magande, Africa Muneti, Mike Juru, Halleluah Chirisa, Winnet Chiganga-Ngarakana, and Kudzai Chitiva (Chapter 16). Moreover, the editors would like to acknowledge the colleagues from CRC Press who have supported the publication of this book, including Tony Moore (Senior Editor) and Andrew Carmount (Editorial Assistant).

During the editing of this book, Cao Wang was supported by the Australian Government through the Australian Research Council's Discovery Early Career Researcher Award (DE240100207). Bilal M. Ayyub was supported by the National Oceanic and Atmospheric Administration, the Federal Railroad Administration, the Qatar National Research Fund, and the Ocean Energy Safety Institute at the Texas

A&M Engineering Experiment Station. These supports are gratefully acknowledged. The results and scientific opinions in this book are those of the authors and are not necessarily that of the sponsors or the United States Government or other governments.

Cao Wang
Bilal M. Ayyub
Michael Beer
September 2025

About the Editors

Cao Wang is a Senior Lecturer and ARC DECRA Fellow at the School of Civil and Environmental Engineering, University of Technology Sydney, Australia. He obtained his PhD degree from the University of Sydney, Australia, in 2019. He is an Associate Editor of three international journals: *ASCE-ASME Journal of Risk and Uncertainty in Engineering Systems, Part A: Civil Engineering*; *Part B: Mechanical Engineering*; and *Engineering Reports*. He is an Editorial Board member of several other journals, including *Structural Safety*. He is the author of two books: *Structural Reliability and Time-Dependent Reliability* (2021) and *Civil Structures and Infrastructure Systems: From Reliability to Resilience* (2025). He is a member of the Committee on Probability and Statistics in the Physical Sciences ($C(PS)^2$) and the Bernoulli Society for Mathematical Statistics and Probability (2021–2025).

Bilal M. Ayyub is a Professor at the University of Maryland, College Park, and the Director of the Center for Technology and Systems Management. He is Visiting Professor, Imperial College London; Senior Economist, NIST; Co-Director of the Center for Resilient Infrastructure at Tongji University; Academician of Georgian National Academy of Sciences; and NOAA Science Advisory Board member. He is a distinguished member of ASCE and an honorary member of ASME. He is a fellow of the Structural Engineering Institute, Risk Society, and the Society of Naval Architects and Marine Engineers. He is a senior member of IEEE. He cochairs the ASCE-NOAA Task Force on Climate Resilience in Engineering Practice. He researches risks in engineering and economics. He has coauthored about 700 publications including 8 textbooks and 15 edited books and has received several awards from American Society of Mechanical Engineers (ASME), American Society of Civil Engineers (ASCE), American Society of Naval Engineers (ASNE), and others.

Michael Beer is a Professor and the Head of the Institute for Risk and Reliability, Leibniz Universität Hannover, Germany. He is also a part-time Professor at the University of Liverpool and a Guest Professor at Tongji University and the University of Science and Technology Beijing, China. He is the Editor-in-Chief of the *ASCE-ASME Journal of Risk and Uncertainty in Engineering Systems, Part A: Civil Engineering* and *Part B: Mechanical Engineering*. He is the Chairman of the European Safety and Reliability Association (ESRA), President-elect of the International Safety and Reliability Association (IASSAR), and Co-Chair of Risk and Resilience Measurements Committee (RRMC), Infrastructure Resilience Division (IRD), ASCE. He is serving on the Executive Board of the European Association of Structural Dynamics (EASD) and on the Board of Directors of the International Association for Probabilistic Safety Assessment and Management (IAPSAM). He is a fellow of the Alexander von Humboldt-Foundation.

1 Introduction

Cao Wang[a], *Bilal M. Ayyub*[b], *and Michael Beer*[c,d,e]

[a]School of Civil and Environmental Engineering, University of Technology Sydney, Ultimo, NSW, Australia

[b]Department of Civil and Environmental Engineering, Center for Technology and Systems Management, University of Maryland, College Park, Maryland; Department of Civil and Environmental Engineering, Imperial College London, London, United Kingdom; International Joint Research Center for Resilient Infrastructure, Tongji University, Shanghai, China

[c]Institute for Risk and Reliability, Leibniz University Hannover, Hannover, Germany

[d]Department of Civil and Environmental Engineering, University of Liverpool, Liverpool, United Kingdom

[e]International Joint Research Center for Resilient Infrastructure & International Joint Research Center for Engineering Reliability and Stochastic Mechanics, Tongji University, Shanghai, China

1.1 RELIABILITY AND RESILIENCE

We are living in a world of uncertainties. For the civil structures and infrastructure systems in our communities, the impacts of environmental factors (e.g., extreme winds, earthquakes, snow loads, and others) on them are random. For example, the Load and Resistance Factor Design (LRFD) method has been widely adopted in national design standards, which takes into account the uncertainties associated with the structural resistance and the load effects (Galambos, 1981). In such a way, the designed structure (or component) has a probability of failure below a target level.

Furthermore, the performance (e.g., stiffness and strength) of a structure degrades with time in a harsh service environment (Ayyub et al., 1989; Ellingwood, 2005; Sanchez-Silva et al., 2011). For instance, dominant mechanisms of deterioration for concrete structures include corrosion, sulfate attack, and diffusion-controlled aging (Mori and Ellingwood, 1993). Hence, the serviceability of structures and infrastructures shall be evaluated in a probabilistic framework, where all the uncertain factors are appropriately incorporated.

Reliability and resilience have been two powerful tools to measure structural responses to disruptive events (Wang, 2025). Structural reliability is defined as the probability that a limit state (e.g., resistance minus a combination of load effects) is not violated. For example, this is the basis for the development of LRFD methods.

DOI: 10.1201/9781003567332-1

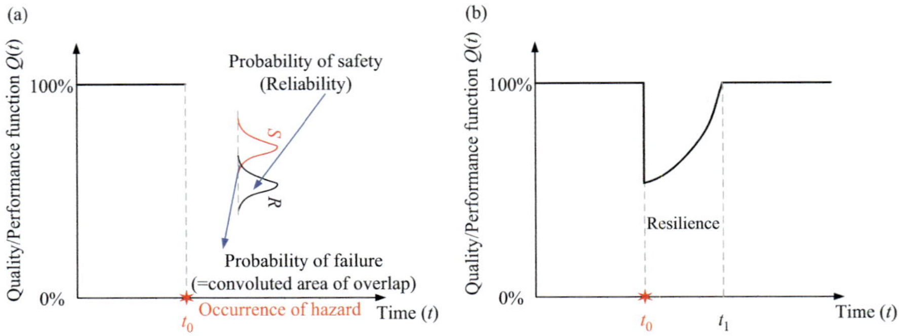

FIGURE 1.1 Comparison between (a) reliability and (b) resilience. (Reproduced from Wang et al., 2022, with permission of reuse from Elsevier.)

The terminology "resilience" was first introduced in ecology, referring to the ability of an ecological system to absorb change and disturbance and still maintain the same relationships between populations or state variables (Holling, 1973). In the context of engineering applications, the resilience assessment of an object (e.g., a structure, an infrastructure system, or a community) has a broader scope compared with reliability. For instance, the National Research Council (2012) defines resilience as "The ability to prepare and plan for, absorb, recover from, or more successfully adapt to actual or potential adverse events." Ayyub (2014; 2015) provided a comparative examination of these definitions for the purpose of quantification and measurement science. This definition was adopted in another NIST report (McAllister, 2013).

A comparison between the concepts of reliability and resilience can be found in Figure 1.1 if considering the occurrence of a single disruptive event. The emphasis is on the different scopes of the two quantities (Wang et al., 2022). The focus of this chapter is on the resilience of infrastructure systems in a changing climate; however, reliability assessment is often beneficial for the purpose of understanding and interpreting infrastructure resilience.

1.2 CLIMATE CHANGE AND CLIMATE-RESILIENT INFRASTRUCTURES

Bruneau et al. (2003) proposed that a resilient system is one showing reduced failure probabilities, reduced consequences from failures, and reduced time to recovery. Hence, a measure for loss of resilience, R, is formulated as follows,

$$R = \int_{t_0}^{t_1} 1 - Q(t)\,dt \tag{1.1}$$

where $Q(t)$ is the performance (percentage) of the system at time t, as illustrated in Figure 1.2. Chapters 9 and 14 will adopt this model for resilience assessment.

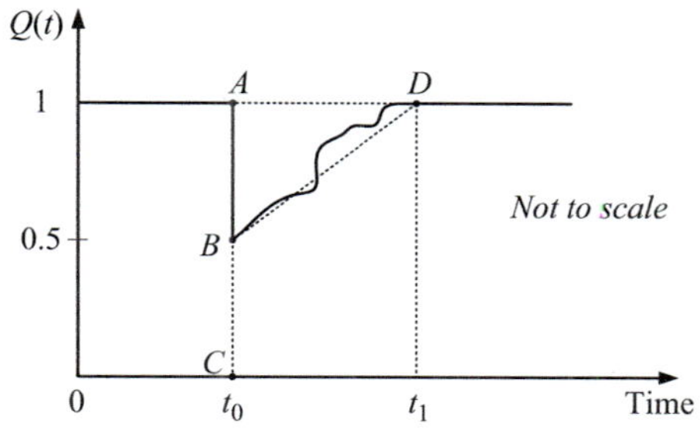

FIGURE 1.2 Resilience quantification. (Redrawn after Bruneau et al., 2003.)

TABLE 1.1
Properties of Resilience Explained through the Resilience Triangle (Ayyub, 2014)

Property	Model (per Figure 1.1)	Units
Robustness	Robustness $= B - C$	Percentage
Redundancy	Not defined	
Resourcefulness	Not defined	
Rapidity	Rapidity $= \dfrac{A-B}{t_1 - t_0}$	Average recovery rate in percentage per time

The resilience can be measured by four properties according to Bruneau et al. (2003), such as robustness, redundancy, resourcefulness, and rapidity (4R). In Figure 1.2, if replacing the recovery trajectory between t_0 and t_1 with a straight line (*BD*), the triangle *ABD* is called "resilience triangle" (Ayyub, 2014), which is a simple model for resilience assessment. The relationship between the properties of resilience and the resilience triangle is explained in Table 1.1.

Note that the term *R* in Eq. (1.1) is not a dimensionless measure. Attoh-Okine et al. (2009) developed a normalized resilience model, denoted by R_s, as follows, by referring to Figure 1.2,

$$R_s = \frac{\int_{t_0}^{t_1} Q(t)\,dt}{t_1 - t_0} \tag{1.2}$$

A similar measure for resilience quantification also appeared in Cimellaro et al. (2010). An alternative version of the model in Eq. (1.2) is that t_1 is replaced by a user-defined time instant t_h (see, e.g., Bocchini et al., 2014). This ensures the resiliences of two objectives to be compared within the same reference period. For example, Chapter 3 will employ the resilience model in Eq. (1.2) to assess infrastructure resilience.

Ayyub (2014) proposed a resilience model (R_e) associated with an event occurring at time t_i as follows,

$$R_e = \frac{T_i + F\Delta T_f + R\Delta T_r}{T_i + \Delta T_f + \Delta T_r} \tag{1.3}$$

where F and R are the failure profile and recovery profile, respectively, which are measured as follows,

$$\text{Failure}(F) = \frac{\int_{t_i}^{t_f} f dt}{\int_{t_i}^{t_f} Q dt} \tag{1.4a}$$

$$\text{Recovery}(R) = \frac{\int_{t_f}^{t_r} r dt}{\int_{t_f}^{t_r} Q dt} \tag{1.4b}$$

where t_f is the end time of the failure event, t_r is the time to recovery, f and r are failure event and recovery event, respectively, as illustrated in Figure 1.3. One can refer to Chapter 2 for an application of Eq. (1.3) in the assessment of critical infrastructure resilience.

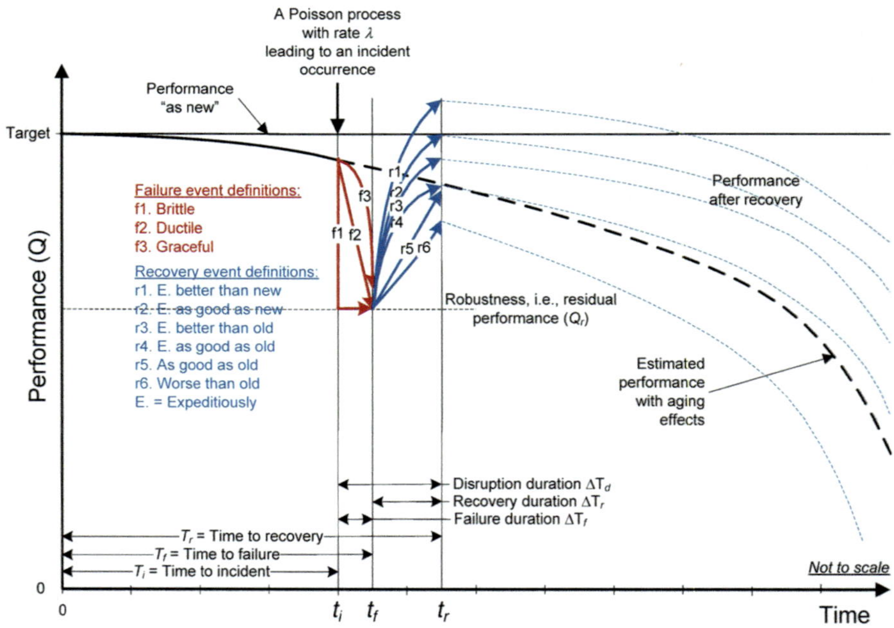

FIGURE 1.3 Definition of resilience by Ayyub (2014). (With reuse permission from Wiley.)

Wang (2023) developed a generalized model for resilience quantification (R_s) as follows,

$$R_e = f^{-1}\left(\frac{1}{t_1 - t_0}\int_{t_0}^{t_1} f\left(Q(t)\right)dt\right) \tag{1.5}$$

where $f(x)$ is known as a "generating function" and f^{-1} is the inverse of $f(x)$.

In Eq. (1.5), if $f(x)=x$, the model reduces to that in Eq. (1.2). Other generating functions include the natural logarithm of x and among others. The choices of the generating function $f(x)$ allow for different resilience models with varying sensitivities to the extremely low/zero performance of the infrastructure under consideration.

The advantage of $f(x)=\ln x$ in Eq. (1.5) compared with the traditional $f(x)=x$ can be shown (Wang et al., 2023; Wang 2024). Other models for resilience quantification are also available in literature (see, e.g., Sun et al. 2020 and Shadabfar et al. 2022).

The impacts of natural hazards on infrastructure resilience shall be properly quantified. However, the statistics of natural hazards (e.g., occurrence rates and magnitudes) will not necessarily be constant in a changing environment (Schreider et al., 2000; Knutson et al., 2010; Di Virgilio et al., 2019). The Intergovernmental Panel on Climate Change (IPCC, 2023) predicted a further increase in temperature until 2100 for different socioeconomic pathway scenarios. Remarkably, the report stated that,

> Urban infrastructure, including transportation, water, sanitation and energy systems have been compromised by extreme and slow-onset events, with resulting economic losses, disruptions of services and impacts to well-being (high confidence)

In the presence of climate change, the "non-stationarity" in natural hazards arises with time-varying trends in their occurrence rates and/or magnitudes. This non-stationarity means a potentially increasing risk to structures and infrastructure systems and hence plays a vital role in the assessment of infrastructure resilience. One may refer to Chapters 7 and 15 for the quantification of hazard non-stationarity in a changing environment. Furthermore, the consequences of climate-related degradation and failure need to be considered in a rigorous risk assessment framework. An overview of recent developments in this direction with a focus on concrete civil infrastructure is provided by Feng et al. (2024).

Hence, we need to ensure the long-term behavior of infrastructure systems with an acceptable level of reliability and resilience through adapting to climate change. This is consistent with the statement by the Committee on Adaptation to a Changing Climate (2018) that,

> Infrastructure designs and plans, as well as institutions, regulations, and codes and standards to which they must adhere, will need to be adapted and even be adaptable to accommodate a range of future climate conditions.

In the above context, this book is themed "Climate-resilient structures and infrastructures." This book addresses how the modern structures and infrastructures can achieve resilience in a changing climate, which offers asset owners/policymakers unique insights into the impacts of climate change, and potential strategies for enhancing climate resilience.

1.3 OVERVIEW OF THIS BOOK

After this chapter, this book contains 15 chapters that explore the climate resilience of infrastructure systems from different perspectives.

In Chapter 2, the sustainable climate resilience strategies for critical energy infrastructure in developing countries are examined. A case study on Sierra Leone is presented, which is recognized by the United Nations as one of the most vulnerable nations to climate change. This chapter also identifies key challenges that hinder the development and maintenance of resilient energy systems.

In Chapter 3, an efficient approach is presented for quantifying the resilience of complex, large, and sub-structured systems subject to monetary and technical restrictions, through expanding and merging a resilience decision-making framework and the concept of survival signature. The efficiency and applicability of the approach are demonstrated through application to two sub-structured systems of increasing complexity, with the second being a sub-structured model of the Berlin U- and S-Bahn system.

Chapter 4 studies the cyber dimensions of industrial water systems: climate change and digitalization. The chapter summarizes the research efforts on cyber resilience management conducted by the Industrial Systems Engineering division at Sapienza University of Rome.

Chapter 5 aims to unpack the "black box" of infrastructure resilience, that is, how the dynamic interdependency of the engaged organizations' capabilities leads to resilient assets. To this end, this chapter initiated cumulative case studies of nine transport infrastructure assets supported by system thinking/dynamics.

In Chapter 6, the emerging role of extreme event attribution (EEA) in improving climate adaptation efforts for civil infrastructure systems is examined. The methodology for incorporating EEA results into probabilistic risk analysis for infrastructure systems and cost-effectiveness analysis for adaptation strategies is presented.

Chapter 7 reviews the methodologies for quantifying basic climatic design values and reliability assessment for infrastructure design under stationary conditions. The concept of climate non-stationarity is also presented, introducing techniques for its detection using historical data and future climate projections. Furthermore, the concept of deep uncertainties in climate modeling for nonstationary future climates is discussed, outlining the engineering design and implementation strategies.

Chapter 8 quantifies the impact of climate change on railroads by developing a methodology that uses a network to represent the system and node/link importance metrics to reveal the relative component importance. A case study of extreme temperatures is used to demonstrate how the methodology may be used to enhance existing research.

In Chapter 9, a methodology is introduced based on a multilayer Monte Carlo simulation to model a two-stage recovery process for residential buildings: functional downtime due to delay and functional downtime due to repair. It is found that short-term post-disaster activities (e.g., post-hazard policies) can effectively guide and facilitate the recovery process.

Chapter 10 introduces a portfolio-level fragility assessment framework integrating data management, machine learning (ML)–based fragility modeling, and probabilistic aggregation techniques predicts. A virtual city consisting of 600 RC frame and 400 RC frame–shear wall buildings is analyzed. It is found that the seismic fragility of building portfolios could be assessed using the Poisson binomial distribution, rather than solely focusing on the damage probability of individual buildings in previous regional-scale research.

Chapters 11 and 12 are “sister chapters.” Chapter 11 evaluates the resistance of toughened glass to hail hazards, by combining experimental observations with both deterministic and probabilistic modeling approaches. This work aims to support the development of more resilient, reliable, and climate-adaptive building envelopes in the face of increasingly extreme weather patterns. In Chapter 12, the vulnerability, damage progression, and recovery performance of aluminum cladding panels are evaluated subjected to hail hazards. A resilience-based approach was developed to evaluate the capacity of the system to maintain, or recover, functionality following a disruption.

Chapter 13 reviews the climate change impacts on maritime transport and port infrastructure resilience, drawing on interdisciplinary research from climate science, transport economics, disaster risk management, and maritime logistics. A regional analysis is conducted to compare how ports in America, Europe, and Asia are addressing climate risks, highlighting disparities in adaptation measures, policy frameworks, and investment priorities.

Chapter 14 develops a model integrating dynamic Bayesian networks (DBNs) with ML to predict critical infrastructure resilience. A dataset was generated by simulating a validated DBN model that represents a pump house within a water distribution system. This dataset was then used to train and evaluate six tree-based ML algorithms. The findings highlight the potential of integrating probabilistic modeling and ML-driven analytics to enhance resilience assessment frameworks.

Chapter 15 focuses on the hurricane evacuation decisions under future climate condition. It is revealed that climate change fundamentally alters the hurricane evacuation challenge, creating conditions where even optimal decision-making cannot prevent substantial degradation in evacuation outcomes. Hence, relying solely on improved evacuation management to address climate risks is inadequate, and communities must therefore pursue multiple adaptation pathways simultaneously.

Chapter 16 examines whether the adoption of green building concepts will result in enhanced infrastructure and superstructure resilience in the Zimbabwean construction industry. The concept stresses aspects including energy generation and conservation, efficient water use (and reuse), choice of building materials, and the environmental aspect. It is found that designing for climate resilience involves various strategies that can significantly reduce a building’s energy consumption and carbon footprint.

REFERENCES

Attoh-Okine, N. O., Cooper, A. T., and Mensah, S. A. (2009). Formulation of resilience index of urban infrastructure using belief functions. *IEEE Systems Journal*, *3*(2), 147–153.

Ayyub, B. M. (2014). Systems resilience for multihazard environments: Definition, metrics, and valuation for decision making. *Risk Analysis*, *34*(2), 340–355.

Ayyub, B. M. (2015). Practical resilience metrics for planning, design, and decision making. *ASCE-ASME Journal of Risk and Uncertainty in Engineering Systems, Part A: Civil Engineering*, *1*(3), 04015008.

Ayyub, B. M., White, G. J., & Purcell, E. S. (1989). Estimation of the structural service life of boats. *Naval Engineers Journal, ASNE, 101*(3), 156–166.

Bocchini, P., Frangopol, D. M., Ummenhofer, T., & Zinke, T. (2014). Resilience and sustainability of civil infrastructure: Toward a unified approach. *Journal of Infrastructure Systems, 20*(2), 04014004.

Bruneau, M., Chang, S. E., Eguchi, R. T., Lee, G. C., O'Rourke, T. D., Reinhorn, A. M., … & Von Winterfeldt, D. (2003). A framework to quantitatively assess and enhance the seismic resilience of communities. *Earthquake Spectra, 19*(4), 733–752.

Cimellaro, G. P., Reinhorn, A. M., & Bruneau, M. (2010). Seismic resilience of a hospital system. *Structure and Infrastructure Engineering, 6*(1–2), 127–144.

Committee on Adaptation to a Changing Climate (2018). Climate-Resilient Infrastructure: Adaptive Design and Risk Management (MOP 140).

Di Virgilio, G., Evans, J. P., Blake, S. A., Armstrong, M., Dowdy, A. J., Sharples, J., & McRae, R. (2019). Climate change increases the potential for extreme wildfires. *Geophysical Research Letters, 46*(14), 8517–8526.

Ellingwood, B. R. (2005). Risk-informed condition assessment of civil infrastructure: State of practice and research issues. *Structure and Infrastructure Engineering, 1*(1), 7–18.

Feng, D. C., Ding, J. Y., Xie, S. C., Li, Y., Akiyama, M., Lu, Y., Beer, M., & Li, J. (2024). Climate change impacts on the risk assessment of concrete civil infrastructures: A state-of-the-art review. *ASCE OPEN: Multidisciplinary Journal of Civil Engineering, 2*(1), 03124001.

Galambos, T. V. (1981). Load and resistance factor design. *Engineering Journal, 18*(3), 74–82.

Holling, C. S. (1973). Resilience and stability of ecological systems. *Annual Review of Ecology and Systematics, 4*, 1–23.

IPCC (2023). Sections. In: *Climate Change 2023: Synthesis Report. Contribution of Working Groups I, II and III to the Sixth Assessment Report of the Intergovernmental Panel on Climate Change* [Core Writing Team, H. Lee and J. Romero (eds.)]. IPCC, Geneva, Switzerland, pp. 35–115. doi: 10.59327/IPCC/AR6-9789291691647

Knutson, T. R., McBride, J. L., Chan, J., Emanuel, K., Holland, G., Landsea, C., … & Sugi, M. (2010). Tropical cyclones and climate change. *Nature Geoscience, 3*(3), 157–163.

McAllister, T. (2013). *Developing Guidelines and Standards for Disaster Resilience of the Built Environment: A Research Needs Assessment.* US Department of Commerce, National Institute of Standards and Technology, Maryland, US.

Mori, Y., & Ellingwood, B. R. (1993). Reliability-based service-life assessment of aging concrete structures. *Journal of Structural Engineering, 119*(5):1600–1621.

National Research Council (2012). *Disaster Resilience: A National Imperative.* The National Academies Press, Washington DC.

Sanchez-Silva, M., Klutke, G. A., & Rosowsky, D. V. (2011). Life-cycle performance of structures subject to multiple deterioration mechanisms. *Structural Safety, 33*(3), 206–217.

Schreider, S. Y., Smith, D. I., & Jakeman, A. J. (2000). Climate change impacts on urban flooding. *Climatic Change, 47*(1), 91–115.

Shadabfar, M., Mahsuli, M., Zhang, Y., Xue, Y., Ayyub, B. M., Huang, H., & Medina, R. A. (2022). Resilience-based design of infrastructure: Review of models, methodologies, and computational tools. *ASCE-ASME Journal of Risk and Uncertainty in Engineering Systems, Part A: Civil Engineering, 8*(1), 03121004.

Sun, W., Bocchini, P., & Davison, B. D. (2020). Resilience metrics and measurement methods for transportation infrastructure: The state of the art. *Sustainable and Resilient Infrastructure, 5*(3), 168–199.

Wang, C. (2023). A generalized index for functionality-sensitive resilience quantification. *Resilient Cities and Structures, 2*(1), 68–75.

Wang, C. (2024). Role of recovery profile dependency in time-dependent resilience. *Engineering Reports, 6*(2), e12716

Wang, C. (2025). *Civil Structures and Infrastructure Systems: From Reliability to Resilience.* Elsevier, Cambridge, MA.

Wang, C., Ayyub, B. M., & Ahmed, A. (2022). Time-dependent reliability and resilience of aging structures exposed to multiple hazards in a changing environment. *Resilient Cities and Structures*, *1*(3), 40–51.

Wang, C., Ayyub, B. M., Zhang, H., & Beer, M. (2023). Time-dependent resilience in the presence of interacting multiple hazards in a changing climate. *ASCE OPEN: Multidisciplinary Journal of Civil Engineering*, *1*(1), 04023006.

2 Sustainable Climate Resilience for Critical Energy Infrastructure in Developing Countries

Needs and Challenges

Ansumana L. M. Swarray[a], *Bilal M. Ayyub*[a], *and Tareq A. Al-Ansari*[b]

[a]Department of Civil and Environmental Engineering, University of Maryland, College Park, Maryland

[b]College of Science and Engineering, Hamad Bin Khalifa University, Doha, Qatar

2.1 BACKGROUND AND METHODOLOGY: CLIMATE VULNERABILITY OF DEVELOPING COUNTRIES

2.1.1 Developing Countries

Critical Energy Infrastructures (CEIs) in developing countries (DCs) must be strengthened for resilience against climate change. This chapter explores CEI vulnerabilities and the strategic needs for sustainability, emphasizing energy infrastructure's role in economic stability, social development, and environmental sustainability. Long-term resilience is crucial for mitigating climate-induced disruptions.

The UN defines DCs as nations with low living standards, weak industrial bases, and moderate or low Human Development Index (HDI) scores (UN 2024a). The Multi-Dimensional Vulnerability Index (MVI), comprising the Human Asset Index and the Economic and Environmental Vulnerability Index (UN 2021), highlights their socioeconomic and environmental risks, underscoring the need for tailored resilience strategies.

DCs, spanning Africa, the Caribbean, South and Central America, and most of Asia, are categorized by the UN based on per capita Gross National Income into upper-middle-income (above \$4,085), lower-middle-income (\$1,036–\$4,085), and low-income groups. Despite a 4.7% Gross Domestic Product (GDP) growth rate in 2023 (UN 2024a), these nations face slow economic development. Least Developed Countries (LDCs), comprising structurally disadvantaged and heavily indebted economies, account for 15% of the global population but contribute only 1.4% to

DOI: 10.1201/9781003567332-2

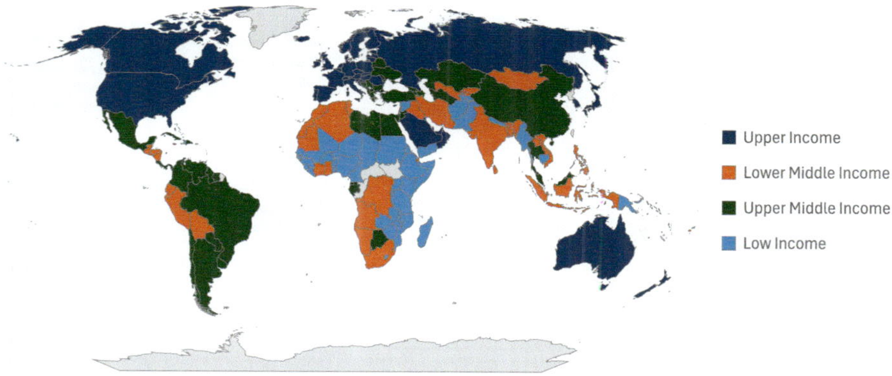

FIGURE 2.1 Map of the world based on UN Gross Domestic Product (GDP) growth rate (UN 2024a).

GDP and 1.1% to global trade (UN 2024b). With 70% of LDCs in Africa, 12 in Asia, and one in Latin America and the Caribbean, these regions face heightened climate vulnerabilities due to weak institutions, financial constraints, and inadequate infrastructure. Addressing these risks requires climate adaptation, policy reforms, international cooperation, and investments in resilient energy systems (Figure 2.1).

2.1.2 Methodology

This chapter presents a methodology for analyzing the development and management of climate-resilient CEI in DCs. It begins with an overview of DCs based on the United Nations classification (UN 2024a), emphasizing vulnerability as a key factor shaping their socioeconomic conditions, particularly for LDCs. Climate projections and the impacts of extreme temperatures on developing nations are defined and characterized, with Sierra Leone serving as a case study.

Next, the methodology defines and analyzes energy infrastructure in DCs, highlighting its vulnerability to climate change, with a focus on Sierra Leone's energy landscape. The study then explores CEI network typology and modeling, focusing on how extreme temperatures affect power grid structures. Infrastructure resilience, its measurement, and the specific vulnerabilities of CEIs in DCs—again using Sierra Leone as an example—are examined in detail, including the impacts of extreme weather events (EWSs). The final section discusses the challenges and requirements for establishing and maintaining sustainable climate-resilient CEIs in Sierra Leone. Figure 2.2 outlines the framework for this methodology.

2.2 CLIMATE HAZARDS AND PROJECTIONS

2.2.1 Climate Hazards

Global climate change is primarily driven by rising ambient temperatures due to the greenhouse effect, caused by both natural and human activities (IPCC 2023).

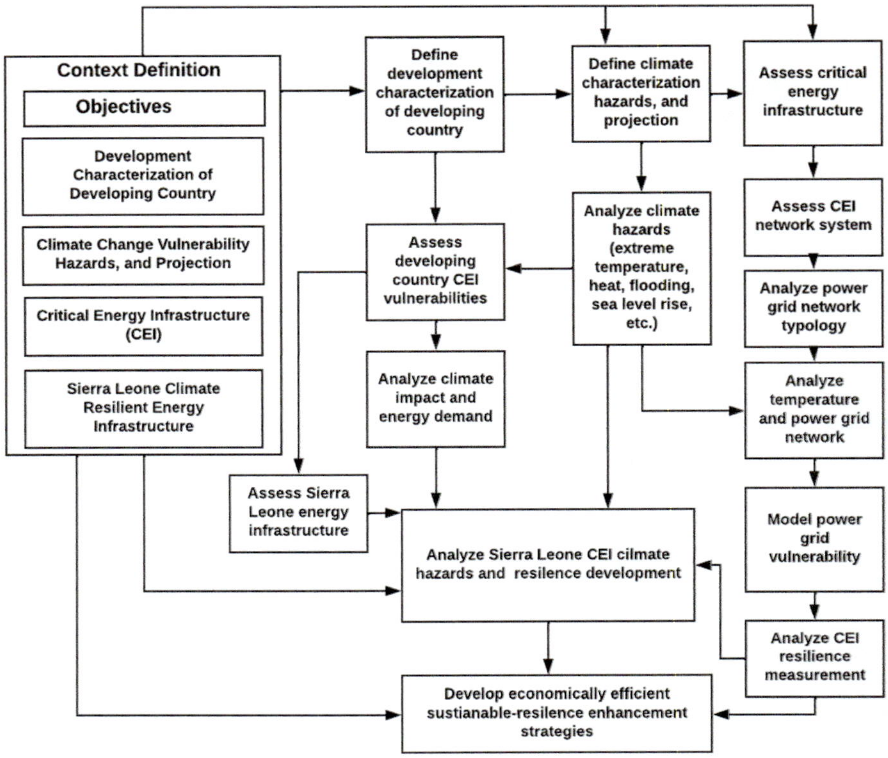

FIGURE 2.2 Methodology for assessing sustainable climate-resilient critical energy infrastructure. (Adapted and significantly revised based on Ayyub & Hill 2019.)

This effect occurs when greenhouse gases (GHGs) such as carbon dioxide (CO_2), methane (CH_4), nitrogen and sulfur oxides, and chlorofluorocarbons trap heat in the atmosphere, leading to global warming and associated environmental conditions like extreme precipitation, flooding, sea level rise, saltwater intrusion, drought, and extreme heat (IPCC 2023). These resultant environmental conditions are sources of harm or loss to the global society and are collectively known as climate hazards, and their occurrence has led to devastating outcomes on different parts of the globe, especially DCs, and continue to exert negative impacts. The vulnerability to these climate hazards, and the degree of vulnerability and occurrence of these hazards are a function of several factors, specific to specific region or country (UN 2024c).

To assess the present and future climate impacts, models such as the Representative Concentration Pathway (RCP) and Shared Socioeconomic Pathway (SSP) have been developed based on the Coupled Model Intercomparison Project (CMIP 5 and 6) that generated five scenarios of potential global climate outcomes with or without mitigation efforts (IPCC 2023). To underscore the critical nature of rising global temperature, a recent European Copernicus Climate Service report, cited by the British Broadcasting Corporation (BBC), states that global average temperatures for

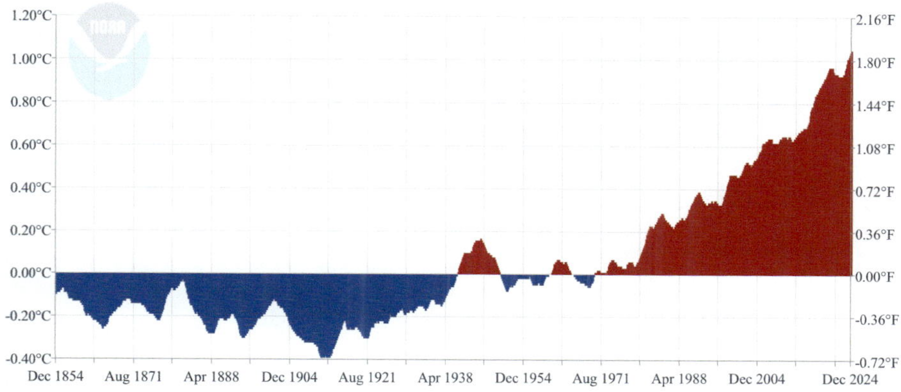

FIGURE 2.3 Global temperature anomaly (NOAA 2025).

2024 were 1.6°C above preindustrial levels, making January 2025 the hottest month on record. UN Secretary-General António Guterres referred to this as evidence of "climate breakdown" (BBC 2025). Figure 2.3 illustrates the continued rise in global temperatures (NOAA 2025).

2.2.2 Infrastructure Vulnerability to Extreme Temperature

Infrastructure vulnerability is defined as the susceptibility of an infrastructure system to disruptions, damage, or failure due to natural hazards, cyber threats, human errors, or systemic weaknesses. It reflects the extent to which infrastructure components or networks are exposed to risks and lack resilience in maintaining functionality under adverse conditions. EWEs, mainly by the rising global temperature such as droughts, floods, rising sea levels, heatwaves, and landslides, significantly threaten economic stability, energy, water, and food security in developing nations, as noted in the IPCC AR6 climate projections. For example, increase in temperature, aridity, and drought have increased the frequency and intensity of wildfires, and the interaction of these wildfires with land-use change (e.g., deforestation) is negatively impacting CEIs (IPCC 2023). These risks are exacerbated by pre-existing vulnerabilities like poverty, inequality, and systemic challenges.

Vulnerability is broadly recognized as a weakness in coping with external shocks (Provitolo 2011). The UN Committee for Development Policy defines it as susceptibility to climatic, social, and political disruptions (UN 2021), while the IPCC AR6 links it to reduced adaptive capacity (IPCC 2023). Frameworks such as the Pressure and Release and Access Models explain disasters as the result of both hazards and social, economic, and environmental deficiencies (Ayyub 2014a, Michellier et al. 2020). Key vulnerability sources include natural hazards, social fragility, economic instability, and environmental degradation (Provitolo 2011; Ranjan 2025).

To measure vulnerability, tools like the MVI assess economic, social, and environmental risks, guiding policymaking and international resource allocation (UN

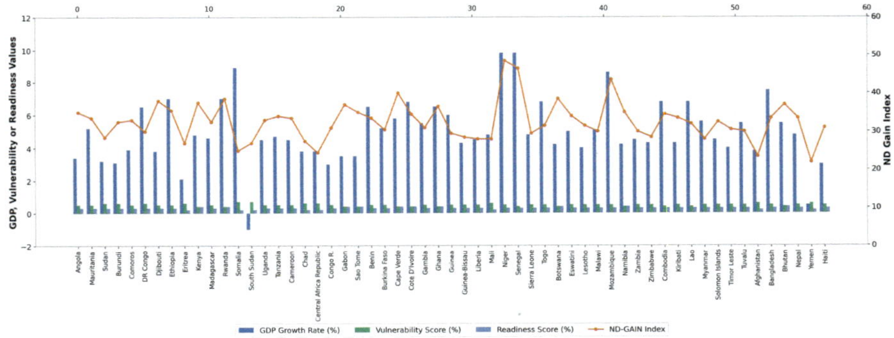

FIGURE 2.4 GDP growth rate and climate vulnerability indices of LDC. (Data from United Nations 2024a, and ND-GAIN 2024.)

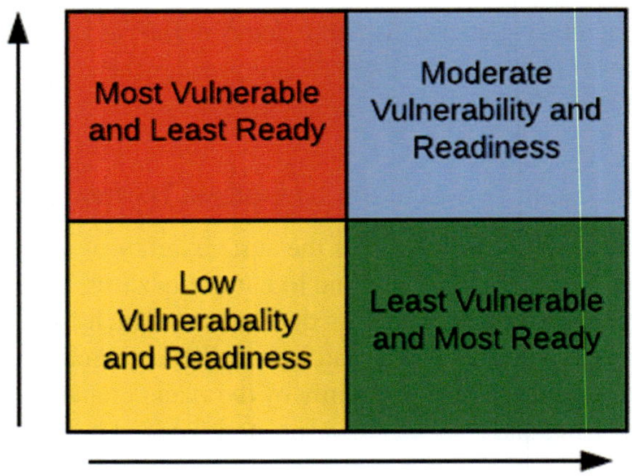

FIGURE 2.5 ND-GAIN climate matrix with vertical arrow indicating increase in vulnerability and horizontal arrow increase in readiness (ND-GAIN 2024).

2021). Another key metric, the ND-GAIN index, evaluates climate vulnerability and readiness, supporting strategic climate policies (ND-GAIN 2024). However, economic growth does not always equate to climate resilience. For example, Cape Verde has the lowest vulnerability (0.428) and highest preparedness (0.450) among 44 LDCs, whereas Chad ranks as the most vulnerable (0.646) with the lowest preparedness (0.189). Niger, despite a 9.9% GDP growth rate, remains highly vulnerable (0.632) and poorly prepared (0.342), highlighting the complex relationship between economic performance and climate resilience.

Figure 2.5 provides the ND-GAIN scale that shows the position of countries with respect to climate vulnerability and readiness to adapt to climate risks. Based on the scores in Figure 2.4, about 75% of the LDCs fall in the upper left quadrant, and the remaining 25% are in the lower left quadrant. This is a strong indication that DCs, especially the LDCs, are very vulnerable to the impacts of climate change with low readiness capacity to manage it. The inherent vulnerabilities and the projected climate challenges due to an increasing global temperature hinder the ability of these nations to achieve their sustainable development objectives (OECD 2021).

2.2.3 Sierra Leone as a Case Study

Sierra Leone is among the LDCs, ranking 182nd out of 189 with a HDI of 0.452 (UNDP 2019). Despite being the fifth most peaceful country in Africa, it remains highly fragile, scoring 81.4 on the Fund for Peace Fragile State Report (2024). With a population of 8.5 million (SSL 2022), the nation ranks as the 11th most vulnerable globally, with an ND-GAIN index score of 35.0 (155th place). It has a vulnerability score of 0.598 (179th) and a readiness score of 0.298 (177th), signaling the urgent need for climate adaptation (ND-GAIN 2024). Although contributing just 0.01% of global emissions, Sierra Leone's CO_2 output rose by 12.5% in 2021 to 6.87 million metric tons (IMF 2024).

The country experiences a tropical monsoon climate, with an average temperature of 26.7°C and annual rainfall of 2,746 mm. The Inter Tropical Convergence Zone (ITCZ) governs seasonal variations, with a wet season from May to November and a dry season, including harmattan winds, from December to May (SL BUR 2021). Climate change is Sierra Leone's most significant threat due to its high vulnerability. CMIP6 model projections indicate moderate conditions until 2070 (SSPs 2–4.5), followed by severe scenarios (SSPs 5–8.5) (NDC 2021). General circulation models predict temperature increases of 1°C–2.5°C by 2060 and up to 5°C by 2100, alongside unpredictable rainfall patterns fluctuating from −27% to +29% (July–September) and −19% to +33% (October–December). This instability heightens risks of floods, landslides, and coastal erosion, making low-lying areas increasingly vulnerable.

Environmental degradation and growing carbon sink deforestation due to planned and unplanned urbanization are the main factors contributing to Sierra Leone's rising temperature. Along with deforestation, energy generation contributes significantly to Sierra Leone's GHG emissions. The emissions are modest compared to other DCs, yet they rise 11% per 5 years (NDC 2021). The country generated 0.01% of global greenhouse emissions and had a 2.6% compound annual growth rate from 1990 to 2021 (Boyle 2024), and emits 0.8 tons of CO_2e per person, ranking 185th out of 191 nations, while other small and DCs like Qatar, Kuwait, Bahrain, and Turkmenistan emit the most per capita. Finally, Sierra Leone's agriculture business produced 4.07m of GHG emissions in 2021, 59.2% of total followed by land-use change (49%), and forestry and garbage (22.1%) (Figure 2.6).

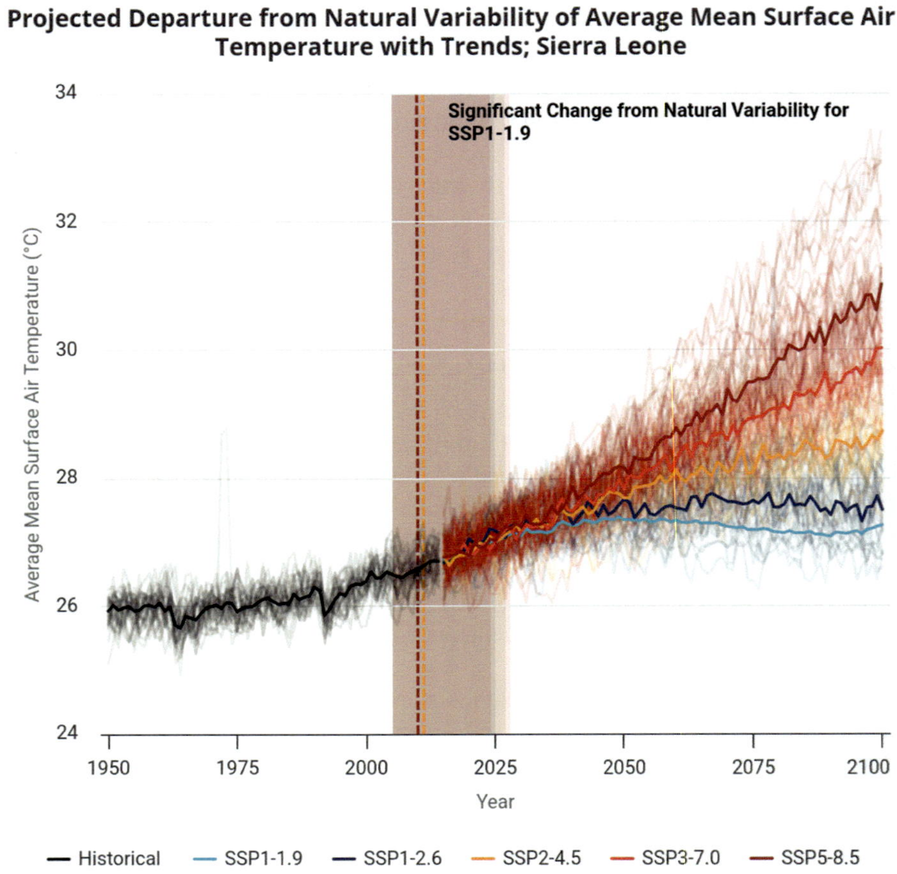

FIGURE 2.6 Projected departure of natural variability of average mean air surface temperature with trend in Sierra Leone, showing Shared Socioeconomic Pathway (SSP) scenarios (World Bank 2024a).

2.3 ENERGY INFRASTRUCTURE IN DEVELOPING COUNTRIES USING SIERRA LEONE AS A CASE STUDY

2.3.1 Infrastructure Definition and Significance

Infrastructure encompasses the essential physical, cyber, human, and organizational systems necessary for societal functions, including buildings, roads, and power networks. It drives socioeconomic progress and sustainable development. Key infrastructure systems, such as transport networks, water supply, energy, and telecommunications, contribute to urban growth and expansion (CDRI 2023). Among various definitions, the UN's is the reference in this context, describing infrastructure as "physical structures, facilities, networks, and assets providing services crucial for a community's social and economic functions" (UNDRR 2022).

Energy is a fundamental and critical infrastructure, and assessing climate change impacts on it requires a clear understanding of energy systems. According to the US Federal Energy Regulatory Commission (FERC), energy infrastructure includes bulk-power systems, whether physical or virtual, whose failure or destruction could compromise national security, economic stability, or public health and safety (FERC 2023).

2.3.2 Energy Infrastructures

Energy infrastructure, which includes power plants (coal, nuclear, and renewables), transmission and distribution (T&D) networks, storage systems, and control mechanisms, is essential for human development (Vaia 2024). However, traditional energy systems have struggled to meet global demand while contributing to climate risks. To ensure future energy security without environmental harm, resilience strategies such as infrastructure reinforcement, asset relocation, climate-resistant technology integration, energy diversification, and policy implementation are necessary.

Developing nations face significant financial and governance challenges in energy infrastructure development. Climate change and inadequate asset management further threaten energy security and economic stability (World Bank 2024b). Many of these countries have infrastructure that is highly vulnerable to disasters such as floods, earthquakes, and tropical cyclones (CDRI 2023), and unreliable energy systems cost developing economies between $391 billion and $647 billion annually due to frequent failures and capacity limitations (Hallegatte et al. 2019). Strengthening infrastructure resilience is critical for long-term socioeconomic growth.

2.3.2.1 Sierra Leone Energy Infrastructure

Sierra Leone has extremely low and insufficient power supply that is far below the demand for households, businesses, and social institutions. Demand for household consumption, industrial (excluding mining), and low-voltage commercial usages is estimated to be around 245 MW and projected to grow up to 485 MW by 2030 (MoE 2020). The total national grid installed capacity is 268 MW, with 96 MW being provided by the government owned, Electricity Generation and Transmission Company (EGTC) and the remainder by Independent Power Producers (IPPs). However, only about 130MW of the national installed capacity is typically operational.

2.3.2.1.1 Source of Energy and Generation Capacity

Sierra Leone has one of the lowest energy supplies globally, with a per capita consumption of 23 kWh (SL Compact 2024). In 2022, the country generated 0.20 TWh of electricity, with hydropower contributing 90% (World in Data 2024). IPPs, including Karpowership and Cote D'Ivoire, Sierra Leone, Liberia, and Guinea power pool, supplied 92 MW, although Karpowership operates below its 126 MW capacity, generating 60 MW in the dry season and 20 MW in the rainy season at 16.8 US cents/kWh (SL Compact 2024). By 2023, renewables accounted for 62% (81.45 MW) of installed capacity, reflecting a 40% rise since 2015. However, fluctuations remain—Bumbuna dam generates 45 MW in the rainy season but drops to 8 MW in the dry season, which necessitating Karpowership support. Electricity generation increased from 247,045.30 MWh to 600,000 MWh in 2022 (Conteh et al. 2023).

Sierra Leone's energy and transport infrastructure is highly vulnerable to climate change. Fossil fuel–based energy expanded by 325% from 43 MW in 2015 to 185 MW in 2022, with projected total generation reaching 770 MW by 2030—90% from renewables (Conteh et al. 2023). However, this remains insufficient as national demand rose 146% from 52.4 MW in 2015 to 125 MW in 2022 and is expected to reach 220 MW by 2030 (MoE 2020), and electricity access is limited, with only 23% of the population connected 51% in urban areas and just 1.5% in rural regions, where 60% of the population resides (SL Compact 2024).

2.3.2.1.2 Transmission and Distribution Network

Sierra Leone's transmission system is inefficient, with two separate networks: a 205 km, 161 kV line from the 50 MW Bumbuna hydro project to Freetown and a 67 km, 33 kV line linking Kenema's 6 MW Goma hydro dam to Bo, and outdated substations on the 161 kV line further limit performance. While the main line can handle 70 MW, instability above 40 MW reduces its effective capacity to 57%. The system also lacks bidirectional power flow, complicating integration beyond Bumbuna, and maintenance work and additional Karpowership loads further destabilize the line, causing outages in Freetown. With transmission losses at 38% in 2020—one of Africa's highest—and tariffs below cost recovery, upgrades are critical (SL Compact 2024), and plans include a $78 million, Indian Exim Bank–funded 225 kV line from Fadugu to Freetown, now delayed to 2025, and a World Bank–funded 66 kV line between Bo and Kenema, also set for completion in 2025 (SL Compact 2024).

The Electricity Distribution and Service Authority (EDSA), on the other hand, manages an unreliable distribution network prone to climate risks and vandalism. Originally designed for 40 MW, Freetown's system has been upgraded to 85 MW with World Bank support. However, overloaded transformers, insufficient substations, and aging infrastructure continue to cause frequent outages, and EDSA's limited resources lead to prolonged service disruptions (World Bank 2024a). Figure 2.7 is Sierra Leone's electricity infrastructure map.

2.3.2.2 Energy Consumption, Access, and Service Metrics

Sierra Leone's energy consumption increased from 1 TWh in 1980 to 4 TWh in 2021, with per capita usage reaching 493 kWh in 2022 but energy intensity declined from 0.59 kWh per dollar in 1980 to 0.37 in 2021 (World in Data 2024). However, electricity access remains among the lowest globally, at 26.31%, with rural connectivity at just 1.5%, compared to 51% in urban areas. Of 1.27 million households, only 19% (240,404) are grid connected, while 46,600 rely on isolated solar microgrids, mainly in Freetown and district capitals. Public services are also affected, with only 25% of hospitals having grid or off-grid solar access, while 90% of elementary schools lack power (SL Compact 2024).

Electricity tariffs, at 18.6 US cents/kWh, are among the highest in the region but remain financially unsustainable. While government subsidies rose from $1.7 million in 2016 to over $14 million in 2020, mostly allocated to EDSA (SL Compact 2024),

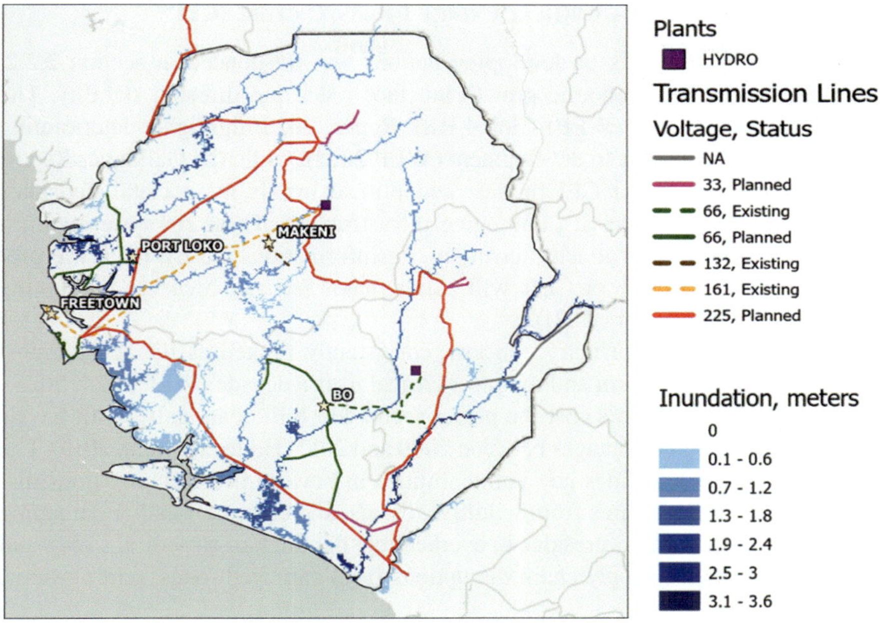

FIGURE 2.7 Energy infrastructure map of Sierra Leone (SL Compact 2024).

the dollar-based Karpowership contract, combined with fluctuating fuel costs and fixed tariffs, strains the sector. Since 2018, EDSA's main grid has provided power for 18 hours daily, but frequent equipment failures cause disruptions (World Bank 2019b), and isolated grids in the provinces experience seasonal fluctuations, leading to prolonged outages. High connection fees and costs also limit access for low-income households, despite a social tariff introduced in 2019 (World Bank 2019b), and rural communities, relying on expensive off-grid solutions, benefit from lower connection fees but still face affordability challenges.

2.3.2.3 Energy and Carbon Efficiency

The main index relating energy and carbon efficiency is the level of GHG emissions such as carbon dioxide (CO_2), methane, and nitrous oxide from all sources including land-use change measured in tons of carbon dioxide-equivalent (CO_2 e) over a 100-year time scale. Sierra Leone is one of the countries with low GHG emissions of 16.23 million metric tons and a per capita of 2 tons in 2023 (World in Data 2024). In Sierra Leone, agriculture and land-use change are the most emitting sectors with a combined value of 7.83 million tons, and electricity and heat accounts for 10,000 tons in 2021. Comparing CO_2 emissions per capita and GDP per capita for 2022, Sierra Leone has a per capita CO_2 emission of 0.15 tons and a per capita GDP of $ 1,690. Another determinant of the carbon intensity of a country is the CO_2 emission per dollar of GDP. In this regard, Sierra Leone had a value of 0.26 kg/$ in 1969 and 0.11 kg/$ in 2023. Finally, with respect to the carbon intensity of energy production, Sierra Leone was 0.19 in 1980 and has slightly increased to 0.30 in 2021 (World in Data 2024).

2.3.3 Vulnerability of Critical Energy Infrastructure (CEI)

The vulnerability of CEIs in developing nations was introduced in section 2.2.2, which are crucial for economic growth but face risks that threaten stability. The World Economic Forum (WEF) Global Risk Report 2023 highlights deteriorating infrastructure as a barrier to development (WEF 2023), while the Gallagher Report 2023 identifies five major CEI threats: geopolitical instability, societal divisions, natural disasters, cyber threats, and aging infrastructure. These risks are manifest as trade disruptions, cyberattacks, terrorism, misinformation, and EWEs. It was predicted that by 2025, 30% of CEIs will suffer major security breaches, disrupting operations (Gartner Report, 2021).

Climate change is a primary concern, consistently ranked as the second-most severe risk in the short term and the most critical over a decade (WEF 2024, 2025), and climate-related hazards cost the power sector $15 billion annually, with EWEs causing 74% of recorded outages between 2000 and 2017 (Hallegatte et al. 2019). The World Bank (2019a) outlines key vulnerabilities in power infrastructure: transmission and distribution failures from wildfires, floods, and extreme weather damaging power lines; generation failures due to overheating, flooding, or structural stress; and fuel and maintenance supply chain disruptions from damaged roads, port closures, and broken pipelines.

Developing nations, particularly in Sub-Saharan Africa (SSA), face heightened risks due to outdated grids, rising energy demand, and poor risk management (Hallegatte et al. 2019; World Bank 2016). Despite contributing less than 3% of global emissions, Africa bears severe climate impacts, with fragile infrastructure frequently exposed to floods, droughts, and landslides (AFDB 2023). These disruptions hinder daily life and economic activities, forcing households and businesses to rely on costly alternatives such as diesel generators and water tanks (World Bank 2019c). For instance, a World Bank study in 2021 found that in SSA, high-voltage transmission lines are vulnerable to flooding and extreme heat, exacerbated by poor maintenance and informal settlements encroaching on power corridors (World Bank 2021).

2.3.4 CEI Resilience Measurement

Resilience measurement and vulnerability assessment are closely linked when assessing the impact of climate change on CEIs, especially in DCs. As noted above, climate change exacerbates the frequency and severity of EWEs, and DCs often lack the institutional capacity, adaptive technology, and financial resources to mitigate these threats, thus making resilience measurements vital for planning and protecting CEIs (IPCC 2014).

On many practical levels, vulnerability serves as input to resilience measurement, as it helps to identify weak points (e.g., aging infrastructure, poor spatial planning, lack of redundancy, etc.), which serves as baseline inputs to resilience assessment by making use of the vulnerability dimensions of exposure, location of infrastructure in climate-prone areas; sensitivity, degree to which energy systems respond to climate stress; and adaptive capacity, the ability to cope with and recover from disruptions (SLECVA 2022; IPCC 2014). Furthermore, resilience assessment evaluates how infrastructural systems for example energy, respond, absorb, adapt to, and recover from vulnerabilities, expressed in resilience metrics such as time to recovery, extent of service restoration, network redundancy, and system adaptability over time (Panteli & Mancarella 2015).

It should also be noted that most resilience models and indexes are probabilistic risk models based on historical risk event frequency. These include the Critical Infrastructure Resilience Index (CIRI), Technical Resilience Analysis, Resilience Measurement Index (RMI), Critical Infrastructure Resilience Evaluation, Resilience Analysis Grid, and Global Infrastructure Risk Model and Resilience Index (GIRI) (CDRI 2023). GIRI, a fully probabilistic risk model, estimates infrastructure asset risk for the most important geology and climatic risks. Ayyub (2014a) suggests quantifying a system's resilience by its functional performance under uncertainty and disruptions, with the performance (Q) or resilience (R) of a system depending on its strength (S) or resistance, load (L), and incident occurrence rate (λ) that is time-dependent using the Poisson process. Figure 2.8 shows a resilience model that measures system performance based

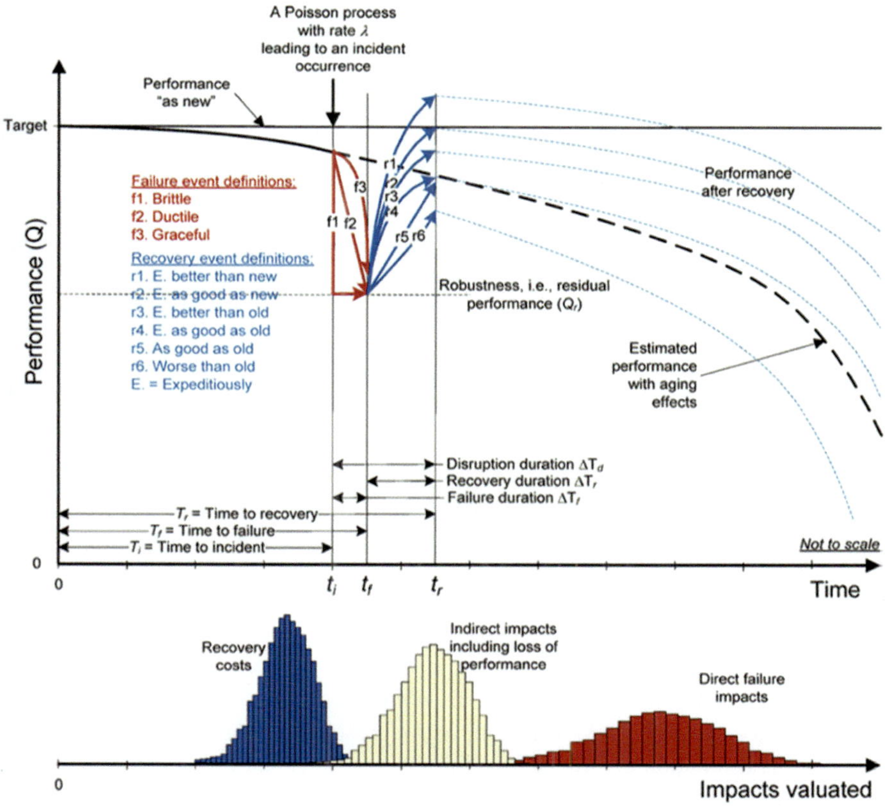

FIGURE 2.8 Schematics of a resilience model (Ayyub 2014a,b).

on disturbance or aging trajectory and recovery trajectory. Based on this, total disruption (*D*) lasts

$$\Delta T_d = \Delta T_f + \Delta T_v \tag{2.1}$$

and the resilience (*R*) is given by the equation:

$$R = \frac{T_i + F\Delta T_f + V\Delta T_v}{T_i + \Delta T_f + \Delta T_v} \tag{2.2}$$

where,

T_i is the start of the failure event
T_f is the final failure event or the time to failure
ΔT_f is the duration of the failure event
ΔT_d is the total disruption
ΔT_v is the duration of the recovery event
F and *V* are the failure profile and recovery profile, respectively.

The failure profile (*F*) considered as a measure of the robustness and redundance of a system is determined from the equation below:

$$F = \frac{\int_{t_i}^{t_f} f\, dt}{\int_{t_i}^{t_f} Z\, dt} \tag{2.3}$$

Whereas the recovery profile (*V*) which measures system's resourcefulness and rapidity is determined from the equation below:

$$V = \frac{\int_{t_f}^{t_v} v\, dt}{\int_{t_f}^{t_v} Z\, dt} \tag{2.4}$$

Based on these equations above, the time to failure (T_f) is characterized by the population density function (f_x) as expressed by the equation below:

$$-\frac{d}{dt}\int_{s=0}^{\infty} \exp\left[1 - \lambda t\left(F_L \alpha(t) s d\tau\right)\right] f_{s0(s)ds} \tag{2.5}$$

where F_L is the cumulative probability distribution function of *L*, and f_s is the probability density function of *S*. The term $\alpha(t)$ represents a degradation or disruption condition as a function of time *t*. It could also represent an improvement condition in the system performance, the unit of which performance may vary depending on the type of system and the objective of the analysis (Ayyub 2014a). It is also noted in (Ayyub 2014b) that for energy or electrical systems, power delivered might be treated as the performance.

2.4 MODELING OF CEI NETWORK

2.4.1 CEI Network System

Managing the vulnerability and resilience of CEI requires an understanding of its components and their interconnections, collectively referred to as networks. This section explores infrastructure networks and power grid modeling. Since CEIs are vital for economic and sustainable development, assessing their vulnerability involves analyzing their physical structure and cyber-human interactions (Duan & Ayyub 2019; Demsar et al. 2005).

CEI networks function as cyber-physical systems, consisting of variables known as nodes (vertices) and the relationships between them, called links (edges) (Newman 2018; Hevey 2018; Hallegatte et al. 2019). Analyzing these networks involves defining their structure using statistical models, evaluating connectivity, and assessing accuracy (Newman 2018; Hevey 2018; Pagani & Aiello 2013). Key network properties include centrality indices, which measure the importance of nodes based on metrics such as degree centrality, node strength, closeness, betweenness, and clustering. According to network theory, networks vary in structure, such as the "small-world" Watts–Strogatz model or the "scale-free" network, where a few highly connected hub nodes dominate (Newman 2018; Hevey 2018).

Additionally, network accuracy is critical and must reflect node relationships, direction, and strength (Duan & Ayyub 2019), with connectivity metrics describing the ratio of edges to nodes, while accessibility metrics assess flow and identify critical points in the network (Newman 2018; Hevey 2018; Pagani & Aiello 2013; Hines 2010). Finally, CEI networks are designed to maintain power grid connectivity, which can be affected by any vulnerability in the system. However, the shape of the network contributes to its coping capacity, with redundancy, and multiple interconnection paths, thus enhancing accessibility and reducing node isolation., and these factors determine CEI resilience against EWEs.

2.4.2 Power Grid Network

Power grid systems exhibit small-world network properties, where nodes are interconnected by short paths that scale logarithmically with network size. This structure enhances resilience to random failures but increases vulnerability to targeted attacks and cascading failures (CFs). Power grids can be classified as complex networks (CNs), multi-agent systems (MAS), or cyber-physical and human systems (CPHS) (Duan & Ayyub 2019; Amani & Jalili 2021). As a CN, a power grid consists of nodes (V) and edges (E), forming a system that follows network or graph theory principles: $G = G(V)$ and $G(E)$, where V represents vertices (nodes) and E represents edges (links) (Pagani & Aiello 2013; Newman 2018; Amani & Jalili 2021; Cockburn & Goodloe 2025). Two adjacent nodes are connected by an edge, and the shortest path between nodes i and j is denoted as d_{ij}. The average shortest path $\bar{S}$ is given by

$$\bar{S} = \frac{1}{N(N-1)} \sum_{i,j \in V} d_{ij} \tag{2.6}$$

where N is the set of nodes and d_{ij} is the shortest path between nodes (Cockburn & Goodloe 2025).

Conversely, a MAS power grid modeling focuses on decentralized power management in micro and smart grids. It involves load shedding, protection, voltage and frequency control, and energy trading using Distributed Energy Resources and communication technologies. The CPHS power grids integrate physical and cyber layers that process modules and control loops to manage distributed control algorithms, ensuring real-time monitoring and optimization. The CPHS layers also has the advantage of ensuring grid dependability that accounts for:

- Physical faults and outages, such as electromagnetic disturbances.
- Cyber vulnerabilities, including internet connectivity issues.
- Human reliability, which impacts inspection and maintenance quality (Amani & Jalili 2021).

Furthermore, in the modeling of power grids, their components are categorized into:

- Generating substations, which produce electricity from fossil fuels, hydro, solar, wind, biomass, geothermal, and nuclear sources.
- Transmission substations, which transfer electricity over high-voltage lines.
- Distribution substations, which reduce voltage levels for local consumers (Cockburn & Goodloe 2025).

2.4.3 Temperature Effect on Power Grid Network

A major consequence of GHG emissions and rising temperatures is the disruption of power services due to overheating and sagging power lines. In power systems, T&D lines act as critical network links, while generators and substations serve as nodes (Pagani & Aiello 2013; Cockburn & Goodloe 2025). These components are highly sensitive to extreme temperatures, which impact their efficiency and reliability.

Like all electrical equipment, power grids are vulnerable to both high and low temperatures, although excessive heat poses a greater risk. Electrical components generate heat, which accelerates wear and shortens system lifespan. Overheated power lines experience increased sagging, reduced capacity, and a higher likelihood of failures (Amani & Jalili 2021). Low temperatures can cause voltage fluctuations, disrupt output regulation, and increase component failure risks (Amani & Jalili 2021). Additionally, in power generation and transmission, dynamic power interactions cause rising temperatures, leakage power, and increased total power, creating a feedback loop known as thermal runaway. This phenomenon complicates energy and power management, leading to equipment failures as ambient temperatures rise (Cockburn & Goodloe 2025).

2.4.4 Modeling Power Grid Climate Vulnerability

As noted in the previous section, power grids are vulnerable to both natural and man-made events, which could be in the form of a widespread flooding, hurricane or tornado outbreak, a targeted attack, and finally a CF. According to Cockburn and Goodloe (2025), and Amani and Jalili (2021), modeling power grid vulnerability or failure involves determining the criticality of the situation. EWEs most often lead to either connectivity loss (CL) or CF or both.

2.4.4.1 Connectivity Loss

When a power grid is functional, the distribution substation can receive power from any of the generation substation or transmission substation. If any of the generation or transmission substation goes offline due to disruptions such as natural hazards, equipment failures, or cyberattacks, the number of substations that will send power to a distribution substation will decrease. This situation is referred to as CL, and it is represented in network-based models as a fraction or number of nodes that become disconnected from energy sources (e.g., substations or generation plants), and reduction in network reachability, often measured in terms of graphs' metrics like giant component size, average path length, or number of unreachable demand nodes (Cockburn and Goodloe 2025; Ouyang 2014). CL is the measure of the average decrease in the number of generating substations connected to the distribution stations, and directly reflects the number of customers who cannot access energy services. For instance, when a substation due to climate-related hazards is disconnected from the transmission grid, all downstream consumers lose access, and this loss of connectivity translates directly to service interruption (Pant et al. 2016; Ouyang 2014). It can be modeled under normal operations (baseline), during failures or cascading outages (e.g., after hurricanes, earthquakes, or cyberattacks), and to compare the impact of mitigation or adaptation measures (Pant et al. 2016). Additionally, as noted by Buldyrev et al. (2010), loss of connectivity is synonymous to structural failures of the network in network-based modeling of energy system with key metrics such as the size of the largest connected components, number of isolated nodes (customer disconnected), and the average node-degree drop after failure. As a result, CL is essential for evaluating system resilience (how quickly the system can restore lost connections) and reliability (how often such disconnections occur) (Panteli & Mancarella 2015).

As noted by Cockburn and Goodloe (2025), CL is expressed as:

$$CL = 1 - \frac{1}{N_d} \sum_{i=1}^{N_d} \frac{N_{g(v_i)}}{N_g} \tag{2.7}$$

where N_g and N_d are the total number of generation and distribution nodes, respectively, and v_i is the load fed into the distribution substations. Figure 2.9 is a simplified schematic of a 9-bus/node system of a power grid to illustrate the concept of CL.

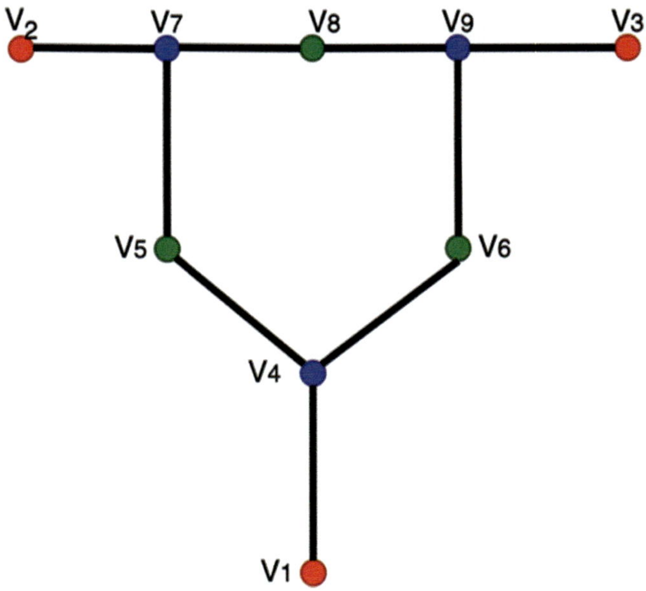

FIGURE 2.9 Graph of a 9-bus/node of power grid system (Cockburn & Goodloe 2025).

From the structure, the generation substations are represented by red nodes of v_1, $v_{2,}$ and v_3; the transmission substations represented by the blue nodes of v_4, $v_{7,}$ and v_9, and the distribution substations represented by the green nodes of v_5, $v_{6,}$ and v_8. As noted earlier, when the grid is functional and intact, the ratio of $N_g(v_i)/N_g$ is one for each of the distribution substation as well as for the averages of them, which produces a CL of zero $(1-1=0)$. Since $N_g=3$ and $N_d=3$, and without any disturbance to the grid, the

$$\mathrm{CL}=1-\frac{1}{N_d}\left[\frac{N_g(v_5)}{N_g}+\frac{N_g(v_6)}{N_g}+\frac{N_g(v_6)}{N_g}\right]=1-\frac{1}{3}\left[\frac{1}{3}+\frac{1}{3}+\frac{1}{3}\right]=1-\frac{1}{3}(3)=0 \tag{2.8}$$

When an EWE occurs and any of the generation nodes are removed or destroyed, an edge incident to the node will also be removed. For instance, if v_2 and v_7 are destroyed, and based on the connectivity loss calculation, about 33.3% of the grid will be affected

$$\left(1-\frac{1}{3}\left[\frac{2}{3}+\frac{2}{3}+\frac{2}{3}\right]=1-\frac{1}{3}(2)=\frac{1}{3}\text{ or }33.3\%\right) \tag{2.9}$$

The CL will increase to 55.5% if a distribution substation, let say v_8 is also destroyed or removed

$$\left(1-\frac{1}{3}\left[\frac{2}{3}+\frac{2}{3}+\frac{0}{3}\right]=1-\frac{1}{3}\left(\frac{4}{3}\right)=\frac{5}{9}\text{ or }55.5\%\right) \tag{2.10}$$

As more nodes are removed or destroyed by during an EWE, the CL increase leading to complete power outage or blackout

$$\left(\frac{1}{3}\left[\frac{0}{3}+\frac{0}{3}+\frac{0}{3}\right]=1-\frac{1}{3}(0)=1\text{ or }100\%\right) \tag{2.11}$$

2.4.4.2 Cascading Failure

CF is a major power grid vulnerability, occurring when a single failure triggers a chain reaction that disrupts interconnected system components, known as the "domino effect" (Amani & Jalili 2021; Cockburn & Goodloe 2025). These failures propagate rapidly through linked systems, creating positive feedback loops that place additional strain on remaining infrastructure. In power grids, a single transformer failure can overload nearby lines, leading to widespread outages. EWEs further amplify CF by disrupting interconnected energy and infrastructure networks.

Modeling EWE-induced CFs can be done using dynamic transient CN models or network-based stochastic models (Amani & Jalili 2021). Since power grids are interlinked with protective measures, localized failures can escalate due to EWEs. For instance, a lightning strike on a transmission line may overload parallel lines (Cockburn & Goodloe 2025). Using CN theory, betweenness centrality can help analyze power grid vulnerabilities, estimating the load impact on substations and failure risks.

According to Cockburn & Goodloe (2025), shutting down one substation increases the load on others, potentially triggering system-wide failures. A 14-bus/node schematic (Figure 2.10) illustrates power grid vulnerability to CFs, with nodes v_2, v_3, and v_6 representing production and distribution substations. Gephi software analysis of failure data identifies node v_4 as having the highest betweenness centrality, making it the most critical point in CF scenarios. From this, critical load (CL) is determined for all nodes, including substations v_2, v_3, and v_6. This study has five generating (red and orange) and eleven distribution (green and orange) substations ($N_g = 5$ and $N_d = 11$). The CL formula is:

$$\text{CL}=1-\frac{1}{11}\left[\frac{N_g(v_2)}{5}+\frac{N_g(v_3)}{5}+\frac{N_g(v_4)}{5}+\frac{N_g(v_5)}{5}+\frac{N_g(v_6)}{5}+\frac{N_g(v_9)}{5}\right.$$
$$\left.+\frac{N_g(v_{10})}{5}+\frac{N_g(v_{11})}{5}+\frac{N_g(v_{12})}{5}+\frac{N_g(v_{13})}{5}+\frac{N_g(v_{14})}{5}\right] \tag{2.12}$$

Based on the betweenness centrality analysis calculated by Cockburn and Goodloe (2025), when a distribution substation is first removed (v_4) at the start of an EWE, the impact will be relatively low, and it will not create an island effect in the grid, as

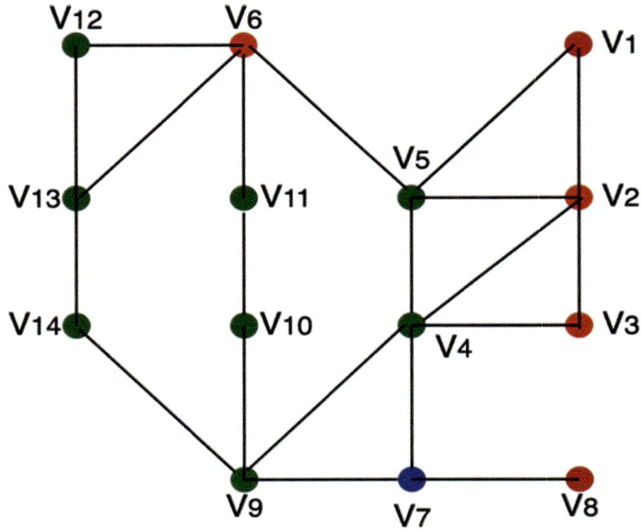

FIGURE 2.10 A graph of a 14-bus power grid system (Cockburn & Goodloe 2025).

the remaining distribution substations will have access to all the generation stations depicted by the low CL below:

$$CL = 1 - \frac{1}{11}\left[\frac{5}{5}+\frac{5}{5}+\frac{0}{5}+\frac{5}{5}+\frac{5}{5}+\frac{5}{5}+\frac{5}{5}+\frac{5}{5}+\frac{5}{5}+\frac{5}{5}+\frac{5}{5}\right] = 1 - \frac{1}{11}\left[\frac{50}{5}\right] = \frac{1}{11} \text{ or } 9.1\% \quad (2.13)$$

As noted in Hallegatte et al. (2019), power grids are built with some redundancy to cope with the disruption of one network, but due to the interdependency and interconnectedness of grid components, the propagated EWE and other inherent vulnerabilities will affect more nodes, increasing the CL until the whole grid is disrupted and a blackout occurs.

According to Cockburn and Goodloe (2025), removing the distribution substations (v_5) and (v_6) will reduce the power load, and v_5, which is both a generation and distribution substation, will leapfrog v_5 and v_9 (both distribution stations) in a betweenness centrality analysis and become the point of disruption (a huge CL of 72.7% and an island effect). As the EWE worsens, this pattern will continue until the grid is completely disrupted.

2.5 DEVELOPING COUNTRIES CEI VULNERABILITIES

2.5.1 Introduction to CEI Vulnerabilities in Developing Countries

Having underscored that CEIs are vulnerable to the impacts of climate on CEI, and the need to critically understand these vulnerabilities to guide in developing a resilient CEI to manage the impacts of climate change. It is also noted in Section 2.3.4 that the link between vulnerability analysis and resilience assessment with vulnerability dimensions of exposure, sensitivity, and adaptive capacity as vital metrics in determining the resiliency of a system. This section is a practical analysis and discussion of the vulnerabilities of CEI in DCs with Sierra Leone as an example, that supports the development of a climate-resilient CEIs in DCs.

As noted under section 2.2.3 above, climate change is affecting weather patterns, water resources, extreme events, land use, and livelihoods in Sierra Leone. Based on climate scenarios generated for Sierra Leone from various studies, the country is experiencing increased average and extreme temperatures and fewer cool days, increased precipitation extremes with more intense heavy precipitation events, increased incidence and intensity of drought, increased intensity of riverine flooding and more frequent landslides, and continued sea level rise (SLR) and damage from storm surge.

Table 2.1 depicts temperature variation in Sierra Leone between 2020 and 2060 based on the findings of an energy sector Climate Vulnerability Assessment (CVA) for Sierra Leone.

The climatic risks above present a framework for assessing climate vulnerability in Sierra Leone. As outlined in section 2.3.2.1 regarding Sierra Leone's energy infrastructure, the principal CEIs of the country include hydropower generation facilities, solar power generation facilities, fossil fuel generation facilities, and electricity T&D assets (substations and lines). These were the focal points of the CVA (SL Energy CVA 2022) that examined their climate exposure, sensitivity, and impact implications for energy assets throughout Sierra Leone. The CVA analysis indicates that energy consumption is expected to rise until the late century due to hotter summers, influenced by variables such as behavioral changes, population growth, alterations in building stock, and the use of cooling systems. Table 2.2 presents an evaluation

TABLE 2.1
Temperature Increases under Climate Change in Sierra Leone (SLECVA 2022)

Variable	Change	Time Horizon	Source
Average annual temperature	+1.0°C–2.5°C (more rapid warming inland)	2060	USAID (2016)
Minimum temperature	+0.4°C–1.0°C	2020–2039	World Bank Group (2022)
	+1.4°C–3.2°C	2060–2079	
Maximum temperature	+0.5°C–0.8°C	2020–2039	Numbers are relative to a 1995–2014 baseline and consider changes across the SSP2-4.5 and SSP5-8.5 scenarios.
	+1.4°C–3.0°C	2060–2079	
Cooling Degree Days (CDD) (number of days above 18.3°C)	+30–50 days	2020–2039	
	+73–173 days	2060–2079	
Very hot days (number of days above 35°C; these occur mostly from December to June)	+1–6 days	2020–2039	

TABLE 2.2
Exposure, Sensitivity, and Adaptive Capacity Ratings for Each of the Impact Metrics (SLECVA 2022)

#	Impact	Exposure Rating	Sensitivity Rating	Adaptive Capacity Rating	Vulnerability Rating
1	Hydropower generation decline from drought	High	High	Medium	High
2	Sedimentation and siltation, flooding/extreme precipitation at hydropower plants	High	High	Low	High
3	Reduction in solar productivity from extreme temperatures	Medium	Low	Low	Medium
4	Transmission and distribution (T&D) infrastructure damaged by riverine flooding and extreme precipitation	Medium	High	Low	High
5	Transmission infrastructure damaged by wildfire	Low	Medium	Low	Medium
6	Transmission line capacity loss due to line sag from increasing temperatures	High	Medium	High	Medium
7	Damage to T&D/solar/hydropower plants' infrastructure from landslides	Low	High	Low	High
8	Damage to T&D infrastructure from sea level rise (SLR)	Low	Low	Low	Low

of exposure, sensitivity, and adaptive ability for the eight impact indicators of the energy industry according to the CVA.

Projections indicate that in a severe climate change scenario, Cooling Degree Days (a metric for estimating temperature-related demand) would rise by around 20%–50% throughout all provinces of Sierra Leone by 2070, with the most significant increases occurring in the Eastern Region. In 2040, demand is anticipated to be double that of 2020, irrespective of climate change. The analysis indicates that Sierra Leone may more successfully manage rising energy demand during climate change through the implementation of appropriate climate-resilient policies and practices, and these strategies are discussed in the next section.

2.5.2 Vulnerability of Generation Infrastructure to Climate Change

Hydropower as the major energy generation in Sierra Leone has huge potential because of the several river basins and high precipitation received annually. The other sources of energy generation in Sierra Leone are solar power, wind power, and biomass. These generation sources are vulnerable to the impacts of floods and droughts.

2.5.2.1 Hydropower Infrastructure

Hydropower infrastructure is vulnerable to the impacts of climate change from their declines due to droughts, siltation and sedimentation from excessive precipitation and degradation of the environment in the upstream of the hydropower facility

2.5.2.1.1 The Impact of Drought on Hydropower Infrastructure

The Bumbuna I dam in the Northwest province and Goma (or Dodo) dam in the southeastern region are two current hydropower plants utilized by the country for electricity generation, and both will be affected by drought from the CVA study. Table 2.3 shows that, under severe climate change (SSP5-8.5), both dams could experience an increase in consecutive dry days (CDD), potentially leading to more intense or longer dry seasons. By the late century, the number of CDD could increase by 14 days, significantly reducing water availability for hydropower generation. Under a severe climate scenario (SSP5-8.5), precipitation may decline by 58%. Recent field

TABLE 2.3
Average Annual Consecutive Dry Days at the Two Hydropower Plant Locations (SLECVA, 2022)

	Drought Metric: Consecutive Dry Days (CDD)		
Hydropower Plant	**Historical Median (1995–2014)**	**SSP5-8.5 (Severe Climate Scenario)**	
		2030	**2070**
Goma dam	121.9	125.0	134.9
Goma dam change	-	+3.1	+13.0
Bumbuna dam	178.5	183.9	194.2
Bumbuna dam change	-	+5.4	+15.7

observations indicate that the dry season is already extending into June, with the most significant precipitation declines expected between April and June.

Sierra Leone's run-of-river hydropower systems are highly sensitive to seasonal water fluctuations (Republic of Sierra Leone 2015). The Bumbuna hydroelectric project (Phase 1), designed for 50 MW, generates only 11 MW in the dry season and 40 MW in the wet season, and this variation results in load shedding and increased reliance on fossil fuels (SL Energy CVA 2022). While the country has a planned hydropower capacity of 1,190 MW, current production remains constrained—Bumbuna could expand to 100 MW, while Goma, operating at 6 MW, delivers just 2 MW. Research by Neumann & Price (2009) indicates that a 1% drop in precipitation reduces hydropower generation by 1% (SL Energy CVA 2022). Under extreme climate scenarios, Bumbuna and Goma's output could decline by up to 60%. However, due to the variability of run-of-river hydropower, actual generation fluctuations require detailed historical flow and precipitation data for accurate modeling (Neumann & Price 2009; SL Energy CVA 2022).

2.5.2.1.2 Flooding and Sedimentation on Hydropower Plants

EWEs, such as floods, contribute to sedimentation and siltation in degraded catchments, impacting hydropower turbine efficiency. Heavy rainfall and flooding have caused sediment accumulation at Bumbuna and Goma dams, as reported by local officials (SLECVA, 2022). Deforestation and land-use changes have further exacerbated sedimentation issues, highlighting the connection between human activity and climate risks for hydropower infrastructure. Under a severe climate scenario (SSP5-8.5), precipitation trends at these dams are projected to shift throughout the 21st century. By 2030, rainfall at both dams is expected to decline by 10%, followed by an 8% increase at Bumbuna Dam and a 2% decrease at Goma Dam by 2070. However, single-day extreme precipitation events are predicted to remain close to historical values, indicating a minor change in flood risk at dam sites. Despite this, World Bank (2018) research suggests that precipitation return periods in Freetown, Makeni (near Bumbuna), and Bo will shorten significantly by 2050. Heavy storms that currently occur once in 500 years could become 1-in-10-year events, increasing flooding and sedimentation risks at hydropower dams.

2.5.2.2 Solar Productivity and Extreme Temperatures and Dust

Extreme temperature and dust may diminish solar production, affecting CEI. Maximum temperatures in warm tropical climates like Sierra Leone reduce solar PV cell efficiency. Sierra Leone's maximum temperatures will rise modestly. Maximum temperatures might rise 10% in each region by 2070, with the Eastern Province seeing the largest increase (Table 2.4). The 96 mini-grid solar systems in Sierra Leone may see more days over 40°C as maximum temperatures rise, and the average annual temperature over 40°C is less than 1 day for all 96 microgrids. In the dry season (December–February), the hot, dry, and dusty harmattan winds may cover solar cells in dust, reducing their effectiveness (SL BUR 2021). Climate change raises the likelihood of more violent droughts, especially in the northern area with most mini-grid solar systems, and dust storms or sandstorms that cover solar photovoltaic cells.

TABLE 2.4
Average Annual Maximum Temperatures across Sierra Leone (SLECVA 2022)

Province	Average Annual Maximum Temperatures (°C)				
	Historical Median (1995–2014)	SSP2-4.5 (Moderate Climate Scenario)		SSP5-8.5 (Severe Climate Scenario)	
		2030	2070	2030	2070
Eastern	31.5	32.6	34.2	32.7	35.8
Northern	31.8	32.8	34.3	32.8	35.7
Southern	31.6	32.6	34.0	32.6	35.4
Western (freetown)	30.4	31.3	32.5	31.2	33.4
Country wide	31.3	32.3	33.8	32.4	35.3

Increased temperatures may lead to lower solar PV efficiency and energy output, reducing energy produced and requiring backup energy options. Solar cell output is usually rated at 25°C and decreases by 0.25% (for amorphous cells) to 0.5% (for most crystalline cells) per 1°C rise in temperature (ICF 2017). In warm climates, cell temperatures for roof-mounted arrays can easily reach 50°C–70°C; at 50°C, output can be 12% below rated output (ADB 2012). However, impacts to solar radiation are projected to be minimal. Given the maximum temperature projections and the above sensitivity thresholds, Sierra Leone provinces could experience approximately 1%–3% decrease in PV output efficiency by 2070. Larger changes in output capacity are likely due to the dust deposits from harmattan winds. High ambient temperatures can also lower the capacity of underground conductors, which can lower overall solar energy production capacity (ADB 2012).

2.5.3 Vulnerability of Transmission and Distribution Infrastructure

T&D lines are affected by extremes of precipitation and temperature associated with coastal and tropical countries. In Sierra Leone, T&D lines are impacted by the climate risks of heat, drought, and precipitation, and these are discussed below.

2.5.3.1 Flooding and Extreme Precipitation on Transmission Infrastructure

Total rainfall over numerous days is important because flooding can undermine T&D infrastructure. Climate models forecast severe precipitation increases in Sierra Leone, including small to significant increases in annual maximum 5-day precipitation along existing and proposed transmission lines. The ensemble median values for this extreme precipitation statistic match historical values in all models, and the top 90th percentile forecasts suggest the number almost doubles by 2070. Independent of climatic scenario, SSP2-4.5 and SSP5-8.5 estimations exhibit large 50th–90th percentile differences. Model magnitude differences, however large, are typical for precipitation variables because they are more sophisticated than temperature variables. Floods and extreme precipitation can flood substations, gouge transmission tower bases, and erode pole foundations, exposing subsurface cabling and perhaps damaging it. Flooding can destroy ground-level substation equipment (ICF 2017), and one tower damage might create local power outages and expensive repairs or asset replacement owing to the lack of backup T&D networks (ADB 2012).

2.5.3.2 Impact of Wildfire and Line Sagging on Transmission Infrastructure

Transmission line extreme temperatures are predicted to rise somewhat in the 21st century. Even in a severe climate scenario (SSP8.5), Sierra Leone's tropical environment cushions major temperature fluctuations. In the 21st century, the transmission line has more very hot days (above 40°C). In a severe climate scenario (SSP5-8.5), this extreme temperature measure increases most in the north and center, where some of the planned transmission lines are to be built. Up to 30–40 more days of temperatures over 40°C might damage network lines in these places. Extreme heat will cause wildfires that destroy transmission cables in the impact corridor.

2.5.3.3 Exposure of Transmission Facilities to Wildfire

Due to rising maximum temperatures and dry circumstances (including an extended dry period from February to June), transmission line wildfire danger increased in the 21st century, and Sierra Leone traditionally has the most wildfires in the northwest and southwest especially in January–March (Fayiah & Xavier 2021). In a catastrophic climate scenario, average maximum temperatures might rise by 4°C (7.2°F) by late century. Table 2.5 shows that relative humidity, a stable climatic characteristic, might fall by 6% by late century, and it fosters wildfires.

Wildfires can damage T&D lines, cause soot deposition on insulators to leak, and produce ionized air and smoke that arcs wires, increasing maintenance costs (ICF 2017). Damage to one tower might cause local power interruption and necessitate

TABLE 2.5
Relative Humidity Values across Sierra Leone through Late Century (SLECVA 2022)

		Relative Humidity Values (%) SSP5-8.5 (Severe Climate Scenario)	
Province	**Historical Median (1995–2014)**	**2030 (Change)**	**2070 (Change)**
Eastern	77.8	74.9 (−2.9)	71.1 (−6.7)
Northern	78.4	76.2 (−2.2)	73.7 (−4.6)
Southern	83.9	80.8 (−3.1)	78.0 (−5.9)
Western (freetown)	83.7	82.1 (−1.6)	80.7 (−3.0)
Country wide	77.6	75.5 (−2.1)	71.7 (−5.9)

expensive repairs or asset replacement due to the lack of backup T&D networks (ADB 2012).

T&D lines can cause wildfires. This is especially true for old timber poles (RMIT University 2012). Heavy wind can bring down poles at high wildfire danger (low relative humidity and high wind), igniting sparks if wires hit dry, combustible vegetation. Safety, the environment, and water quality around T&D lines are at danger. Poor upkeep of aged or degraded infrastructure increases wildfire risk and susceptibility. Low relative humidity and high severe temperatures across the country make wildfires more likely to start and spread.

2.5.3.4 The Impact of Sea Level Rise to Transmission and Distribution Infrastructure

SLR from heavy precipitation, especially around the coast, will erode and deteriorate coastal facilities like T&D lines. The T&D system in Sierra Leone is now under little danger from SLR. Globally, sea levels will rise 0.1–0.56 m by 2100 (USAID 2016). Sierra Leone's entire coastline is low- to low-medium coastal flood risk, and there are no transmission lines (planned or existing) in SLR-prone locations. However, climate change will intensify storms in the 21st century. High emission scenarios (SSP5-8.5) may enhance coastal erosion in Sierra Leone due to increasing sea levels. Since none of the mapped assets are exposed, distribution lines, poles, substations, and other infrastructure, especially in the Freetown peninsula, should be assessed for SLR and storm surge risks in the Western Area. SLR might harm infrastructure due to poor maintenance and repair (USAID 2018). Aging and decaying infrastructure reduces power sector dependability and climate vulnerability. Degraded or aging infrastructure may be more vulnerable to SLR owing to poor maintenance (Ouedraogo 2019).

SLR can scour transmission tower bases and increase saltwater exposure, which can corrode electrical components of T&D infrastructure, requiring increased maintenance. Again, lack of redundant T&D networks could lead to local power loss and expensive repairs or replacement of assets if inundation damages one tower (ADB 2012). Although asset damage from SLR may be significant, the above exposure section and current planning efforts show it is unlikely to cause many issues within Sierra Leone's T&D infrastructure. Thus, Sierra Leone is at low overall risk to this impact.

2.5.4 Impact of Landslides to CEI Infrastructure

West African countries with heavy rainy seasons are experiencing more frequent landslides due to climate change. Landslides can damage power generation infrastructure and access roads, disrupting service, increasing repair costs, and hindering maintenance efforts. They also contribute to sedimentation, reducing hydropower storage capacity, damaging turbines, and blocking dam outlets, leading to higher maintenance costs and safety concerns (ICF 2017). Older infrastructure is more vulnerable to landslides and climate risks, with poor maintenance worsening structural weaknesses (USAID 2018). Additionally, landslides can erode transmission tower bases, expose underground cables, and cause breakages, leading to power outages and expensive repairs due to the lack of network redundancy (ADB 2012).

However, prior landslide occurrences and increased extreme precipitation could heighten future risks. Bumbuna I and Goma dams are in a moderate (3) risk zone, while some transmission lines are at high or very high risk. Around 50% of the India Exim Extension project line and 20% of the Bumbuna to Yiben line fall within these high-risk categories. As extreme precipitation is the primary driver of landslides, 10%–20% increases in rainfall by late century could significantly raise landslide risks along key transmission routes, including Bumbuna to Yiben, the India Exim Extension, and Fadugu-Kabala lines.

2.5.5 Climate Impacts on Energy Demand

As said, climate change affects energy sources and infrastructure. Sierra Leone's per capita energy consumption is low since many people use charcoal and firewood (MoE, 2020). As the government expands access to the electrical grid, demand is likely to rise, and climate change is also affecting demand. The 2020–2040 Sierra Leone Integrated Resource Plan (IRP) projected future electricity demand based on stakeholder consultations and government policies like energy efficiency and universal electricity access, but it did not account for climate change. The IRP estimated demand based on 32% of homes having power in 2020 and 82% in 2030. The IRP created a "low case," "base case," and "high case" to assess demand possibilities.

Table 2.6 presents load forecast modeling results, indicating that Sierra Leone's energy consumption will increase 1.8 times by 2025 and double by 2040, even without climate change. This highlights the urgent need for energy infrastructure investment (Figure 2.11).

The Western region will have the highest demand (14,458 GWh), followed by the North and North-West (8,957 GWh), East (6,432 GWh), and South (5,847 GWh) (MoE 2020). Climate change will further escalate demand due to higher cooling needs during hotter summers and increased irrigation requirements in droughts. Other factors influencing future demand include population growth, technological

TABLE 2.6

Projected Growth in Energy Demand for Sierra Leone from 2020 to 2040 (MoE 2020)

Scenario	Demand	Annual Average Rate of Increase	System Peak Demand
Low case	657 GWh in 2020 2183 GWh in 2040	6.3%	150 MW in 2020 499 MW in 2040
Base case	696 GWh in 2020 1239 GWh in 2025 2193 GWh in 2035 2800 GWh in 2040	7.2%	159 MW in 2020 639 MW in 2040
High case	745 GWh in 2020 4651 GWh in 2040	9.6%	170 MW in 2020 1062 MW in 2040

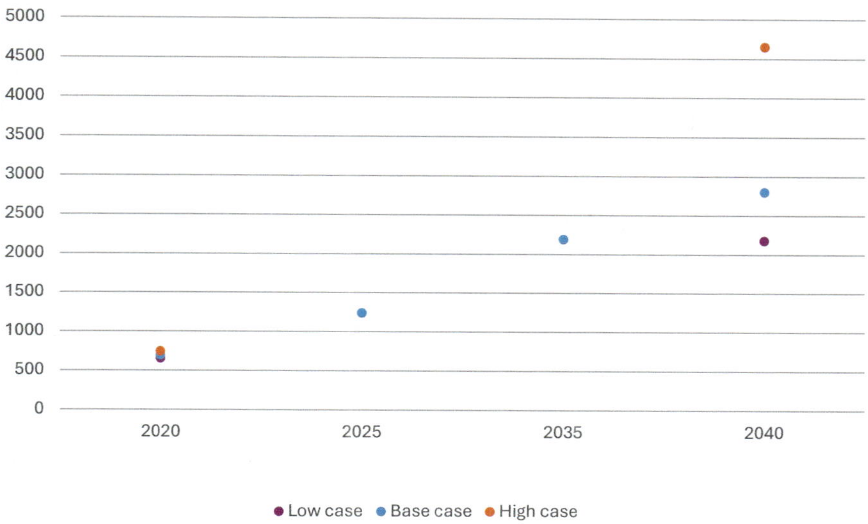

FIGURE 2.11 Projected energy demand in Sierra Leone from 2020 to 2040 (MoE 2020).

advancements (e.g., air conditioning and irrigation), and behavioral changes. Energy sector vulnerabilities could impact health, economy, and food security.

Sierra Leone's agriculture sector is highly vulnerable to climate extremes like heatwaves, droughts, and floods, which reduce yields and cause food spoilage. Energy shortages disrupt food processing, refrigeration, and milling (SEforALL 2021). Rising temperatures will increase cooling demand, further straining the energy sector. Unreliable electricity may lead to greater use of polluting fuels like kerosene and wood, worsening indoor air pollution and respiratory health risks (Arku et al. 2020; WHO 2015). Unless renewable energy solutions are adopted, reliance on diesel and biomass will increase GHG emissions (WHO 2015).

Expanding electricity access can help reduce deforestation and natural resource depletion. Reliable refrigeration minimizes food spoilage, improving fishing sustainability and profitability. In rural areas, wood is the primary cooking fuel, but improved energy availability could curb deforestation. A weak power sector also exacerbates coastal and forest degradation by forcing communities to overexploit resources. Although irrigation currently has low energy demand, future droughts and rising temperatures may increase the need for water pumping.

Power sector vulnerability hinders community socioeconomic progress and health center and school services. Unreliable energy in rural Sierra Leone inhibits economic growth and quality of life (UNOPS 2020) because outages might disrupt hospital and household systems (ICF 2017). A crucial healthcare system without a stable power supply puts the population at danger everyday (e.g., enough illumination to see) and in severe crises (e.g., pandemic resources). Since improved energy availability permits medication and vaccine storage, refrigeration is essential, yet unstable energy systems and outages might destroy vital medicine (Mobarak et al. 2019). Health centers could provide services and outreach to their town and nearby areas with stable supply (UNOPS 2020). Health clinics and communities need enough illumination to promote quality of life. Elevating access and reliability of electricity, especially in rural areas, can help schools with lighting, communications tools, and technology (e.g., radio, phones, and computers), solar-powered water treatment, food and medical supply refrigeration, and heating and cooling (Welland 2017).

2.6 CLIMATE RESILIENCE DEVELOPMENT IN SIERRA LEONE: NEEDS AND CHALLENGES

2.6.1 Climate Governance and Regulatory Landscape

Strengthening climate resilience in Sierra Leone's power sector requires addressing institutional, financial, technical, and policy challenges. Fiscal capacity is key to financing climate adaptation and energy sustainability. Recognizing the link between climate vulnerability and development, Sierra Leone has integrated climate action into national strategies. The 2019–2023 Medium-Term National Development Plan emphasized low-carbon development, while the 2024–2030 plan builds on this to establish a resilient green economy by 2034. The recently established Presidential Initiative on Climate Change, Renewable Energy, and Food Security supports low-carbon transition with goals such as: developing a Just Energy Transition Plan by 2024, expanding grid capacity from 200 MW to 1 GW by 2034, establishing 200 small grids by 2030 to improve access, and ensuring 20% of households use sustainable cooking fuels by 2030.

Sierra Leone also participates in global climate agreements and has introduced national policies and structures such as the establishment of the Ministry of Environment and Climate Change (MoE 2020), Nationally Determined Contribution (NDC, 2021), and National Climate Policy (2021). These frameworks focus on reducing vulnerabilities, adaptation, and low-carbon development. However, challenges persist, including weak coordination and institutional capacity to integrate climate risks into key sectors such as water, energy, agriculture, and disaster management. Poor alignment among ministries, environmental agencies, and local governments creates policy gaps, weakening climate action. A shortage of climate experts and technical professionals further limits progress, and although awareness of climate change is increasing, many lack specialized skills to implement effective policies. This limited understanding of climate interactions and long-term impacts further complicates resilience planning, making climate management especially difficult for developing nations.

2.6.2 Energy Governance and Regulatory Regime

Sierra Leone has introduced several energy programs to enhance supply and support the transition, including the National Energy Strategy Plan (2010–2025), Renewable Energy Policy (2016), Off-Grid Solar Energy Strategy (2016), National Renewable Energy Action Plan (2015), Energy Efficiency Policy, Electricity Sector Roadmap (2017–2030), National Electricity Act (2011), and Integrated Resource Plan (IRP 2020–2040). The National Energy Strategy Plan sets short-, medium-, and long-term supply targets (SL Compact 2024), and the National Electricity Act (2011) replaced the 1982 law, creating two state-owned entities—EGTC, which oversees power generation and high-voltage transmission, and the Electricity Distribution and Supply Authority (EDSA), which handles 66 kV and low-voltage distribution and sales. Additionally, the Energy and Water Regulatory Commission was established in 2011 to act as an independent regulator. However, governance and inefficiencies have weakened EDSA's financial stability, and a Pricewater Coopers financial review reported losses of Le 42.3 billion in 2017, following Le 13.3 billion (2015) and Le 66.6 billion (2016) to such an extent that EDSA owes EGTC $40 million despite a Power Purchase Agreement. High electricity tariffs and low electrification rates reduce Sierra Leone's regional competitiveness. The sector also struggles with seasonal hydropower variability, inefficiencies in transmission/distribution, and weak institutional frameworks. While efforts to expand solar, wind, hydro, and biomass in rural and industrial areas continue, financial and institutional barriers slow progress. Goals to increase grid connections to 42% by 2025 and expand mini-grids (27%) and solar systems (10%) by 2030 remain challenging. Another key issue is poor coordination between the Ministry of Energy's Renewable Energy Directorate and the Climate Change Secretariat (EPA Sierra Leone), delaying the Nationally Appropriate Mitigation Action plan.

To improve resilience, the government should upgrade the IRP into an Integrated Resource and Resilience Plan (IRRP), incorporating climate adaptation. The IRRP would strengthen power planning and stakeholder engagement through methodologies such as demand forecasting models (econometric, sectoral, and end-use assessments), energy efficiency evaluations (load monitoring and economic impact analysis), transmission and distribution planning (grid integration, loss reduction, and network optimization), and financial modeling (capital investment and revenue analysis). These strategies would improve long-term energy resilience, ensuring sustainable investments and sector expansion.

2.6.3 Technical Capacity of the Energy Sector

Despite efforts by government and development partners such as the World Bank, UN, AfDB, and EU, Sierra Leone's energy sector stakeholders struggle to apply technical knowledge and data to mitigate climate impacts and enhance power sector resilience. Various initiatives have supported climate-related technical studies, data collection, and informational platforms. The Sierra Leone Meteorological Agency and other institutions utilize this data for climate resilience planning. However, many weather stations are nonfunctional, and available data is limited to a few locations with restricted coverage. Expanding hydrological and meteorological monitoring networks would enhance understanding of climate risks to existing and future power infrastructure, improving hydropower investment decisions.

A major challenge is the lack of real-time weather and asset data due to poor climate monitoring and forecasting. Effective hydrometeorological data collection is crucial for early warning systems to predict windstorms, floods, and droughts, helping utilities, farmers, and insurers manage risks. Limited risk assessment capacity and inadequate downscaled climate models hinder climate resilience planning for CEIs.

Most cities and towns lack technology to integrate climate data into decision-making. The government remains unprepared for disasters and lacks contingency planning for floods and droughts. Additionally, climate information is poorly integrated into power sector planning, emergency response, and infrastructure development.

2.6.4 Financial Need and Flow

Adequate funding is essential for implementing Sierra Leone's NDC 2021 and 2024–2030 NMTDP climate strategies to achieve a green economy. The country has received international climate finance for mitigation and adaptation. In 2010, Sierra Leone secured USD 11.1 million for adaptation and USD 10.5 million for mitigation. By 2015, climate funding rose to USD 36.7 million, with USD 23.5 million allocated to adaptation. In 2020, funding significantly increased to USD 178 million, with USD 123 million for adaptation and USD 73 million for mitigation. Between 2016 and 2020, Sierra Leone received USD 910 million in climate financing, with USD 490 million for adaptation and USD 420 million for mitigation.

However, a major funding gap remains for achieving a low-carbon economy by 2034. The NDC 2021 estimates that USD 3.47 billion (74% of 2020 GDP) is required between 2020 and 2030. Projected expenditures include USD 1.38 billion each for mitigation and adaptation, with additional costs for climate-related damages. The estimated annual fiscal cost is USD 347 million (8.3% of GDP), or USD 276 million per government estimates. To meet these targets, Sierra Leone requires USD 84.7–131.2 million per year in climate financing, averaging USD 108 million annually. Adaptation needs range from USD 31–49 million per year, while loss and damage costs are estimated at USD 50–78 million annually.

2.7 CONCLUSION AND RECOMMENDATION

As global temperatures rise, EWEs are becoming more frequent, disproportionately impacting DCs due to their socioeconomic vulnerabilities. These nations must adapt and build resilience to ensure sustainability. CEIs are vital for economic growth, but climate change poses increasing risks. To mitigate these threats, developing nations need climate-resilient energy systems and sustainable development strategies. GHG emissions and environmental degradation accelerate climate change, worsening flooding, heatwaves, droughts, rising sea levels, and landslides. These risks are particularly severe in LDCs, where high temperatures weaken power grids, causing equipment failures, sagging power lines, and reduced efficiency, which threaten energy reliability.

This chapter analyzes CEI vulnerabilities in DCs, focusing on Sierra Leone. With a HDI of 0.452 (ranked 182nd of 189) (UNDP 2019) and the 11th most climate-vulnerable country on the ND-GAIN index, Sierra Leone also has a fragility score of 81.4 (Fund for Peace 2024). Its economy depends on natural resources and rain-fed agriculture, making it highly climate-sensitive.

Sierra Leone's small energy sector is mainly renewable based, with significant hydro and solar potential. While the country aims to expand energy access and transition to a low-carbon economy, challenges such as climate risks, technology gaps, and institutional weaknesses slow its progress. Integrated climate and energy policies are essential, but financing gaps, poor infrastructure, and weak adaptation measures hinder many developing nations.

Climate Change poses significant threats to CEI, particularly in DCs, and from the weak point, an urgent action is needed to reduce their vulnerability and build and strengthen their resilience

- DCs should prioritize huge investment in climate-resilient energy systems to ensure energy security, economic development, and well-being.
- There is a need for strong and sustained collaborative efforts such as the sharing of knowledge, technology, expertise best practices, and resources amongst DCs, and between developing and the developed countries to build climate-resilient energy systems worldwide.
- DCs with support from developed countries must accelerate the transition to a more sustainable and resilient energy future by implementing innovative solutions and strengthening global partnerships.

ACKNOWLEDGMENTS

The authors acknowledge the support provided by Hamad Bin Khalifa University, a member of Qatar Foundation, located in Education City, Doha, Qatar. This research was funded by the Qatar National Research Fund under grant number NPRP14C-0909-210008.

REFERENCES

African Development Bank (AFDB). 2023. Climate Resilience and a Just Energy Transition in Africa. https://www.afdb.org/sites/default/files/2022/05/25/aeo22_chapter2_eng.pdf (Accessed, 12/19/2024).

Amani, A. M., and Jalili, M. 2021. Power grid as complex networks: Resilience and reliability analysis. *IEEE Xplore*. https://ieeexplore.ieee.org (Accessed, 01/21/2025).

Arku, R., Michael, D., MyLinh, B., Li, W., Bo, H., Lap, A. T., Prem, K., and Mony, P. V. M. L. 2020. Adverse health impacts of cooking with kerosene: A multi-country analysis within the prospective urban and rural epidemiology study. *Environmental Research* 188: 109851. https://doi.org/10.1016/j.envres.2020.109851 (Accessed, 02/16/2022).

Asian Development Bank (ADB). 2012. Climate Risk and Adaptation in the Electric Power Sector. https://www.adb.org/documents/adb-annual-report-2012 (Accessed, 02/16/2022)

Ayyub, B. M. 2014a. *Risk Analysis in Engineering and Economics*. 2nd Edition. Chapman and Hall/CRC, New York.

Ayyub, B. M. 2014b. Systems resilience for multihazard environments: Definition, metrics, and valuation for decision making. *Risk Analysis* 34(2). https://doi.org/10.1111/risa.12093

Ayyub, B. M, and Hill, A. C. 2019. Climate-resilient infrastructure: Engineering and policy perspectives. *The Bridge Articles* 49(2). https://www.nae.edu/212185/ClimateResilient-Infrastructure-Engineering-and-Policy-Perspectives

Boyle, R. 2024. Greenhouse Gas Emissions in Sierra Leone. Overview of greenhouse gases and emissions per capita in Sierra Leone. Are they prepared to meet net zero targets and invest in the energy transition. https://www.emission-index.com/countries/sierra-leone (Accessed, 11/14/2024).

British Broadcasting Corporation (BBC). 2024. New Cast. https://www.bbc.com/ (Accessed, 01/16/2025).

Buldyrev, S. V., Parshani, R., Paul, G., Stanley, H. E., and Havlin, S. 2010. Catastrophic cascade of failures in interdependent networks. *Nature* 464(7291): 1025–1028. https://doi.org/10.1038/nature08932 (Accessed, 07/04/2025).

Coalition for Disaster Resilient Infrastructure (CDRI). 2023. Global Infrastructure Resilience Capturing the Resilience Dividend. A Biennial Report from the Coalition for Disaster Resilient Infrastructure (CDRI). https://www.cdri.world/biennial-report (Accessed, 01/18/2025).

Cockburn, S., and Goodloe, M. 2025. Graphical Measures of Power Grid Vulnerability Command, Control, and Interoperability Center for Advanced Data Analysis. Department of Homeland Security. https://ccicada.org (Accessed, 01/21/2025).

Conteh, F., Masahiro, F., Shriram, S. R., Edward, R. C., Conteh, M. A., Rashwan, A., and Senjyu, T. 2023. A LEAP model application for sustainable power generation system. *Sustainability* 15: 11838. https://doi.org/ 10.3390/su151511838

Demsar, U., Spatenkova, O., and Virrantaus, K. 2005. Centrality measures and vulnerability of spatial networks. ISCRAM 2005 Conference Proceedings. Researchgate Publication. https://www.researchgate.net/profile/Kirsi-Virrantaus/publication/228672324_Centrality_measures_and_vulnerability_of_spatial_networks/links/00b4952cc62958f70a000000/Centrality-measures-and-vulnerability-of-spatial-networks.pdf (Accessed 12/10/2024)

Duan, S., and Ayyub, B. M. 2019. Assessment methods of network resilience for cyber-human-physical systems. *ASCE-ASME. Journal of Risk Uncertainty Engineering System, Part A: Civil Engineering* 6(1): 03119001. https://ascelibrary.org/doi/10.1061/AJRUA6.0001021

Fayiah, M., and Roberto X. 2021. Seasonal Wildfire Outbreak Trend and Its Consequences on Forest Biodiversity and the Environment: A Case Study of Sierra Leone. https://ideas.repec.org/a/pkp/ijseer/v10y2021i2p69-84id2156.html (Accessed, 02/11/2022).

Federal Energy Regulatory Commission. 2023. Critical Energy/Electric Infrastructure Information (CEII). https://www.ferc.gov/ceii (Accessed, 11/19/2024).

Fund for Peace. 2024. Fragile States Index. https://fragilestatesindex.org/ (Accessed, 12/18/2024).

Gallagher Report. 2023. Four Key Threats to Critical Infrastructure. Building Resilience in an Increasing Interconnected World. https://www.ajg.com (Accessed, 01/16/2025).

Gartner Report. 2021. Press Release on Critical Infrastructure and Risk. www.gartner.com/en/newsroom (Accessed, 01/18/2025).

Hallegatte, S., Rentschler, J., and Rozenberg, J. 2019. *Lifelines: The Resilient Infrastructure Opportunity. Sustainable Infrastructure*. ©World Bank, Washington, DC. https://hdl.handle.net/10986/31805 License: CC BY 3.0 IGO.

Hevey, D. 2018. Network analysis: A brief overview and tutorial. *Health Psychology and Behavioral Medicine* 6(1): 301–328. https://doi.org/10.1080/21642850.2018.1521283 (Accessed, 01/18/2025).

Hines, P. D. H. 2010. Do Typological Models Provide Good Information about Vulnerability in Electric Power Networks? University of Vermont. https://www.researchgate.net/publicatiopn/47297656 (Accessed, 01/18/2025).

ICF. 2017. Ghana Integrated Resource and Resilience Planning Program: Risks and Resilience in Ghana's Electric Power System. USAID. https://ndcpartnership.org/knowledge-portal/good-practice-database/ghanas-integrated-resources-and-resilience-planning-program

Intergovernmental Panel on Climate Change (IPCC). 2014. AR5: Climate Change 2014: Impacts, Adaptation, and Vulnerability – Technical Summary. https://www.ipcc.ch/report/ar5/wg2/ (Accessed, 07/05/2025).

Intergovernmental Panel on Climate Change (IPCC). 2023. Sixth Assessment Report (AR). https://www.ipcc.ch/assessment-report/ar6/ (Accessed, 11/15/2024).

International Monitoring Fund (IMF). 2024. Sierra Leone Climate Public Investment Institutional Design. Technical Assistance Report. https://www.elibrary.imf.org/view/journals/019/2024/062/article-A001-en.xml (Accessed, 11/19/2024).

Michellier, C., Pigeon, P., Paillet, A., Trefon, T., Dewitte, O., and Kervyn, F. 2020. The challenging place of natural hazards in disaster risk reduction conceptual models: Insight from Central Africa and the European Alps. *International Journal of Disaster Risk Science* 11: 316–332. https://link.springer.com/article/10.1007/s13753-020-00273-y (Accessed, 01/21/2025).

Ministry of Energy (MoE). 2020. *Integrated Resource Plan (IRP).* Ministry of Energy Publication, Freetown, Sierra Leone.

Mobarak, A., Kevin, G., Niccoló, M., and Maarten, V. 2019. Electrifying Growth in Sierra Leone: Energy Access for Productive Use. https://www.theigc.org/project/electrifying-growth-in-sierra-leone-energy-access-for-productive-use/ (Accessed, 02/14/2022).

National Oceanic and Atmospheric Administration (NOAA). 2025. National Centers for Environmental information, Climate at a Glance: Global Time Series, published February 2025, retrieved on February 17, 2025 https://www.ncei.noaa.gov/access/monitoring/climate-at-a-glance/global/time-series (Accessed, 02/17/2025).

NDC. 2021. Sierra Leone Nationally Determined Contribution (NDC) Report to the United Nations Framework Convention on Climate Change (UNFCCC), Environment Protection Agency (EPA), Freetown, Sierra Leone.

ND-GAIN. 2024. Notre Dame Global Adaptation Initiative. Country Index Scores. https://gain.nd.edu/our-work/country-index/ (Accessed, 12/01/2024).

Neumann, J. E., and Price, J. C. 2009. Adapting to Climate Change: The Public Policy Response – Public Infrastructure. Resource for the Future. https://rosap.ntl.bts.gov/view/dot/17294 (Accessed, 02/19/2022).

Newman, M. 2018. *Networks*. 2nd Edition. University of Michigan, Oxford University Press, Oxford, UK.

Organization for Economic Cooperation and Development (OECD). 2021. Green Transition: Strengthening Climate Resilience in Developing Countries. What Are the Priorities? https://oecd-development-matters.org/2021/03/26/strengthening-climate-resilience-in-developing-countries-what-are-the-priorities/ (Accessed, 11/16/2024).

Ouedraogo, N.S. 2019. "Opportunities, Barriers and Issues with Renewable Energy Development in Africa: a Comprehensible Review." *Current Sustainable/Renewable Energy Reports* 6: 52-60. doi:https://doi.org/10.1007/s40518-019-00130-7

Our World in Data. 2024. Communicating the Best Research and Data to Understand Big Global Problems and Make Progress Against Them. https://ourworldindata.org/search?q=CO2%20emissions%20%20Sierra%20Leone (Accessed, 12/07/2024).

Ouyang, M. 2014. Review on modeling and simulation of interdependent critical infrastructure systems. *Reliability Engineering and System Safety* 121: 43–60. https://doi.org/10.1016/j.ress.2013.06.040 (Accessed, 07/02/2025).

Pagani, G. A. and Aiello, M. 2013. The power grid as a complex network. A survey. *Physica A* 392: 2688–2700. Elsevier. www.elsevier.com (Accessed, 01/17/2025).

Pant, M., Hall, J. W., and Blainey, S. P. 2016. Vulnerability assessment framework for interdependent critical infrastructures: Case-study for Great Britain's rail and electricity networks. *European Journal of Operational Research* 248(1): 273–286. https://doi.org/10.1016/j.ejor.2015.07.044 (Accessed, 07/03/2025).

Panteli, M., and Mancarella, P. 2015. Influence of extreme weather and climate change on the resilience of power systems: Impacts and possible mitigation strategies. *Electric Power Systems Research* 127: 259–270 https://www.researchgate.net/publication/279313514_Influence_of_extreme_weather_and_climate_change_on_the_resilience_of_power_systems_Impacts_and_possible_mitigation_strategies (Accessed, 07/04/2025).

Provitolo, D. 2011. Resiliencery Vulnerability Notion: Looking in Another Direction to Study Complex and Systemic Risks and Disasters. ResearchGate. https://www.researchgate.net/publication/280750163_Resiliencery_Vulnerability_Notion_Looking_In_Another_Direction_To_Study_Complex_And_Systemic_Risks_and_Disasters (Accessed, 01/15/2025).

Ranjan, R. 2025. Hazard, Vulnerability and Risk. Geography of Natural hazards and Disaster Management. https://ebooks.inflibnet.ac.in/geop15/chapter/hazard-vulnerability-and-risks/ (Accessed, 01/19/2025).

Republic of Sierra Leone. 2015. "National Renewable Energy Action Plan." /https://www.se4all-africa.org/fileadmin/uploads/se4all/Documents/Country_PANER/Sierra_Leone_National_Renewable_Energy_Action_Plans.pdf (Accessed 12/07/2024)

RMIT University. 2012. Ageing Wooden Power Poles Increase Risk of Fire. https://phys.org/news/2012-03-ageing-wooden-power-poles.html

SEforALL. 2021. Increasing Energy Access in Sierra Leone. https://www.seforall.org/system/files/2021-05/Energy-Access-SierraLeone-SEforALL.pdf

Sierra Leone Biennial Update Report (BUR). 2021. Environment Protection Agency (EPA) Sierra Leone Report, Freetown, Sierra Leone.

Sierra Leone Energy Climate Vulnerability Assessment (SLECVA). 2022. Sierra Leone Energy Climate Change Vulnerability Assessment and Applied Political Economy Analysis (APEA). Millennium Challenge Corporation (MCC), Washington DC, USA.

Sierra Leone Energy Compact. 2024. Project Document. Millennium Challenge Corporation (MCC), Washington DC, USA

Statistics Sierra Leone (SSL). 2022. Sierra Leone 2021 Mid-term Population and Housing Census. Thematic Report on Population Statistics Sierra Leone (SSL) https://www.statistics.sl/images/StatisticsSL/Documents/Census/MTPHC_Preliminary_Report/Final_Preliminary_Report_2021_MTPHC.pdf (Accessed, 11/10/ 2024).

United Nations Development Program (UNDP). 2019. Human Development Report 2019. Beyond Income, Beyond Averages, Beyond Today: Inequalities in Human Development in the 21st Century. https://hdr.undp.org/content/human-development-report-2019 (Accessed, 11/09/2024).

United Nations Office for Project Services (UNOPS). 2020. Access to Energy: Giving Sierra Leone the Power to Change Lives. https://www.unops.org/news-and-stories/stories/access-to-energy-giving-sierra-leone-the-power-to-change (Accessed, 02/19/2022).

United Nations Office of Disaster Risk Reduction (UNDRR). 2022. Principles for Resilient Infrastructure. https://doi.org/10.1163/26662531_00301_017 (Accessed, 11/09/2024). (Accessed, 11/09/2024).

United Nations. 2021. Multidimensional Vulnerability Index: Potential Development and Uses. Analysis and Recommendations. Committee for Policy Development (CPD), United Nations. https://www.un.org/development/desa/dpad/publication/world-economic-situation-and-prospects-2024/ (Accessed, 01/18/2025).

United Nations. 2024a. World Economic Situation and Progress (WESP). Statistical Annex. Department of Economic and Social Affairs (DESA), United Nations Publication. https://www.un.org/development/desa/dpad/publication/world-economic-situation-and-prospects-2024/ (Accessed, 01/18/2025).

United Nations. 2024b. Handbook on the Least Developed (LDC) Country. Committee for Development Policy (CDP) United Nations Publication. https://www.un.org/development/desa/dpad/publication/world-economic-situation-and-prospects-2024/ (Accessed, 01/18/2025).

United Nations. 2024c. Global Set of Climate Change Statistics and Indicators. Implementation Guidelines. Studies in Methods. Series M No. 11. Statistic Division, Department of Economic and Social Affairs, United Nations, New York.

United States Agency for International Development (USAID). 2018. Power Africa Transmission Roadmap to 2030: A Practical Approach to Unlocking Electricity Trade. https://policycommons.net/artifacts/18358909/power-africa-transmission-roadmap-to-2030/19259381/ (Accessed, 02/14/2022).

Vaia. 2024. Understanding Energy Infrastructure. https://www.vaia.com/en-us/explanations/environmental-science/energy-resources/energy-infrastructure/ (Accessed, 11/19/2024).

Welland, A. 2017. Education and the electrification of rural schools. *Smart Villages Initiative*, 1–64. https://e4sv.org/wp-content/uploads/2017/05/TR13-Education-and-the-Electrification-of-Rural-Schools_web-1.pdf (Accessed, 02/11/2022).

World Bank, 2016. Investing in Urban Resilience- *Protecting and promoting Development in a Changing World. https://www.worldbank.org/en/news/video/2016/10/13/investing-in-urban-resilience-protecting-and-promoting-development-in-a-changing-world.* (Accessed, 11/07/2024)

World Bank. 2018. Sierra Leone Multi-City Hazard Review and Risk Assessment Final Report (Volume 1 of 5): Technical Methodology and Summary of Results. World Bank. 68. https://documents1.worldbank.org/curated/en/821161549318730387/pdf/130797-v1-FInal-Report-Volume-1-of-5-Technical-Methodology-and-SoR.pdf (Accessed, 02/11/2022).

World Bank. 2019a. Resilience and Critical Power Infrastructure – Lessons learned from Natural Disasters and Future Research Needs. World Bank Climate Change Group. Global Facility for Disaster Reduction and Recovery, Policy Research Working Paper 8900 (Accessed, 01/19/2025).

World Bank. 2019b. Vulnerabilities of Networked Energy Infrastructure. A Primer. Lifeline. The Resilient Infrastructure Opportunities. Background Paper. Policy Research Working Paper 8901. www.worldbank.org (Accessed, 01/19/2025).

World Bank. 2019c. Sierra Leone Multi-City Hazard Review and Risk Assessment. https://www.worldbank.org/en/topic/disasterriskmanagement/brief/sierra-leone-multi-city-hazard-review-and-risk-assessment (Accessed, 01/19/2025).

World Bank. 2021. *Measuring the Climate Resilience of the Power Sector: Harmonization, Not Homogenization*. Washington, DC: World Bank. Retrieved from https://openknowledge.worldbank.org/bitstreams/e0a89f45-f205-4a5d-8321-d7dfdddd7c1e/download. (Accessed, 07/05/2025)

World Bank. 2024a. World Bank Climate Change Knowledge Portal (CCKP). https://climate knowledgeportal.worldbank.org/. (Accessed, 11/15/2024).

World Bank. 2024b. Sierra Leone: Beyond Connections Energy Access Diagnostic Report Based on the Multi-Tier Framework. Energy Sector Management Program (ESMAP). www.worldbank.org (Accessed, 12/01/2024).

World Economic Forum (WEF), 2023. Global Risk Report- Insight Report. https://www.weforum.org/report/global-risks-report (Accessed, 12/15/2024)

World Economic Forum (WEF), 2024. Global Risk Report- Insight Report. https://www.weforum.org/report/global-risks-report (Accessed, 01/19/2025)

World Economic Forum (WEF). 2025. Global Risk Report – Insight Report. https://www.weforum.org/report/global-risks-report (Accessed, 01/19/2025).

World Health Organization (WHO). 2015. Residential Heating with Wood and Coal: Health Impacts and Policy Options in Europe and North America, 1–50. The WHO Regional Office for Europe, Copenhagen. https://www.who.int/europe/publications/i/item/residential-heating-with-wood-and-coal--health-impacts-and-policy-options-in-europe-and-north-america Accessed, 02/11/2022)

3 Resilience Decision-Making under Climate Hazards for Complex and Substructured Infrastructure Systems

Julian Salomon[a], *Niklas R. Winnewisser*[a], *and Michael Beer*[a,b]

[a]Institute for Risk and Reliability, Leibniz University Hannover, Hannover, Germany

[b]Institute for Risk and Uncertainty, University of Liverpool, Liverpool, United Kingdom; International Joint Research Center for Resilient Infrastructure & International Joint Research Center for Engineering Reliability and Stochastic Mechanics, Tongji University, Shanghai, China

Infrastructure networks represent complex systems essential for maintaining modern societal functions. However, these critical systems increasingly face severe threats due to changing climatic conditions, including extreme weather events and long-term environmental shifts. Recent research emphasizes not only enhancing system robustness and reliability but also improving the capability of infrastructure to recover from disruptions. The holistic approach to addressing these combined challenges is captured by the concept of resilience. Yet, resilience assessment for today's complex infrastructures has become notably challenging. Due to their increased complexity, individual components often possess intricate internal structures, necessitating their treatment as interconnected subsystems. Consequently, innovative methods for efficient resilience analysis are required to effectively address these emerging challenges under evolving climatic conditions.

This work is based on Salomon et al. (2020) and Salomon et al. (2022), presenting an efficient resilience decision-making procedure for complex and substructured systems. A methodology is presented by bringing together two methods from the fields of reliability analysis and modern resilience assessment. A resilience decision-making framework and the concept of survival signature are extended and merged, providing an efficient approach for quantifying the resilience of complex, large, and

 DOI: 10.1201/9781003567332-3

substructured systems subject to monetary and technical restrictions. The approach combines both of the advantageous characteristics of its two original components: a direct comparison between various resilience-enhancing options from a multidimensional search space, leading to an optimal trade-off with respect to the system resilience and a significant reducing computational effort due to the separation property of the survival signature. Once a subsystem structure has been computed, any possible characterization of the probabilistic part can be validated with no need to recompute the structure.

The presented methods are applied to two substructured systems of increasing complexity, with the second being a substructured model of the Berlin's U- and S-Bahn system. In both cases, the methods provide accurate results and demonstrate their efficiency and general applicability.

3.1 INTRODUCTION

In today's highly developed societies, complex systems such as infrastructure networks, industrial plants, and jet engines are ubiquitous and of paramount importance to societal functioning. It is evident that these systems are exposed to a variety of harmful influences of natural, technical, and anthropogenic origin. Among these threats, climate-related hazards have become particularly significant, driven by extreme weather events and long-term environmental shifts, which increasingly challenge the robustness and reliability of infrastructure systems (Field 2012; Pörtner et al. 2022). At the same time, Punzo et al. (2020) highlighted that "It is an undeniable fact that modern day systems are more integrated, more interdependent, evolve at faster pace and, in a word, are more complex than the systems of the previous century [...]." Given this growing complexity combined with escalating climatic pressures, it becomes impractical to detect and prevent all potential adverse impacts proactively.

Therefore, new developments in engineering must emphasize not only monitoring and improving system robustness and reliability but also enhancing their ability to recover effectively from disruptions after adverse climatic events (Tran et al. 2017; Linkov et al. 2014). The concept of resilience encapsulates precisely these aspects, integrating analyses and optimization strategies for robustness, reliability, and recovery from both technical and economic perspectives (Cimellaro et al. 2010; Ayyub 2015; Fang et al. 2015). Applying resilience thinking to engineered systems thus leads to a paradigm shift: secure systems cannot rely solely on preventive strategies but must explicitly incorporate efficient recovery capabilities, particularly in response to climate-driven disturbances (Cimellaro 2016; Pant et al. 2018; Sharifi 2020).

In engineering, the concept of resilience has steadily gained popularity in recent years (Bergström et al. 2015; Patriarca et al. 2018; Punzo et al. 2020). The notion of "resilience" appears in various fields such as ecology, economics, and psychology, as well as in the context of mechanical systems and is derived from the Latin word "resilire," which means "to bounce back." The concept of resilience was first introduced by Holling in the field of ecological systems (Holling 1973). Although several other definitions by various scientists followed, most of them have certain key aspects in common that were already captured by Holling's early definition (Fiksel

2003; Little 2003; Hollnagel et al. 2006; Bruneau and Reinhorn 2007; Youn et al. 2011). Ayyub (2014) provided a literature review and developed a comprehensive definition of resilience in the context of complex systems based on the content of the Presidential Policy Directive (PPD) on critical infrastructure security and resilience (Presidential Policy Directive (PPD) 2013). His widely accepted definition provides a solid foundation for quantifying resilience.

Numerous options exist for improving the resilience of complex systems. However, resources are not unlimited, and resilience cannot be increased at will. Therefore, it is essential not only to be able to distinguish and weigh between a variety of different resilience-enhancing measures but also to consider their monetary aspects (Gilbert and Ayyub 2016; Fang and Sansavini 2017). Salomon et al. (2020) presented a method for identifying the most cost-effective allocation of resilience-enhancing investments by merging the resilience metric of Ouyang et al. (2012) and an adaptation of the systemic risk measure of Feinstein et al. (2017). Their approach allows for a direct comparison of the effects of heterogeneous controls on the resilience of a system over an arbitrary time period in a two-dimensional parameter space.

Additionally, current research in the context of resilience focuses on improved resilience quantification measures, as proposed in Singh et al. (2022), and overarching frameworks for stakeholder decision-making, for example, for transportation networks in the presence of seismic hazards (Alipour and Shafei 2022). For a comprehensive literature review on resilience assessment frameworks that balance resources and performance, see Capacci et al. (2022). Other researchers recently studied the complexity of realistic infrastructure systems, failure consequences, recovery sequences, and varying external effects. In Duan and Ayyub (2020), for example, the authors revealed the vast complexity of modern critical infrastructures and their multifactorial nature as cyber–human–physical systems and studied appropriate modeling and resilience analysis approaches. Furthermore, the works performed by Emanuel and Ayyub (2019) and Saadat et al. (2020) are concerned with the effects on decision-making when considering stakeholder preferences or enhancement and recovery strategies. External effects and challenges arising from climate change were studied in the context of resilience, for example, in Zhang et al. (2022).

Various technical and infrastructural systems in today's society are large and complex in nature. In particular, when system components have such complexity that they themselves need to be modeled as systems, so-called systems of systems (Nielsen et al. 2015; Jamshidi 2017), resulting in a significantly high number of components. This is in accordance to Batty, who highlights "A very simple definition of a complex system is a system that is composed of complex systems" (Batty 2016). As each of the subsystems affects the top-level system under consideration, this causes a significant increase in computational effort for system analysis and constitutes a major challenge (Kolowrocki 2014; Negi and Singh 2015). Therefore, it is particularly important to have tools capable of efficiently assessing all three resilience phases. Typically, the reliability phase involves the most system evaluations, in particular when various different system configurations and (environmental) scenarios need to be assessed that have an impact on the probability structure of the subsystems and thus on the overall system. Therefore, a particularly efficient analysis approach is required for this phase.

An efficient approach to modeling the reliability of systems with multiple component types is provided by the concept of survival signature, introduced and discussed in Coolen and Coolen-Maturi (2013, 2016). Its major benefit over conventional approaches is the separation of the system structure from the probabilistic properties of the system components. Once the system structure has been analyzed, any possible probabilistic characterization can be tested without having to reevaluate any system states. Consequently, this approach reduces the computational cost of repeated model evaluations typically required in design and maintenance processes (Patelli et al. 2017). Current research is focused on multistate components (Patelli et al. 2017), common cause failures (Feng et al. 2018), multiple failure modes and dependent failures (George-Williams et al. 2019), approximation techniques for large systems (Behrensdorf et al. 2021), and reliability analysis in consideration of imprecision (Salomon et al. 2021).

The challenges described above are addressed in Salomon et al. (2022) by (1) extending the resilience decision-making approach initially proposed in Salomon et al. (2020) to multidimensional parameter spaces and (2) establishing a comprehensive methodological framework. This framework comprises two central components: the extended resilience decision-making method and the survival signature concept. Together, these facilitate efficient and multidimensional resilience analyses of large-scale, complex, and substructured infrastructure systems. In this chapter, the methodology is demonstrated using two substructured systems of increasing complexity, the second of which is a detailed substructured representation of Berlin's U-Bahn and S-Bahn network.

3.2 RESILIENCE DECISION-MAKING

Assessing the resilience of complex systems under technical or economic constraints demands a sophisticated methodology capable of efficiently guiding optimal decision-making. Salomon et al. (2020) proposed such a versatile methodology, comprising three core elements: a dedicated metric for resilience quantification, an adapted systemic risk measure, and a computationally efficient grid search algorithm.

3.2.1 RESILIENCE QUANTIFICATION

A suitable quantitative measure of resilience constitutes a fundamental prerequisite for the resilience assessment of engineered systems. Comprehensive overviews on resilience metrics within a systemic context are provided by the authors in Bergström et al. (2015); Hosseini et al. (2016); Sun et al. (2020). While Bergström et al. (2015) emphasized resilience as an essential link between the increasing complexity of modern systems and their inherent risks, Sun et al. (2020) specifically addressed infrastructure resilience, highlighting its close association with functionality and performance measures. Hosseini et al. (2016) proposed a general classification scheme for resilience quantification approaches. According to their analyses, performance-based metrics constitute the most frequently utilized category. These metrics evaluate resilience by comparing system performance before and after disruptive events, distinguishing further subcategories depending on their time-dependency and characterization as deterministic or probabilistic processes.

According to Henry and Ramirez-Marquez (2012) and Hosseini et al. (2016), performance-based and time-dependent metrics explicitly consider the following system and transition states associated with a disruptive event:

1. The original stable state, which persists until a disruptive event occurs, primarily governed by the system's inherent reliability.
2. The vulnerability of the system, represented by performance losses following a disruptive event, and its robustness, counteracting these vulnerabilities and mitigating the performance losses. Both aspects depend on the degradation characteristics of system components.
3. The recoverability, characterized by the disrupted state of the system and the subsequent recovery toward a new stable state.

An illustration of these phases and transitions is provided in Figure 3.1. It should be noted that the performance level of the recovered stable state might differ from that of the original state.

The performance loss area between the original and the recovered stable state shown in Figure 3.1 refers to the well-established principle of the "*resilience triangle*," originally introduced by Bruneau et al. (2003) and depicted in Figure 3.2. In their seminal work, Bruneau et al. (2003) proposed a time-dependent, performance-based, and deterministic metric to quantify resilience losses experienced by a community after seismic events. Denoting by t_0 the time at which a disruptive event occurs, and by t_1 the time at which recovery is completed, and defining $Q(t)$ as the

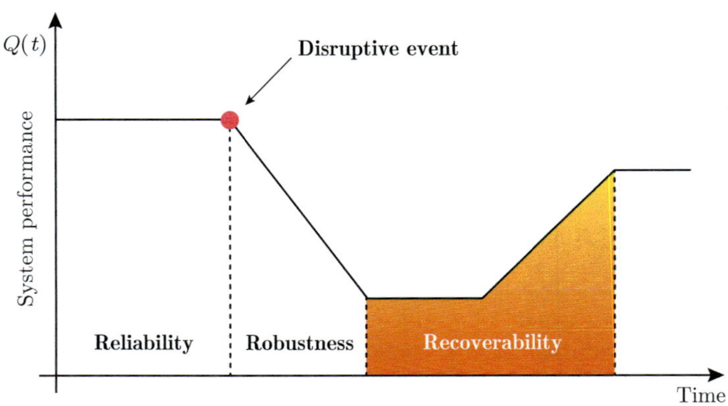

FIGURE 3.1 In the evolution of a system before and after the impact of a disruptive event, distinct phases can be identified: (i) original stable state, (ii) disruptive impact, vulnerability, robustness, and (iii) disrupted state and recovery; adapted from Henry and Ramirez-Marquez (2012).

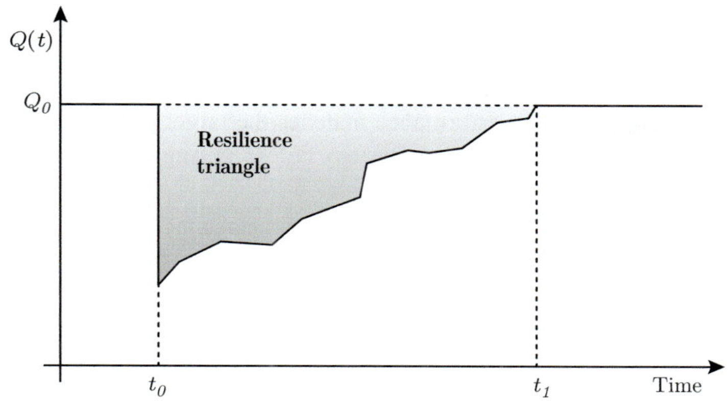

FIGURE 3.2 Resilience triangle. (Adapted from Bruneau et al., 2003.)

quality (or performance) of the community infrastructure at time t, the resilience loss metric is expressed as:

$$R_{Br} = \int_{t_0}^{t_1} [100 - Q(t)]\,dt. \tag{3.1}$$

Here, system performance is evaluated against a time-independent ideal performance level of 100 within the specified performance loss interval. This foundational concept has significantly influenced several subsequent resilience metrics developed across various contexts (see e.g., Zobel 2011; Adams et al. 2012; Sahebjamnia et al. 2015).

Salomon et al. (2020) utilized the probabilistic, time-dependent resilience metric developed by Ouyang et al. (2012). This metric quantifies resilience as the expected ratio between the integral of the actual system performance $Q(t)$ over the time interval from 0 to a given time T and the integral of a corresponding target system performance $\mathrm{T}Q(t)$ across the same interval:

$$\mathrm{Res} = E[Y] \tag{3.2}$$

where

$$Y = \frac{\int_0^T Q(t)\,dt}{\int_0^T \mathrm{T}Q(t)\,dt} \tag{3.3}$$

In this formulation, the actual system performance $Q(t)$ is treated as a stochastic process. Generally, the target performance $\mathrm{T}Q(t)$ can also be modeled as a stochastic process; however, for simplicity, it is often assumed to be a constant, nonrandom value T$Q(t)$. Under the assumption that actual system performance does not exceed the target, the resilience metric ranges from 0 to 1. A metric value Res = 1 indicates that the actual system performance matches the target performance perfectly, whereas Res = 0 signifies a complete system failure during the entire observation period.

3.2.2 Adapted Systemic Risk Measure

Feinstein et al. (2017) initially proposed a general approach for measuring systemic risk, widely utilized, for example, in financial contexts (Weber and Weske 2017). In Salomon et al. (2020), this systemic risk measure was adapted and further extended for engineering applications, as summarized in the following section. The adapted measure comprises two main elements: a descriptive input–output model and a clearly defined acceptance criterion that reflects normative resilience standards set by regulatory authorities.

Consider a given system composed of m components $i \in \{1,\ldots,m\}$, each associated with a specific component type $k_i \in \{1,2,\ldots,K\} \subseteq \mathbb{N}$, and characterized by e properties influencing the overall system performance $Q(t)$. These properties, hereafter referred to as "endowment properties," directly impact system resilience and may be enhanced through appropriate capital allocation. Each component i is thus represented by:

$$(\mathbf{a}_i;k_i) = (\eta_{i1},\eta_{i2},\ldots,\eta_{ie};k_i) \in \mathbb{R}^{(1\times e)} \times \mathbb{N}, \tag{3.4}$$

where the vector $(\eta_{i1},\eta_{i2},\ldots,\eta_{ie})$ contains the numerical values of the respective endowment properties. Consequently, the entire system configuration can be described by a tuple combining the matrix $\mathbf{A} \in \mathbb{R}^{(m\times e)}$ and the column vector $\mathbf{z} \in \mathbb{N}^m$ representing the component types:

$$(\mathbf{A};\mathbf{z}) = \begin{pmatrix} \eta_{11} & \eta_{12} & \cdots & \eta_{1e}; & z_1 \\ \eta_{21} & \eta_{22} & \cdots & \eta_{2e}; & z_2 \\ \vdots & \vdots & & \vdots & \vdots \\ \eta_{m1} & \eta_{m2} & \cdots & \eta_{me}; & z_m \end{pmatrix} \tag{3.5}$$

The system under consideration is defined through a descriptive, non-decreasing input–output model $Y = Y_{(\mathbf{A};\mathbf{z})}$, which directly relates endowment properties to the system's overall performance. With reference to Eq. (3.2), the model output Y explicitly depends on the current allocation of endowment properties $(\mathbf{A};\mathbf{z})$.

Additionally, the following specific acceptance set is defined:

$$A = \{X \in X \mid E[X] \geq \alpha\} \tag{3.6}$$

where X represents a normalized model output, $E[X]$ is its expected value, and $\alpha \in [0,1]$. Correspondingly, the adapted systemic risk measure is formulated as:

$$R(Y) = \left\{ \mathbf{A} \in \mathbb{R}^{m \times e} \mid Y_{(\mathbf{A};\mathbf{z})} \in A \right\} \tag{3.7}$$

representing the set of all endowment property allocations **A** for which the system achieves or exceeds the resilience threshold α.

In practical applications, structural constraints may need to be imposed on the matrix in Eq. (3.5). For instance, it may be necessary that components of a specific type share identical configurations, implying that the respective row vectors $\mathbf{a}_i$ must be equal. Feinstein et al. (2017) addressed such constraints through monotonically increasing mapping functions $g_z : \mathbb{R}^p \to \mathbb{R}^{(m \times e)}$, $a' \mapsto (\mathbf{A};\mathbf{z})$, where $\mathbf{z} \in \mathbb{R}^m$ denotes the component types. These functions allow parameter sets $a' \in \mathbb{R}^p$ of lower dimension to be efficiently mapped onto the complete system description as shown in Eq. (3.5).

3.2.3 Grid Search Algorithm and the Curse of Dimensionality

Feinstein et al. (2017) and Salomon et al. (2020) proposed to evaluate the systemic risk measure using a combination of grid search algorithms and stochastic simulations. The grid search algorithm systematically explores the space of possible endowment properties, whereas stochastic simulations quantify the resilience corresponding to each endowment allocation generated by the grid search. Specifically, the probabilistic resilience metric (Eqs. 3.2 and 3.3) is estimated through Monte Carlo simulation. According to Feinstein et al. (2017), the grid search algorithm operates in two primary phases:

1. A search conducted along the main diagonal of the endowment property space until the first allocation meeting the acceptance criterion defined by the adapted systemic risk measure is identified.
2. Determination of the Pareto front between the set of acceptable endowments $R(Y)$ and its complementary set $R(Y)^c$, starting from the initial acceptable allocation identified.

The algorithm efficiently computes the full extent of $R(Y)$, substantially reducing computational costs due to the monotonicity assumption inherent to the input–output model $Y_{(\mathbf{A};\mathbf{z})}$, as described in Section 3.2.2. A comprehensive description of the two-dimensional implementation of this algorithm can be found in Chapter 4 of Feinstein et al. (2017).

In Salomon et al. (2020), this algorithm was integrated into the resilience decision-making methodology and successfully applied to case studies involving two-dimensional parameter spaces. Feinstein et al. (2017) explicitly note that although their grid search algorithm can theoretically be extended to higher-dimensional problems, doing so entails significantly increased computational time and memory requirements. However, practical resilience analyses of real-world technical systems typically involve numerous influencing factors, inherently resulting in high-dimensional parameter spaces. To address this issue, Salomon et al. (2022) developed an extension of the previously proposed methodology, enabling resilience decision-making for n-dimensional parameter spaces. This advancement constitutes an integral part of the methodological framework described in Section 3.4.

3.3 CONCEPT OF SURVIVAL SIGNATURE

The concept of survival signature, introduced by Coolen and Coolen-Maturi (2013), provides an efficient method to compute the survival function of complex systems comprising multiple component types. Over the past decade, the survival signature has attracted considerable interest due to its notable advantages. A key merit lies in its computational efficiency during repeated system evaluations, achieved by clearly separating topological reliability from the probabilistic structure of component failures. Simultaneously, this concept drastically condenses information on system topology. Components belong to the same type if their failure times are assumed independent and identically distributed (*iid*) or exchangeable; a distinction particularly relevant when modeling dependent failures (Coolen and Coolen-Maturi 2016). A concise overview of the survival signature concept is provided below; detailed derivations and further applications can be found in Coolen and Coolen-Maturi (2013, 2016); Feng et al. (2016).

3.3.1 Structure Function

Consider a system consisting of m components initially assumed of a single type. The state of each component i is represented by x_i, where $x_i = 1$ indicates functioning and $x_i = 0$ non-functioning. Thus, the state vector is defined as $\mathbf{x} = (x_1, x_2, \ldots, x_m) \in \{0,1\}^m$. The structure function $\phi : \{0,1\}^m \to \{0,1\}$ maps the state vector to the overall system status, where $\phi(\mathbf{x}) = 1$ indicates system functionality, and $\phi(\mathbf{x}) = 0$ represents system failure.

If the system comprises multiple component types ($K \geq 2$), the total number of components becomes $m = \sum_{k=1}^{K} m_k$, with m_k denoting the number of components of type k. Accordingly, the state vector for each component type k is expressed as $\mathbf{x}^k = \left(x_1^k, x_2^k, \ldots, x_{m_k}^k\right)$.

3.3.2 Survival Signature

The survival signature summarizes the probability of system functionality based solely on the number of functioning components l_k per component type $k \in \{1, 2, \ldots, K\}$.

Assuming the component failure times within each type are *iid* or exchangeable, the survival signature is formally defined as:

$$\Phi(l_1,l_2,\ldots,l_K)=\left[\prod_{k=1}^{K}\binom{m_k}{l_k}\right]^{-1}\times\sum_{\mathbf{x}\in S_{l_1,l_2,\ldots,l_K}}\phi(\mathbf{x}) \tag{3.8}$$

where $\binom{m_k}{l_k}$ is the total number of state vectors $\mathbf{x}^k$ for type k, and $S_{l_1,l_2,\ldots,l_K}$ denotes the set of all system state vectors satisfying $l_k=\sum_{i=1}^{m_k}x_i^k$. Consequently, the survival signature encapsulates only the topological reliability, independent of the specific time-dependent failure behaviors of components described in Section 3.3.3. Further discussions about the validity and practical implications of component exchangeability can be found in Coolen and Coolen-Maturi (2016); Salomon et al. (2021).

3.3.3 Probability Structure

The probability structure defines the probability of having exactly l_k functioning components of each type k at time t. Let $C_k(t)\in\{0,1,\ldots,m_k\}$ represent the number of functioning components of type k at time t, and let $F_k(t)$ denote the cumulative distribution function (CDF) describing the failure times of components of type k. Then, the probability structure of the system, independent of its topology, is given by:

$$\begin{aligned}P\left(\bigcap_{k=1}^{K}\{C_k(t)=l_k\}\right)&=\prod_{k=1}^{K}P(C_k(t)=l_k)\\&=\prod_{k=1}^{K}\binom{m_k}{l_k}[F_k(t)]^{m_k-l_k}[1-F_k(t)]^{l_k}\end{aligned} \tag{3.9}$$

3.3.4 Survival Function

The survival function describes the probability of the system functioning at a given time t. Integrating the results from Sections 3.3.2 and 3.3.3, the survival function is defined as:

$$P(T_s>t)=\sum_{l_1=0}^{m_1}\cdots\sum_{l_K=0}^{m_K}\Phi(l_1,l_2,\ldots,l_K)\times P\left(\bigcap_{k=1}^{K}\{C_k(t)=l_k\}\right) \tag{3.10}$$

where T_s denotes the random time to system failure. Notably, the survival signature distinctly separates the time-independent topological information from the time-dependent probability structure. Hence, after initial calculation in a preprocessing step, the survival signature can be efficiently reused to evaluate the survival function across various scenarios involving different probabilistic configurations with unchanged system topology. This survival signature is typically stored as a matrix, thus avoiding repeated, computationally costly evaluations of the structure function. These attributes confer significant computational advantages over traditional methods, particularly when repeated simulations of system configurations are necessary (Patelli et al. 2017). For large-scale systems, Monte Carlo simulation provides an efficient approximation method for computing the survival signature (Behrensdorf et al. 2021).

3.4 RESILIENCE DECISION-MAKING FOR SUBSTRUCTURED SYSTEMS

This section illustrates the resilience decision-making methodology for substructured systems proposed in Salomon et al. (2022). The approach integrates the survival signature concept described in Section 3.3 into the resilience decision-making framework summarized in Section 3.2. First, the system preparation via a formalized substructuring approach is introduced. Second, the augmented methodology is outlined.

3.4.1 Definition of Substructured Systems

Consider a substructured system S composed of subsystems and components, structured across $L \geq 1$ hierarchical levels, where only components exist at level $L+1$. Components are directly associated with probability distributions describing their time-dependent behavior. Level 1 represents the overall system. The substructuring approach is illustrated in Figure 3.3.

At each level $v = 1, 2, \ldots, L$, the system consists of n^v subsystems S_j^v, $j = 1, \ldots, n^v$, and m^v components C_i^v, $i = 1, \ldots, m^v$. Information regarding component states propagates from level $L+1$ upwards to determine the overall system state s^0, corresponding to the system performance $Q(t)$ underlying the resilience measure Res in Eq. (3.2). The (sub)system state s_j^v at level v is related to the state vector $\mathbf{x}_j^v$ via the structure

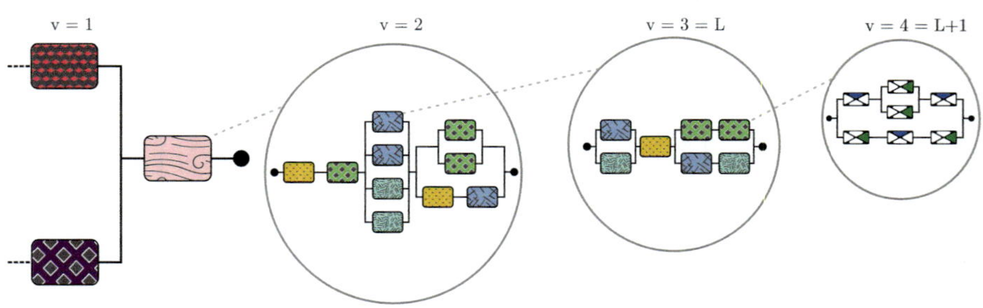

FIGURE 3.3 Illustration of the substructuring concept.

function $s_j^v = \phi_j^v(\mathbf{x}_j^v) \in \{0,1\}$. The vector $\mathbf{x}_j^v$ includes subsystem and component states from the next lower level $w = v+1$:

$$\mathbf{x}_j^v = \left(s_1^w, s_2^w, \ldots, s_{n_j^w}^w, c_1^w, c_2^w, \ldots, c_{m_j^w}^w\right)$$

where subsystem states s_p^w and component states c_q^w are binary $(\{0,1\})$. At the lowest subsystem level L, state vectors contain only component states.

Component failure probabilities are described by known CDF $F_k(t)$, per component type k, as in Section 3.3.3. Binary states $(\{0,1\})$ follow directly from the binary-state formulation of the survival signature; ongoing research explores extensions to multistate scenarios and continuous-state scenarios (Eryilmaz and Tuncel 2016; Liu et al. 2018; Yi et al. 2021; Qin and Coolen 2022; Winnewisser et al. 2023).

3.4.2 Extension of the Adapted Systemic Risk Measure

In substructured system resilience analysis, it is beneficial to assign endowments not only to individual components but also to subsystems. Therefore, the adapted systemic risk measure from Section 3.2.2 is extended.

Consider a system with m components and n subsystems, each subsystem j characterized by d endowment properties:

$$(S_j; b_j) = (\xi_{j1}, \xi_{j2}, \ldots, \xi_{jd}; b_j) \in \mathbb{R}^{(1\times d)} \times \mathbb{N}$$

where b_j specifies the subsystem type. The subsystem configuration is captured by the tuple $(\mathbf{D}; \mathbf{h})$, with matrix $\mathbf{D} \in \mathbb{R}^{(n\times d)}$ and vector $\mathbf{h} \in \mathbb{N}^n$:

$$(\mathbf{D}; \mathbf{h}) = \begin{pmatrix} \xi_{11} & \xi_{12} & \cdots & \xi_{1d}; & h_1 \\ \xi_{21} & \xi_{22} & \cdots & \xi_{2d}; & h_2 \\ \vdots & \vdots & & \vdots & \vdots \\ \xi_{n1} & \xi_{n2} & \cdots & \xi_{nd}; & h_n \end{pmatrix}$$

The overall system is thus defined by an input–output model $Y = Y_{(\mathbf{A};\mathbf{z}),(\mathbf{D};\mathbf{h})}$, dependent on component and subsystem endowment configurations. Using acceptance set A from Eq. (3.6), the extended systemic risk measure becomes:

$$R(Y) = \left\{\mathbf{A} \in \mathbb{R}^{m\times e}, \mathbf{D} \in \mathbb{R}^{(n\times d)} \mid Y_{(\mathbf{A};\mathbf{z}),(\mathbf{D};\mathbf{h})} \in A\right\}$$

This allows incorporating diverse performance-influencing endowments into resilience analyses.

3.4.3 Expansion of Resilience Analysis

The system resilience *Res* depends on reliability, robustness, and recoverability. Reliability typically imposes significant computational demands; thus, efficiency is enhanced by incorporating the survival signature and substructuring.

In a preprocessing step, survival signatures for each subsystem are computed from structure functions (Section 3.3.2), capturing topological reliability independent of probability structures. This greatly facilitates repeated online evaluations required by resilience assessments across varying endowment scenarios.

The computational complexity of resilience analyses depends on repeatedly evaluating the model $Y_{(\mathbf{A};\mathbf{z}),(\mathbf{D};\mathbf{h})}$ via stochastic simulations and grid search (Section 3.2.3). Utilizing survival signatures, the evaluation of subsystem reliability significantly reduces complexity by avoiding direct computation of numerous component-level structure functions.

Subsystem reliabilities propagate upwards via survival functions (Eq. 3.10), using failure rates derived from cumulative hazard functions:

$$\lambda^{s_j^1}(t) = -\frac{d \ln P(T_{s_j^1} > t)}{dt} \tag{3.11}$$

Thus, online evaluations of resilience become computationally tractable, requiring simulation of significantly fewer elements. Importantly, subsystems of identical types share survival signatures, further enhancing efficiency.

This structured resilience analysis, augmented by survival signatures and substructuring, enables assessing realistically complex systems with substantially reduced computational resources. To illustrate its practical efficiency and general applicability, this approach is applied to a complex illustrative example in Section 3.5 and to Berlin's metro system in Section 3.6.

The detailed algorithmic implementation is outlined below:

Algorithm 3.1

Step A Compute survival signatures for all subsystems S_j^v, $v = 1,\ldots, L$,.

Step B Identify the Pareto front via the grid search algorithm:

Step B1 Compute subsystem and component failure rate matrices based on Eq. (3.11).

Step B2 Perform N stochastic simulations for $t_h = 0$:

a. Evaluate system performance $Q(t_h)$.
b. Sample subsystem and component failures based on failure rate matrices.
c. Check recovery status and reset failure rates as necessary.
d. Increment t_h until reaching the final time T.

Step B3 Compute *Res* via Eqs. (3.2) and (3.3).

The presented methodology has been implemented in the publicly accessible Julia package *ResilienceDecisionMaking.jl* (Behrensdorf et al. 2022), proving its efficiency and practical applicability. Illustrative applications to a complex hypothetical system and the Berlin metro system are discussed in Sections 3.5 and 3.6.

3.5 ARBITRARY COMPLEX SYSTEM

In Patelli et al. (2017) and Salomon et al. (2021), the authors demonstrated their proposed reliability simulation methods using an arbitrary complex system. To illustrate the broad applicability and efficiency of the methodology presented herein, this system is adapted through substructuring and analyzed for resilience decision-making.

3.5.1 Model

The considered complex system comprises $n = 14$ subsystems assigned to $B = 6$ subsystem types. Figure 3.4 illustrates the system configuration and subsystem-to-type assignments. The overall system is deemed functional if a continuous path from the start node to the end node exists; otherwise, it is considered nonfunctional. This binary definition of functionality formally represents a reliability block diagram (RBD) and defines the system performance $Q(t)$, serving as the basis for resilience decision-making. For clarity and simplicity, it is assumed that only one

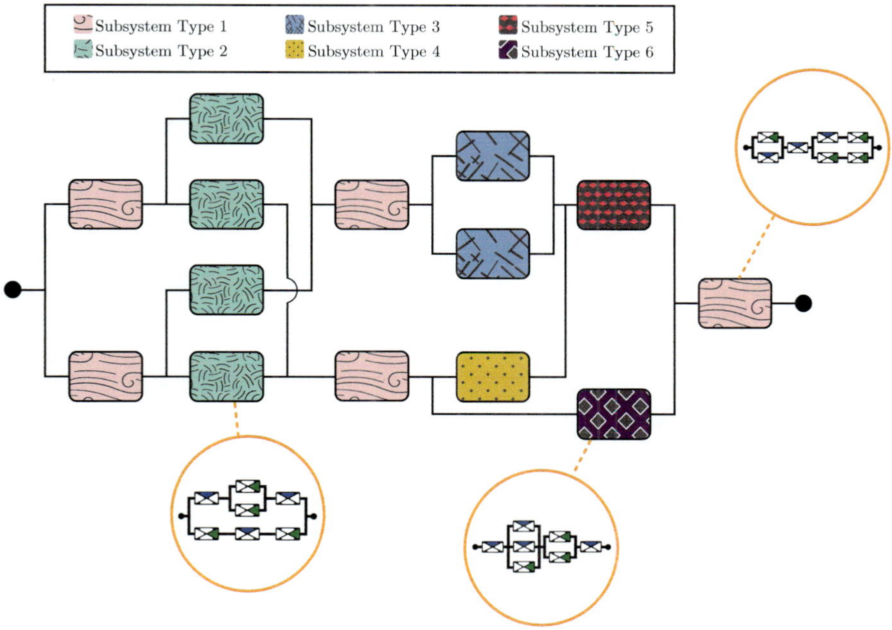

FIGURE 3.4 Representation of the complex system with 14 subsystems. (Adapted from Patelli et al., 2017.)

level of subsystems is present $(l = L = 1)$, hence $\mathbf{x}^s = (s_1, s_2, \ldots, s_{14})$, $S_j^1 = S_j$, and $\lambda^{s_j^1}(t) = \lambda^{s_j}(t)$.

The structure of the six subsystem types, also represented as RBDs, is depicted in Figure 3.5. Each subsystem of the same type shares identical structure. A subsystem S_j is regarded as operational if a continuous path between the start and end points exists; otherwise, it is nonoperational, that is, $s_j \in \{0,1\}\ \forall j \in \{1,\ldots,14\}$. Depending on their type, the subsystems consist of seven to ten components, resulting in a total of $m = 106$ individual components.

Components are classified into $K = 2$ types, with 50 components of type 1 and 56 components of type 2. For resilience analysis, each component is characterized by a single endowment property, the reliability improvement *rel*, fully describing a component as:

$$(a_i; k_i) = (\mathrm{rel}_i; k_i).$$

The reliability improvement rel_i solely depends on the component type, thus for components of the same type, $\mathrm{rel}_i = \mathrm{rel}_{i'}$ if $k_i = k_{i'}$.

Moreover, each component type is characterized by a specific time-dependent failure behavior. In practice, the underlying probability distribution functions describing this behavior must be derived from operational data, which can be challenging due

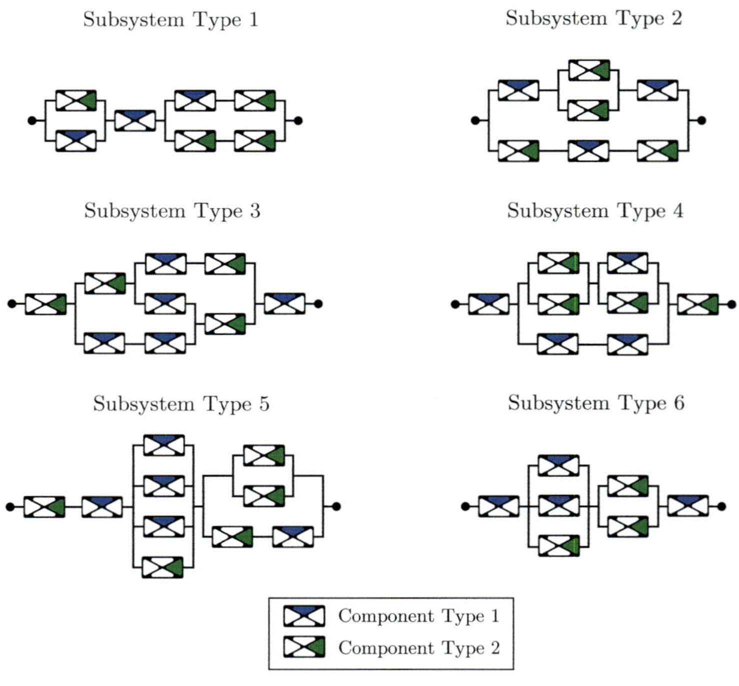

FIGURE 3.5 Structures of the $B = 6$ subsystem types within the complex system.

to data scarcity, measurement uncertainty, or subjective expert judgments. Recent advances, such as those presented in Salomon et al. (2021), facilitate efficient propagation of uncertainty within the survival signature framework, which will be integrated into the methodology in future work. However, for this illustrative case, exponential distributions are considered for both component types:

$$F_i\left(t;\lambda_i\left(\text{rel}_i\right)\right)=1-e^{-\lambda_i(\text{rel}_i)t}\quad \text{for } t\geq 0, \tag{3.12}$$

with the component-specific failure rate:

$$\lambda_i\left(\text{rel}_i\right)=\lambda_{i,\max}-\Delta\lambda_i\cdot\text{rel}_i \tag{3.13}$$

where $\lambda_{i,\max}$ denotes the maximum failure rate of component i of type k, and $\Delta\lambda_i$ represents the constant decrease in failure rate per reliability improvement rel_i.

The simulation procedure can be summarized as follows: After computing the system performance at time t_h, each subsystem S_j may fail randomly according to the extracted time-dependent failure rate $\lambda^{s_j}\left(t_h\right)$, see Eq. (3.1). A failed subsystem no longer contributes to system performance until it has fully recovered. The probability of subsystem S_j failing within the interval $\left(t_h,t_{h+1}\right)$ is:

$$P\left\{S_j \text{ fails during } \left(t_h,t_{h+1}\right)\right\}=\Delta t\cdot\lambda^{s_j}\left(t_h\right) \tag{3.14}$$

Upon failure, subsystems undergo immediate and complete restoration after r time steps, corresponding to a one-step recovery profile. It is assumed that repaired subsystems, along with their components, return to their original "as-new" condition. This assumption simplifies the demonstration but can differ in practical applications, where further endowment properties might influence recovery duration and quality. After recovery, the subsystem's survival function is reset, resulting in a failure rate per simulation step $\lambda^{s_j}\left(t_h\right)$ that again matches that of a new subsystem.

3.5.2 Costs of Endowment Properties

Optimal endowment properties are closely related to the quality of system components, and an increase in production quality typically results in rising costs. This economic aspect is essential to consider in resilience decision-making. According to Mettas (2000), increasing component reliability in complex networks often leads to an exponential increase in associated costs. Enhancing the endowment property *reliability improvement* effectively reduces the failure rate of individual components and, consequently, the corresponding subsystems. To model this relationship, an exponential cost function is assumed. Thus, the total costs associated with the reliability improvements of the components are defined as:

$$\text{cost}_{(\mathbf{A};\mathbf{z})}=\text{cost}^{\text{rel}}=\sum_{i=1}^{106}\text{price}^{\text{rel}}_{(\text{rel}_i;k_i)}\cdot 1.2^{(\text{rel}_i-1)} \tag{3.15}$$

where $(\text{rel}_i;k_i)$ represents the reliability improvement value of component i, k_i denotes its corresponding component type, and $\text{price}^{\text{rel}}_{(\text{rel}_i;k_i)}$ is an arbitrary but fixed base price assigned per component type.

3.5.3 Scenario

Table 3.1 provides the detailed scenario settings, encompassing model and simulation parameters used to apply the proposed resilience decision-making approach to the arbitrary complex system presented in Figure 3.4. The recovery period is assumed constant at $r = 20$ time steps for all subsystems, independent of subsystem type. Reliability improvement levels rel_i for each component are explored across the discrete range $\text{rel}_i \in \{1,\ldots,10\}$ for both component types.

In a preliminary step, the survival signatures of all 14 subsystems are computed. Table 3.2 provides, exemplarily, the survival signature for subsystem type $b_j = 5$. For clarity, only nontrivial survival signature values (i.e., those differing from zero and one) are listed. Subsequently, the resilience analysis proceeds in two steps: Firstly, the set of all acceptable endowment configurations $R(Y)$, satisfying a resilience criterion of $\text{Res} \geq 0.90$, is determined via Algorithm 3.1. Secondly, due to the

TABLE 3.1
Parameter Values for the Resilience Decision-Making Analysis of the Arbitrary Complex System

Parameter	Scenario
Acceptance threshold α	0.90
Number of time steps u	200
Length of time step Δt	0.05
Maximum analysis time T	10
Maximum failure rate $\lambda_{i,\max}$	0.15 for component type $k_i = 1$
	0.20 for component type $k_i = 2$
Failure rate reduction $\Delta\lambda_i$	0.014 for component type $k_i = 1$
	0.019 for component type $k_i = 2$
Reliability improvement rel_i	$\{1,2,\ldots,10\}$ for component types $k_i \in \{1,2\}$
Reliability improvement base price $\text{price}^{\text{rel}}_{(\text{rel}_i;k_i)}$	1,000€ for component type $k_i = 1$
	2,000€ for component type $k_i = 2$
Recovery time steps r	20
Number of stochastic simulations N	500

TABLE 3.2
Nontrivial Survival Signature Values $\Phi(l_1, l_2)$ for Subsystems of Type $b_j = 5$, as Depicted in Figures 3.4 and 3.5

l_1	l_2	$\Phi(l_1,l_2)$
2	4	1/25
3	3	3/50
2	5	3/25
4	3	3/20
3	4	9/50
2	6	1/5
5	3	11/50
6	3	3/10
3	5	3/10
4	4	33/100
3	6	2/5
5	4	23/50
4	5	12/25
6	4	3/5
4	6	3/5
5	5	16/25
6	5	4/5
5	6	4/5

associated costs of improving component quality, the most cost-effective acceptable configuration $\hat{\mathbf{A}}$ within $R(Y)$ is identified.

Figure 3.6 visualizes the grid search results, indicating reliability improvement configurations within the set $R(Y)$ that achieve a satisfying resilience performance. The results clearly reveal that reliability improvements for components of type $k_i = 1$ exert a greater influence on overall system resilience compared to improvements for type $k_i = 2$. For instance, at maximum improvement levels for type 1 components ($\text{rel}_i = 10$), minimal improvements ($\text{rel}_i = 2$) for type 2 components are sufficient to fulfill the acceptance criterion. Conversely, even maximal improvement for type 2 ($\text{rel}_i = 10$) requires at least moderate improvements ($\text{rel}_i = 4$) for type 1 components to satisfy the criterion.

The plausibility of these outcomes arises from the subsystem topologies (see Figure 3.5). Type 1 components occupy significantly more bottleneck positions within subsystems than type 2, resulting in a higher impact on subsystem and overall system resilience. Although type 2 components benefit from greater failure rate reductions per improvement step, their lower topological importance limits their overall influence.

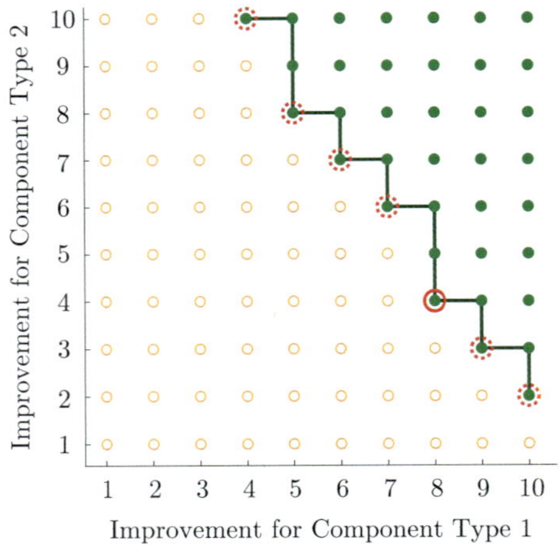

FIGURE 3.6 Results of the 2D grid search for reliability improvement values in the complex system.

Incorporating economic constraints, the most cost-efficient acceptable endowment configuration $\hat{\mathbf{A}}$ consists of reliability improvements $\text{rel}_i = 8$ for type 1 and $\text{rel}_i = 4$ for type 2, with total costs calculated by Eq. (3.15) as $\text{cost}_{(\hat{\mathbf{A}};\mathbf{z})} = 372{,}695\,€$. Due to the computational efficiency gained through the proposed method and the grid search, only 23% of all possible configurations had to be evaluated, significantly reducing computational efforts, and demonstrating scalability for more complex systems.

3.6 *U-BAHN* AND *S-BAHN* SYSTEM OF BERLIN

Berlin's subway (*U-Bahn*) and suburban rail (*S-Bahn*) systems transport about two-thirds of the city's 1.5 billion annual passengers (Berliner Verkehrsbetriebe 2018; S-Bahn Berlin 2018). Consequently, these transportation systems constitute the backbone of public mobility in Berlin, significantly influencing social and economic activities within the German capital. Given their critical role, ensuring the resilience of these infrastructure networks – especially in the face of increasing climatic threats such as extreme weather events, flooding, heatwaves, and storms due to climate change – is of paramount importance. To demonstrate the versatility and efficiency of the presented resilience decision-making methodology, it is applied to a detailed model of the combined Berlin U-Bahn and S-Bahn network, hereafter referred to as the "metro system." The goal of this application is to identify suitable resilience-enhancing endowment properties for each station within the system under given budgetary constraints. This approach systematically determines acceptable endowments that optimize system resilience with respect to reliability, robustness, and recoverability, considering anticipated increases in climate-related disruptions. The presented method can be implemented during any life-cycle phase, including the planning and design phases, to proactively enhance the resilience of both existing and future transportation infrastructure under changing climatic conditions.

3.6.1 Model

Berlin's U-Bahn and S-Bahn constitute a highly interconnected network unified through numerous shared stations. Following the modeling strategy proposed by Zuev et al. (2015) and further utilized in Salomon et al. (2020), these networks are jointly considered as a single integrated metro system.

Zhang et al. (2013) introduced a methodology for mapping metro networks onto topological graphs, which is utilized here. Accordingly, the Berlin metro system is represented by a topological network composed of 306 nodes (representing metro stations) interconnected by 350 edges (connections between stations). For simplicity and computational manageability, parallel connections are represented as single undirected edges. This simplified network representation is shown in Figure 3.7.

In complex infrastructures, system-level performance critically depends on the operational states of its components, which themselves often depend on the states of further subcomponents. Consequently, an essential modeling challenge is determining the appropriate granularity of system decomposition.

The proposed resilience decision-making framework explicitly addresses this issue through a systematic, substructuring-based propagation of reliability characteristics. In this case study, each metro station is modeled as an individual subsystem

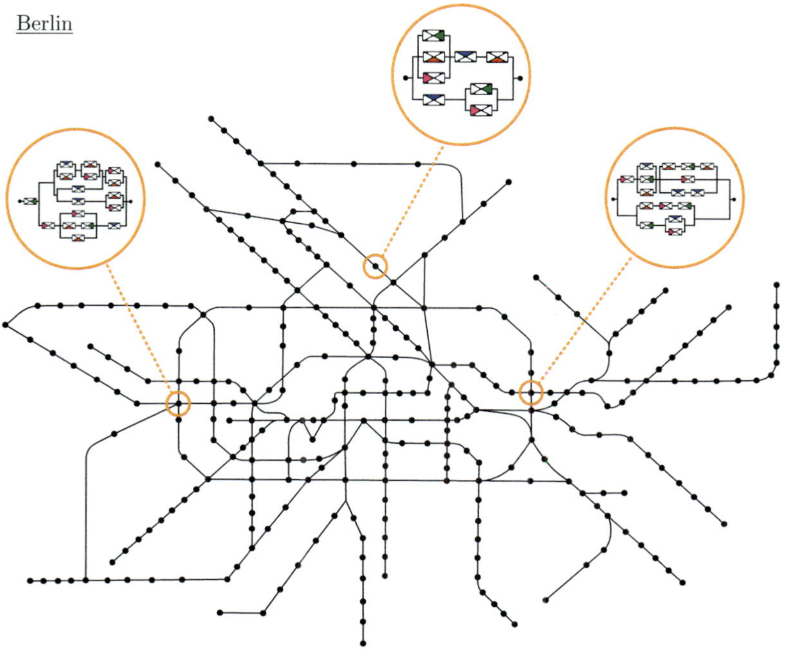

FIGURE 3.7 Topological network representation of the Berlin metro system.

characterized by its own reliability structure, see Figure 3.7. For clarity and illustrative purposes, a single subsystem level is assumed ($L = 1$), yielding subsystem definitions $S_j^1 = S_j$, state vector $\mathbf{x}^s = (s_1, s_2, \ldots, s_{306})$, and corresponding subsystem-specific failure rates $\lambda^{s_j^1}(t) = \lambda^{s_j}(t)$, where $n = 306$ subsystems (stations) are considered.

In practice, a metro station's functionality depends on multiple subcomponents, including structural elements (e.g., stairs, platforms, columns, and rails) and electrical systems (e.g., power supply, escalators, elevators, lighting, and communication systems). These can be further subdivided based on functionality relevant to rail operations and passenger accessibility. For demonstration purposes, the case study restricts its analysis to reliability modeling at the subsystem (station) level. Each subsystem's reliability structure is represented by a RBD. Consequently, station functionality is binary, where subsystem S_j is considered functional if a continuous path exists between subsystem start and end nodes, otherwise nonfunctional, that is, $s_j \in \{0,1\}$, $\forall j \in \{1,\ldots,306\}$. Figure 3.8 illustrates the reliability structures for all six distinct subsystem types considered. The classification of metro stations into $B = 6$ types depends on their connectivity to neighboring stations (node degree), with subsystem type 1 corresponding to stations with one connection, subsystem type 2 for two connections, and so forth. Figure 3.9 provides the distribution of the 306 stations among these types. Subsystem complexity increases with node degree, resulting in subsystems consisting of 4–21 components. Considering all subsystem types and the total number of stations, the metro system comprises $m = 2{,}776$ individual components.

These components are classified into $K = 4$ distinct types $\left(k_i \in \{1,2,3,4\}\right)$, each characterized by a specific endowment property: *reliability improvement* (rel_i). Reliability improvements are considered type-dependent, meaning components of the same type k_i share the same rel_i. In practice, reliability improvements can translate to physical enhancements, such as reinforced structural elements, installation of advanced monitoring systems, better water drainage, or improved flood protection measures, which are particularly critical in light of increasing climate-induced hazards.

The time-dependent failure behaviors of component types 1 and 3 $(k_i = 1,3)$ follow exponential distributions according to Eqs. (3.12) and (3.13). Components of types 2 and 4 $(k_i = 2,4)$ follow gamma distributions described by the CDF:

$$F\left(t;\alpha_i,\lambda_i(\mathrm{rel}_i)\right) = \int_0^t \frac{u^{\alpha_i - 1}\lambda_i(\mathrm{rel}_i)^{\alpha_i} e^{-\lambda_i(\mathrm{rel}_i)u}}{\Gamma(\alpha_i)} du \tag{3.16}$$

with shape parameter α_i, rate parameter $\lambda_i(\mathrm{rel}_i)$, and the gamma function $\Gamma(\alpha_i)$. Again, $\lambda_i(\mathrm{rel}_i)$ depends linearly on the reliability improvement rel_i, as expressed in Eq. (3.13).

To quantify the resilience performance of the Berlin metro system, the network efficiency E_f, introduced in Latora and Marchiori (2001), is adopted as

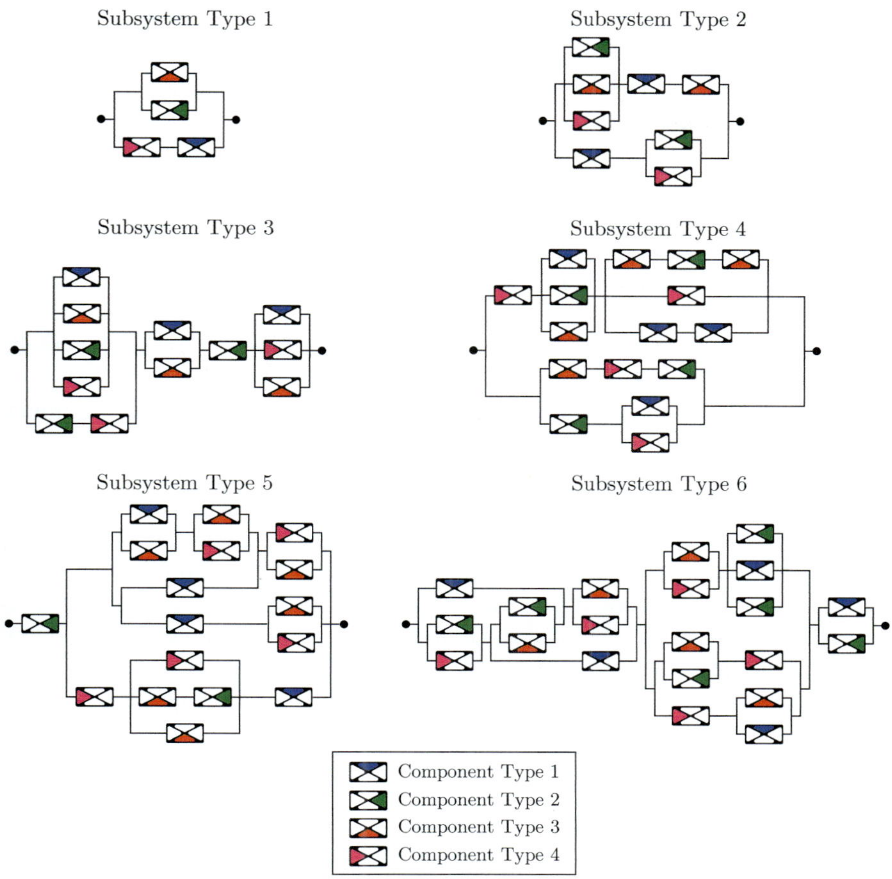

FIGURE 3.8 RBDs representing the $B = 6$ subsystem types of Berlin metro stations.

FIGURE 3.9 Number of individual metro stations per subsystem type.

performance measure $Q(t)$. Network efficiency evaluates system connectivity, reflecting the metro system's fundamental operational requirement of station-to-station accessibility:

$$E_f(t) = \frac{1}{n(n-1)} \sum_{u \neq v} \frac{1}{d_{uv}(t)} \tag{3.17}$$

where n is the number of stations, and $d_{uv}(t)$ denotes the shortest path length between stations u and v. Algorithmic methods for computing $d_{uv}(t)$, including Floyd–Warshall, Dijkstra, and Bellman–Ford algorithms, are extensively described in Dreyfus (1969); Zhan and Noon (1998).

Subsystem failure probabilities within each simulation step (t_h, t_{h+1}) follow Eq. (3.14). Failed stations remain in the system with zero node degree, meaning all direct connections to neighboring stations are temporarily removed. This assumption maintains node count consistency, essential for accurately evaluating network efficiency.

Subsystem recovery occurs fully after a station-specific duration r_j, calculated as:

$$r_j = r_{\max} - 2 \cdot \text{rec}_j, \quad \text{with} \quad \text{rec}_j < r_{\max} \tag{3.18}$$

where rec_j is a station-specific recovery improvement, and $r_{\max}$ is the maximum recovery duration. In practice, recovery improvements could correspond to having larger or specially trained emergency response teams, enhanced logistical preparedness, greater availability of replacement components, or an increased budget allocation specifically for recovery operations, especially important under the anticipated increased frequency of climate-related disruptions. After recovery, the subsystem's original connections are reinstated if the adjacent stations are operational. Components within repaired stations are restored to their initial (as-new) condition. The subsystem's survival function is then reset, ensuring subsequent subsystem reliability evolution matches that of an initially new station. This recovery approach illustrates a simple yet representative scenario. Alternative, more sophisticated recovery profiles can easily be integrated into the proposed framework.

3.6.2 Costs of Endowment Properties

Improving endowment properties such as "reliability improvement" and "recovery improvement" inevitably involves additional costs. Given that resilience measures aim at mitigating potential losses caused by extreme climatic events – such as heavy rainfall, flooding, heatwaves, or storms, whose frequency and severity are expected to increase due to climate change – it is essential to model these costs accurately. Following the discussion in Mettas (2000), an exponential relationship between improvements and costs is assumed, reflecting the higher marginal costs for substantial enhancements in system resilience.

Thus, the total costs associated with reliability improvements are defined as:

$$\text{cost}^{\text{rel}} = \sum_{i=1}^{2,776} \text{price}^{\text{rel}}_{(\text{rel}_i;k_i)} \cdot 1.2^{(\text{rel}_i - 1)} \tag{3.19}$$

where rel_i represents the reliability improvement level of component i, k_i its corresponding type, and $\text{price}^{\text{rel}}_{(\text{rel}_i;k_i)}$ denotes the predefined basic price. Analogously, an exponential relationship is assumed for the total costs associated with recovery improvements:

$$\text{cost}^{\text{rec}} = \sum_{j=1}^{306} \text{price}^{\text{rec}}_{(\text{rec}_j;b_j)} \cdot 1.2^{(\text{rec}_j - 1)} \tag{3.20}$$

where rec_j denotes the recovery improvement level of station j, b_j its type, and $\text{price}^{\text{rec}}_{(\text{rec}_j;b_j)}$ is the corresponding basic price. The total cost $\text{cost}_{(\mathbf{A};\mathbf{z}),(\mathbf{D};\mathbf{h})}$ for a given endowment configuration is then obtained by summing these two cost components:

$$\text{cost}_{(\mathbf{A};\mathbf{z}),(\mathbf{D};\mathbf{h})} = \text{cost}^{\text{rel}} + \text{cost}^{\text{rec}} \tag{3.21}$$

In practice, economic considerations must comprehensively incorporate both direct and indirect costs arising from climate-related disruptions and associated recovery measures. Direct costs encompass explicit investments in reliability and recovery enhancements, while indirect costs reflect economic losses to individuals and businesses due to reduced system performance or service interruptions of the investigated metro system. As highlighted in Saadat et al. (2020), incorporating these indirect effects is crucial for an accurate assessment of resilience strategies. Furthermore, subjective preferences of stakeholders and decision-makers, as discussed in Emanuel and Ayyub (2019), can substantially influence resilience decisions. These aspects, particularly in view of escalating climate risks, should be integrated into the proposed methodology in future research by including additional cost components, discount rates, and stakeholder-specific weighting for performance deterioration and recovery processes.

3.6.3 Scenario

To demonstrate the application of the resilience decision-making method to the Berlin metro system, illustrated in Figure 3.7, the model and simulation parameters detailed in Table 3.3 are considered. The recovery improvement rec_j is explored within the range $\text{rec}_j \in \{1,\ldots,10\}$, assuming equal improvement levels for each station, independent of their subsystem type b_j. Similarly, the reliability improvement rel_i for components is explored within $\text{rel}_i \in \{1,\ldots,10\}$ for all component types $k_i \in \{1,\ldots,4\}$.

In a preprocessing step, the survival signatures for all 306 metro stations are determined. Table 3.4 presents an illustrative example of the nontrivial survival signature values, that is, values $\Phi(l_1,\ldots,l_4)$ that differ from 0 and 1, for station type 2. Subsequently, the set of acceptable endowment configurations $R(Y)$ satisfying a resilience criterion of $\text{Res} \geq 0.99$ over the considered time period is identified using Algorithm 3.1. Given the associated costs of improving component and station endowments, the most cost-effective acceptable endowment, denoted by $(\hat{A}, \hat{D})$, is determined.

TABLE 3.3
Parameter Values for Resilience Decision-Making Applied to the Berlin Metro System

Parameter	Scenario
Acceptance threshold α	0.99
Length of time step Δt	0.05
Number of time steps u	200
Maximum time T	10
Shape parameter gamma distribution α_i	$\alpha_i = 1.2$ for $k_i = 2$
	$\alpha_i = 2.6$ for $k_i = 4$
Maximum failure rate $\lambda_{i,\max}$	$\lambda_{i,\max} = 0.34$ for $k_i = 1$
	$\lambda_{i,\max} = 0.43$ for $k_i = 2$
	$\lambda_{i,\max} = 0.36$ for $k_i = 3$
	$\lambda_{i,\max} = 0.66$ for $k_i = 4$
Failure rate reduction $\Delta\lambda_i$	$\Delta\lambda_i = 0.03$ for $k_i = 1$
	$\Delta\lambda_i = 0.04$ for $k_i = 2$
	$\Delta\lambda_i = 0.034$ for $k_i = 3$
	$\Delta\lambda_i = 0.051$ for $k_i = 4$
Reliability improvement rel_i	$\mathrm{rel}_i \in \{1,\ldots,10\}$ for $k_i \in \{1,\ldots,4\}$
Reliability improvement price $\mathrm{price}^{\mathrm{rel}}_{(\mathrm{rel}_i;k_i)}$	$\mathrm{price}^{\mathrm{rel}}_{(\mathrm{rel}_i;1)} = 100€$
	$\mathrm{price}^{\mathrm{rel}}_{(\mathrm{rel}_i;2)} = 200€$
	$\mathrm{price}^{\mathrm{rel}}_{(\mathrm{rel}_i;3)} = 200€$
	$\mathrm{price}^{\mathrm{rel}}_{(\mathrm{rel}_i;4)} = 400€$
Maximum recovery time $r_{\max}$	22
Recovery improvement rec_j	$\mathrm{rec}_j \in \{1,\ldots,10\}$
Recovery improvement price $\mathrm{price}^{\mathrm{rec}}_{(\mathrm{rec}_j;b_i)}$	$\mathrm{price}^{\mathrm{rec}}_{(\mathrm{rec}_j;1)} = 100€$
	$\mathrm{price}^{\mathrm{rec}}_{(\mathrm{rec}_j;2)} = 200€$
	$\mathrm{price}^{\mathrm{rec}}_{(\mathrm{rec}_j;3)} = 300€$
	$\mathrm{price}^{\mathrm{rec}}_{(\mathrm{rec}_j;4)} = 400€$
	$\mathrm{price}^{\mathrm{rec}}_{(\mathrm{rec}_j;5)} = 500€$
	$\mathrm{price}^{\mathrm{rec}}_{(\mathrm{rec}_j;6)} = 600€$
Sample size N	500

TABLE 3.4
Nontrivial Survival Signature Values for Stations of Type $b_j = 2$ within the Berlin Metro System (cf. Figure 3.8)

l_1	l_2	l_3	l_4	$\Phi(l_1,\ldots,l_4)$
2	2	1	1	1/4
2	1	1	2	1/4
2	2	2	1	3/8
2	2	1	2	3/8
2	1	2	2	3/8
3	2	1	1	1/2
2	3	1	1	1/2
2	1	3	1	1/2
3	1	1	2	1/2
2	3	1	2	1/2
2	1	1	3	1/2
2	2	1	3	1/2
2	3	1	3	1/2
2	2	2	2	9/16
3	2	2	1	3/4
2	3	2	1	3/4
2	2	3	1	3/4
3	2	1	2	3/4
3	1	2	2	3/4
2	3	2	2	3/4
2	1	3	2	3/4
2	1	2	3	3/4
2	2	2	3	3/4
2	3	2	3	3/4
3	2	2	2	7/8
2	2	3	2	7/8

In Figure 3.10, the results of the grid search algorithm are presented. It illustrates the accepted endowments contained in $R(Y)$, that is, all combinations leading to a satisfactory resilience of the metro system. Clearly, type 1 components as well as the recovery improvement of metro stations exert the greatest influence, and thus exhibit the highest importance, on overall system resilience. Only configurations with a reliability improvement of at least $\text{rel}_i = 8$ for type 1 components and a recovery improvement of at least $\text{rec}_j = 8$ for all metro stations fulfill the acceptance criterion. Additionally, type 2 components are significantly relevant, with minimum acceptable reliability improvements starting at $\text{rel}_i = 6$. Reliability improvements for type 3 and 4 components, however, have comparatively less impact. Consequently, numerous acceptable configurations include minimal reliability improvements for at least one of these two component types.

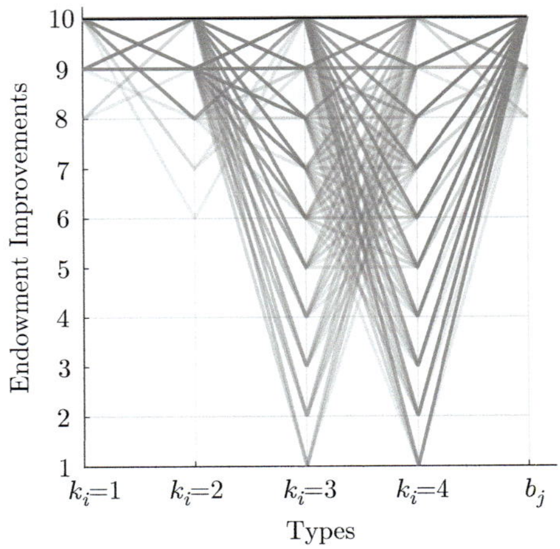

FIGURE 3.10 The set of all accepted endowments $R(Y)$ evaluated via the 5D grid search algorithm for the Berlin metro system with explored reliability improvement and recovery improvement values.

Upon closer examination of the topological structures of the metro system and its subsystems, these results are plausible, as was the case in the previous analyses. Several *U-Bahn* and *S-Bahn* lines start and end in long sequences of directly interconnected type 2 stations, as depicted in Figure 3.7. The resilience analysis of the Berlin metro system conducted in Salomon et al. (2020) showed that interruptions in these station chains severely degrade the network efficiency and thus system resilience. Consequently, the criticality of type 2 stations arises not only from their high frequency within the network but also from their vital topological role in maintaining connectivity. Therefore, components within these stations significantly affect the overall resilience. Inspection of the subsystem structure for type 2 stations (see Figure 3.8) further confirms that type 1 components occupy predominant positions, as the simultaneous failure of these two components alone leads to the failure of the entire station subsystem. No other single component type exerts such influence within type 2 stations.

Furthermore, the pronounced relevance of type 2 components is clarified by examining subsystem structures for station types 3 and 5 (cf. Figure 3.8). These station types uniquely contain critical "bottleneck" positions, where a single component failure disrupts station functionality. Notably, type 2 components occupy these crucial positions. Given that these station types interconnect three and five adjacent stations, respectively, they contribute significantly to the network's connectivity and resilience, thereby highlighting the strategic importance of type 2 components within the overall metro system.

Conversely, type 3 and 4 components do not occupy similarly critical positions within station subsystem structures, which explains their relatively limited impact on overall system resilience. The substantial influence of recovery improvements can be intuitively understood. Given that resilience is defined as the integral over actual system performance, rapid station recovery immediately enhances network efficiency, directly contributing to increased system resilience.

Considering the arbitrary base prices listed in Table 3.3, the most cost-efficient acceptable endowment $\left(\hat{A},\hat{D}\right)$ consists of a reliability improvement configuration with $\text{rel}_i = 10$ for $k_i = 1$, $\text{rel}_i = 9$ for $k_i = 2$, $\text{rel}_i = 7$ for $k_i = 3$, and $\text{rel}_i = 2$ for $k_i = 4$, as well as maximum recovery improvement $\text{rec}_j = 10$ for all metro stations $\left(b_j \in \{1,2,3,4,5,6\}\right)$. Figure 3.11 highlights this optimal configuration. Due to the monotonicity properties of the input–output model and the increasing nature of endowment costs, only configurations located on the dominant vertices of the Pareto front must be evaluated to identify the most cost-efficient solution. Consequently, Figure 3.11 exclusively illustrates these Pareto front configurations. The resulting total costs for the optimal configuration, calculated according to Eqs. (3.19–3.21), amount to $\text{cost}_{\left(\hat{A},\hat{D}\right)} = 1,700,829€ + 361,185€ = 2,062,014€$.

Employing the grid search algorithm significantly reduced the computational effort for this case study, requiring evaluation of only 0.159% of all possible endowment configurations. Thus, the presented methodology allowed reducing the complexity of the resilience analysis from 2,776 individual components to 306 top-level subsystems (stations), substantially lowering the computational burden while ensuring that all component influences were accurately captured. Analogous to the previous case studies, not only the most cost-effective configuration but also feasible configurations constrained by specific budget limits can readily be assessed.

Finally, it is important to note that the methods time-step-based simulation enables the straightforward integration of further complexity scenarios, such as cascading failures Crucitti et al. (2004); Dueñas-Osorio and Vemuru (2009); Buldyrev et al. (2010). In infrastructure networks, this capability allows consideration of increasingly frequent natural hazards, such as localized flooding events, affecting

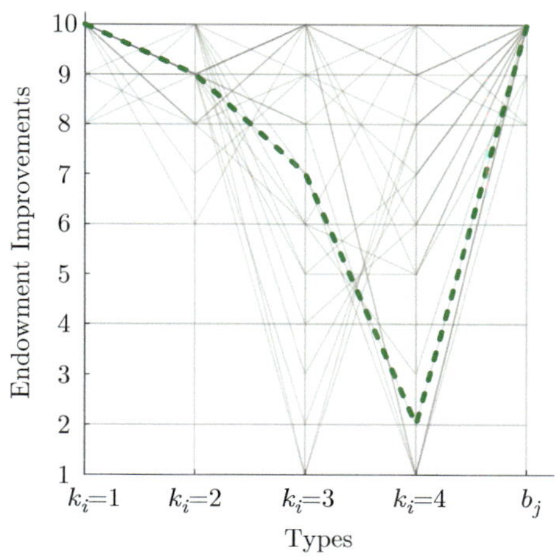

FIGURE 3.11 Dominant Pareto front endowments identified via the 5D grid search algorithm for the Berlin metro system with explored reliability and recovery improvement values, highlighting the most cost-efficient endowment $\left(\hat{A},\hat{D}\right)$.

spatially adjacent stations. The successful inclusion of such scenarios in resilience decision-making analyses for infrastructure systems has already been demonstrated in Salomon et al. (2020).

3.7 CONCLUSION AND OUTLOOK

This work addresses the critical challenge of efficient multidimensional decision-making for enhancing the resilience of complex, substructured systems under increasingly prevalent climatic threats, based on the frameworks proposed in Salomon et al. (2020) and Salomon et al. (2022). The presented approach enables systematic comparison of various heterogeneous controls influencing resilience, including preventive measures against failures (e.g., flood protection and reinforced infrastructure) and recovery improvement strategies (e.g., increased emergency response capacities, expanded repair teams, and resource allocation), applicable both in the design phase and during the entire life cycle of existing complex infrastructure systems.

Thanks to the time-step-accurate simulation of system performance at the system level, the developed methodology seamlessly incorporates complexity extensions, such as cascading failures and intricate dependency structures, which are particularly relevant in the context of climate-induced risks and extreme weather events. The approach features high numerical efficiency, significantly simplifying analyses involving numerous variations in endowment properties that continuously alter the probability structure of system components. This efficiency is primarily achieved through leveraging the separation property of the survival signature.

A key part is the integration of a robust substructuring strategy suitable for large-scale, complex infrastructures. The application of survival signatures facilitates propagating subsystem reliability characteristics efficiently through multiple hierarchy levels up to the system top-level. This notably reduces computational demands, enabling resilience analyses for infrastructures characterized by large numbers of interconnected components.

Economic constraints and budgetary restrictions can be explicitly and flexibly integrated into the resilience assessment. Thus, the methodology not only identifies the most cost-efficient acceptable endowments but also facilitates the evaluation of configurations within predefined financial thresholds. This capability enables tailored decision-making aligned with realistic economic frameworks and resource availability.

The broad applicability and generalizability of the proposed methodology are demonstrated by two fundamentally distinct systems: an arbitrary large-scale system composed of numerous subsystems and components, and an extensive substructured model of the Berlin metro system. The results achieved are coherent and plausible within the context of the selected model parameters, highlighting the approach's adaptability across diverse domains and contexts, including critical urban infrastructures facing escalating climate hazards.

While the presented methodology introduces simplifications for illustrative clarity, the authors are confident that the developed framework constitutes a meaningful foundational advancement. For practical applications to complex, multifaceted systems - such as cyber-human-physical systems in an era of climate change - this foundational framework should be combined with further developments. Integrating existing and emerging methodologies will ensure comprehensive, reality-based resilience assessments that fully capture both technical complexities and socioeconomic factors crucial to modern, climate-conscious societies.

Future research directions include incorporating advanced extensions of the survival signature approach, such as handling epistemic uncertainties, propagating uncertainty toward imprecise system resilience quantifications, and exploring multistate or continuously varying component functionalities. Moreover, to address computational limitations inherent to high-dimensional parameter spaces, research will explore sophisticated sampling methods, such as Subset Simulation (see Au and Wang 2014), with the goal of further enhancing numerical efficiency, scalability, and applicability to even more extensive and climate-sensitive resilience decision-making problems.

REFERENCES

Adams, T. M., Bekkem, K. R., and Toledo-Durán, E. J. (2012). Freight resilience measures. *Journal of Transportation Engineering*, 138(11):1403–1409.

Alipour, A. and Shafei, B. (2022). An overarching framework to assess the life-time resilience of deteriorating transportation networks in seismic-prone regions. *Resilient Cities and Structures*, 1(2):87–96.

Au, S.-K. and Wang, Y. (2014). *Engineering risk assessment with subset simulation*. John Wiley & Sons.

Ayyub, B. M. (2014). Systems resilience for multihazard environments: Definition, metrics, and valuation for decision making. *Risk Analysis*, 34(2):340–355.

Ayyub, B. M. (2015). Practical resilience metrics for planning, design, and decision making. *ASCE-ASME Journal of Risk and Uncertainty in Engineering Systems, Part A: Civil Engineering*, 1(3):04015008.

Batty, M. (2016). Complexity in city systems: Understanding, evolution, and design. In G. de Roo & E. A. Silva (Eds.), *A planner's encounter with complexity*, pp. 99–122. Routledge.

Behrensdorf, J., Regenhardt, T.-E., Broggi, M., and Beer, M. (2021). Numerically efficient computation of the survival signature for the reliability analysis of large networks. *Reliability Engineering & System Safety*, 216:107935.

Behrensdorf, J., Salomon, J., and Winnewisser, N. (2022). ResilienceDecisionMaking.jl. doi:10.5281/zenodo.7034998

Bergström, J., Van Winsen, R., and Henriqson, E. (2015). On the rationale of resilience in the domain of safety: A literature review. *Reliability Engineering and System Safety*, 141:131–141.

Berliner Verkehrsbetriebe (2018). Lagebericht und Jahresabschluss 2017. http://unternehmen.bvg.de/index.php?section=downloads%7B%5C&%7Ddownload=3128(AccessedOct.5, 2018).

Bruneau, M., Chang, S. E., Eguchi, R. T., Lee, G. C., O'Rourke, T. D., Reinhorn, A. M., Shinozuka, M., Tierney, K., Wallace, W. A., and Von Winterfeldt, D. (2003). A framework to quantitatively assess and enhance the seismic resilience of communities. *Earthquake Spectra*, 19(4):733–752.

Bruneau, M. and Reinhorn, A. (2007). Exploring the concept of seismic resilience for acute care facilities. *Earthquake Spectra*, 23(1):41–62.

Buldyrev, S. V., Parshani, R., Paul, G., Stanley, H. E., and Havlin, S. (2010). Catastrophic cascade of failures in interdependent networks. *Nature*, 464(7291):1025–1028.

Capacci, L., Biondini, F., and Frangopol, D. M. (2022). Resilience of aging structures and infrastructure systems with emphasis on seismic resilience of bridges and road networks: Review. *Resilient Cities and Structures*, 1(2):23–41.

Cimellaro, G. P. (2016). Urban resilience for emergency response and recovery. *Fundamental Concepts and Applications*.

Cimellaro, G. P., Reinhorn, A. M., and Bruneau, M. (2010). Framework for analytical quantification of disaster resilience. *Engineering Structures*, 32(11):3639–3649.

Coolen, F. P. and Coolen-Maturi, T. (2013). Generalizing the signature to systems with multiple types of components. In W. Zamojski, J. Mazurkiewicz, J. Sugier, T. Walkowiak und J. Kacprzyk (Eds.), *Complex systems and dependability*, pp. 115–130. Springer.

Coolen, F. P. and Coolen-Maturi, T. (2016). The structure function for system reliability as predictive (imprecise) probability. *Reliability Engineering & System Safety*, 154:180–187.

Crucitti, P., Latora, V., and Marchiori, M. (2004). Model for cascading failures in complex networks. *Physical Review E - Statistical Physics, Plasmas, Fluids, and Related Interdisciplinary Topics*, 69(4):4.

Dreyfus, S. E. (1969). An Appraisal of Some Shortest-Path Algorithms. *Operations Research*, 17(3):395–412.

Duan, S. and Ayyub, B. M. (2020). Assessment methods of network resilience for cyber-human-physical systems. *ASCE-ASME Journal of Risk and Uncertainty in Engineering Systems, Part A: Civil Engineering*, 6(1):03119001.

Dueñas-Osorio, L. and Vemuru, S. M. (2009). Cascading failures in complex infrastructure systems. *Structural Safety*, 31(2):157–167.

Emanuel, R. and Ayyub, B. (2019). Assessing resilience model responsiveness in the context of stakeholder preferences in decision support systems. *ASCE-ASME Journal of Risk and Uncertainty in Engineering Systems, Part A: Civil Engineering*, 5(2):04019005.

Eryilmaz, S. and Tuncel, A. (2016). Generalizing the survival signature to unrepairable homogeneous multi-state systems. *Naval Research Logistics (NRL)*, 63(8):593–599.

Fang, Y., Pedroni, N., and Zio, E. (2015). Optimization of cascade-resilient electrical infrastructures and its validation by power flow modeling. *Risk Analysis*, 35(4):594–607.

Fang, Y. and Sansavini, G. (2017). Optimizing power system investments and resilience against attacks. *Reliability Engineering and System Safety*, 159:161–173.

Feinstein, Z., Rudloff, B., and Weber, S. (2017). Measures of systemic risk. *SIAM Journal on Financial Mathematics*, 8(1):672–708.

Feng, G., George-Williams, H., Patelli, E., Coolen, F., and Beer, M. (2018). An efficient reliability analysis on complex non-repairable systems with common-cause failures. In S. Haugen, A. Barros, C. van Gulijk, T. Kongsvik & J. E. Vinnem (Eds.), *Safety and reliability–safe societies in a changing world*, pp. 2531–2537. CRC Press.

Feng, G., Patelli, E., Beer, M., and Coolen, F. P. (2016). Imprecise system reliability and component importance based on survival signature. *Reliability Engineering & System Safety*, 150:116–125.

Field, C. B. (2012). *Managing the risks of extreme events and disasters to advance climate change adaptation: Special report of the intergovernmental panel on climate change.* Cambridge University Press.

Fiksel, J. (2003). Designing resilient, sustainable systems. *Environmental Science and Technology*, 37(23):5330–5339.

George-Williams, H., Feng, G., Coolen, F. P., Beer, M., and Patelli, E. (2019). Extending the survival signature paradigm to complex systems with non-repairable dependent failures. *Proceedings of the Institution of Mechanical Engineers, Part O: Journal of Risk and Reliability*, 233(4):505–519.

Gilbert, S. and Ayyub, B. M. (2016). Models for the economics of resilience. *ASCE-ASME Journal of Risk and Uncertainty in Engineering Systems, Part A: Civil Engineering*, 2, Article 04016003, pp. 1–11.

Henry, D. and Ramirez-Marquez, J. E. (2012). Generic metrics and quantitative approaches for system resilience as a function of time. *Reliability Engineering and System Safety*, 99:114–122.

Holling, C. S. (1973). Resilience and stability of ecological systems. *Annual Review of Ecology and Systematics*, 4:1–23.

Hollnagel, E., Woods, D. E., and Levensen, N. (2006). *Resilience engineering: Concepts and precepts*. Ashgate Publishing.

Hosseini, S., Barker, K., and Ramirez-Marquez, J. E. (2016). A review of definitions and measures of system resilience. *Reliability Engineering & System Safety*, 145:47–61.

Jamshidi, M. (2017). *Systems of systems engineering: Principles and applications*. CRC Press.

Kolowrocki, K. (2014). *Reliability of large and complex systems*. Elsevier.

Latora, V. and Marchiori, M. (2001). Efficient behavior of small-world networks. *Physical Review Letters*, 87(19):198701-1–198701-4.

Linkov, I., Bridges, T., Creutzig, F., Decker, J., Fox-Lent, C., Kröger, W., Lambert, J. H., Levermann, A., Montreuil, B., Nathwani, J., et al. (2014). Changing the resilience paradigm. *Nature Climate Change*, 4(6):407–409.

Little, R. G. (2003). Toward more robust infrastructure: Observations on improving the resilience and reliability of critical systems. *Proceedings of the 36th Annual Hawaii International Conference on System Sciences, HICSS 2003*, Big Island, Hawaii, USA.

Liu, Y., Shi, Y., Bai, X., and Liu, B. (2018). Stress–strength reliability analysis of multi-state system based on generalized survival signature. *Journal of Computational and Applied Mathematics*, 342:274–291.

Mettas, A. (2000). Reliability allocation and optimization for complex systems. *Reliability and Maintainability Symposium, 2000. Proceedings. Annual Reliability and Maintainability Symposium*, Los Angeles, CA, USA, pp. 216–221.

Negi, S. and Singh, S. (2015). Reliability analysis of non-repairable complex system with weighted subsystems connected in series. *Applied Mathematics and Computation*, 262:79–89.

Nielsen, C. B., Larsen, P. G., Fitzgerald, J., Woodcock, J., and Peleska, J. (2015). Systems of systems engineering: Basic concepts, model-based techniques, and research directions. *ACM Computing Surveys (CSUR)*, 48(2):1–41.

Ouyang, M., Dueñas-Osorio, L., and Min, X. (2012). A three-stage resilience analysis framework for urban infrastructure systems. *Structural Safety*, 36–37:23–31.

Pant, R., Thacker, S., Hall, J. W., Alderson, D., and Barr, S. (2018). Critical infrastructure impact assessment due to flood exposure. *Journal of Flood Risk Management*, 11(1):22–33.

Patelli, E., Feng, G., Coolen, F. P., and Coolen-Maturi, T. (2017). Simulation methods for system reliability using the survival signature. *Reliability Engineering & System Safety*, 167:327–337.

Patriarca, R., Bergström, J., Di Gravio, G., and Costantino, F. (2018). Resilience engineering: Current status of the research and future challenges. *Safety Science*, 102:79–100.

Pörtner, H.-O., Roberts, D. C., Tignor, M., Poloczanska, E. S., Mintenbeck, K., Alegría, A., Craig, M., Langsdorf, S., Löschke, S., Möller, V., Okem, A., and Rama, B., editors (2022). *Climate Change 2022: Impacts, Adaptation, and Vulnerability. Contribution of Working Group II to the Sixth Assessment Report of the Intergovernmental Panel on Climate Change*. Cambridge University Press, Cambridge, UK and New York, NY, USA.

Presidential Policy Directive (PPD) (2013). Critical infrastructure security and resilience. https://www.dhs.gov/presidential-policy-directive-8-national-preparedness (Accessed Nov. 9, 2018).

Punzo, G., Tewari, A., Butans, E., Vasile, M., Purvis, A., Mayfield, M., and Varga, L. (2020). Engineering resilient complex systems: The necessary shift toward complexity science. *IEEE Systems Journal*, 14(3):3865–3874.

Qin, J. and Coolen, F. P. (2022). Survival signature for reliability evaluation of a multi-state system with multi-state components. *Reliability Engineering & System Safety*, 218:108129.

S-Bahn Berlin (2018). https://sbahn.berlin/das-unternehmen/unternehmensprofil/s-bahn-berlin-auf-einen-blick/. (Accessed Oct. 5, 2018).

Saadat, Y., Ayyub, B. M., Zhang, Y., Zhang, D., and Huang, H. (2020). Resilience-based strategies for topology enhancement and recovery of metrorail transit networks. *ASCE-ASME Journal of Risk and Uncertainty in Engineering Systems, Part A: Civil Engineering*, 6(2):04020017.

Sahebjamnia, N., Torabi, S. A., and Mansouri, S. A. (2015). Integrated business continuity and disaster recovery planning: Towards organizational resilience. *European Journal of Operational Research*, 242(1):261–273.

Salomon, J., Behrensdorf, J., Winnewisser, N., Broggi, M., and Beer, M. (2022). Multidimensional resilience decision-making for complex and substructured systems. *Resilient Cities and Structures*, 1(3):61–78.

Salomon, J., Broggi, M., Kruse, S., Weber, S., and Beer, M. (2020). Resilience decision-making for complex systems. *ASCE-ASME Journal of Risk and Uncertainty in Engineering Systems, Part B: Mechanical Engineering*, 6(2), Article 020901: 1–11.

Salomon, J., Winnewisser, N., Wei, P., Broggi, M., and Beer, M. (2021). Efficient reliability analysis of complex systems in consideration of imprecision. *Reliability Engineering & System Safety*, 216:107972.

Sharifi, A. (2020). Urban resilience assessment: Mapping knowledge structure and trends. *Sustainability*, 12(15):5918.

Singh, R. R., Bruneau, M., Stavridis, A., and Sett, K. (2022). Resilience deficit index for quantification of resilience. *Resilient Cities and Structures*, 1(2):1–9.

Sun, W., Bocchini, P., and Davison, B. D. (2020). Resilience metrics and measurement methods for transportation infrastructure: The state of the art. *Sustainable and Resilient Infrastructure*, 5(3):168–199.

Tran, H. T., Balchanos, M., Domercant, J. C., and Mavris, D. N. (2017). A framework for the quantitative assessment of performance-based system resilience. *Reliability Engineering & System Safety*, 158:73–84.

Weber, S. and Weske, K. (2017). The joint impact of bankruptcy costs, fire sales and cross-holdings on systemic risk in financial networks. *Probability, Uncertainty and Quantitative Risk*, 2(1), Article 9, pp. 1-38.

Winnewisser, N. R., Salomon, J., Broggi, M., and Beer, M. (2023). The concept of diagonal approximated signature: New surrogate modeling approach for continuous-state systems in the context of resilience optimization. *Disaster Prevention and Resilience*, 2(2).

Yi, H., Cui, L., and Balakrishnan, N. (2021). Computation of survival signatures for multi-state consecutive-k systems. *Reliability Engineering & System Safety*, 208:107429.

Youn, B. D., Hu, C., and Wang, P. (2011). Resilience-driven system design of complex engineered systems. *Journal of Mechanical Design*, 133(10):101011.

Zhan, F. B. and Noon, C. E. (1998). Shortest path algorithms: An evaluation using real road networks. *Transportation Science*, 32:65–73.

Zhang, J., Zhao, M., Liu, H., and Xu, X. (2013). Networked characteristics of the urban rail transit networks. *Physica A: Statistical Mechanics and Its Applications*, 392(6):1538–1546.

Zhang, Y., Ayyub, B. M., and Fung, J. F. (2022). Projections of corrosion and deterioration of infrastructure in United States coasts under a changing climate. *Resilient Cities and Structures*, 1(1):98–109.

Zobel, C. W. (2011). Representing perceived tradeoffs in defining disaster resilience. *Decision Support Systems*, 50(2):394–403.

Zuev, K. M., Wu, S., and Beck, J. L. (2015). General network reliability problem and its efficient solution by subset simulation. *Probabilistic Engineering Mechanics*, 40:25–35.

4 Navigating New Waters
Ensuring Cyber and Climate Resilience in Industrial Water Systems

Francesco Simone and Riccardo Patriarca
Department of Mechanical and Aerospace Engineering, Sapienza University of Rome, Rome, Italy

NOTATION

AI	Artificial intelligence
BI	Business intelligence
CPS	Cyber-physical system
CSTS	Cyber-socio-technical system
HHIL	Human-hardware in the loop
IoT	Internet of things
IT	Information technology
KPI	Key performance indicator
KRI	Key result indicator
OT	Operations technology
PI	Performance indicator
RI	Result indicator
STAMP	Systems-Theoretic Accident Model and Processes
STPA-Sec	System-theoretic process analysis for security
STPA-Sec/S	System-theoretic process analysis for security with simulation
SWaT	Secure water treatment
UCA	Unsafe control action

4.1 INTRODUCTION

Historically, industrial development has always had a dual impact. On one hand, it drove economic progress enhancing societal well-being, while on the other, it contributed significantly to climate change through the extensive use of natural resources – particularly fossil fuels – leading to their depletion, and to environmental pollution. Over the centuries, industries have – and continue to – emitted, a vast amount of greenhouse gases, accelerating global warming, and potentially increasing the frequency of climate-related disruptions, such as floods and heatwaves. While the strength of this correlation remains debated, it is undeniable that such a connection exists[1].

DOI: 10.1201/9781003567332-4

The quantity of emissions reflects the evolution and revolutions, of industrial systems: the first one began in the late 18th century, driven by mechanization and steam power; the second revolution started in the late 19th century, characterized by the mass adoption of electricity; the third industrial revolution, beginning in the mid-20th century, featured widespread electronics usage, automation, and digital technologies such as computers and the internet. In recent years, a fourth industrial revolution took place (i.e., Industry 4.0), characterized by the widespread merging of physical and digital technologies by means of AI and IoT, among others.

While being a causal factor for larger emissions with an increase in consumption, the latter also offers an increasingly usable way to address the growing challenges posed by climate change and its related hazards. Indeed, as industries faced mounting pressure to reduce their environmental impact, and to adapt to an increasingly volatile climate, the role of advanced digital technologies gained growing importance. They enable industries to optimize their operations, improving resource utilization efficiency, and eventually managing climate-related risks. As a result, the principles of Industry 4.0 have been widely adopted across industrial sectors, with nearly every competitive industrial organization in the world embracing these innovations.

4.1.1 The Advent of Digitalization

However, even if digitalization became instrumental to better manage climate change and its risks, its affirmation in technology posed new challenges for systems' safety (Humayed et al., 2017). It has undeniably made operations more efficient, but it also drove systems toward additional complexity, introducing a range of new vulnerabilities which expose modern systems to cyber threats. This is true both for everyday life and industrial operations, by means of the so-called cyber-physical systems (CPSs). CPSs refer to integrative systems consisting of interconnected computing and control devices interacting with a physical infrastructure via sensors and actuators (Rajkumar et al., 2010). A CPS is capable of continuously interacting with the physical system in which it operates through its capacity to compute, communicate, and control the physical environment. A big increase in the use of CPSs has occurred in recent years ranging from smart grids to smart factories, industrial control systems, and robotics. As such, modern systems can be denoted as cyber-socio-technical systems (CSTSs) due to the tight interaction among cyber-technical artifacts, with humans, and organizations (Patriarca et al., 2021). CSTSs are open to a series of completely new disruption scenarios with a potential disastrous impact. In this context, a cybersecurity issue does not only refer to data or information hacking, but it can lead to a modification of physical-world processes causing tangible damage and breakages with even societal impact. Even though systems are becoming more and more software-centric to better adapt to climate change, attacks could nowadays have physical consequences, potentially circling back on the environment, too.

In this increasingly complex scenario, traditional risk management approaches, which deal with disruptive events trying to reduce both their probability of occurrence and their impact, demand for updates. The goal of securing systems, which have been usually achieved through risk mitigation in the form of prevention and protection, is still challenging. Even if these strategies are critical to prevent extremely negative consequences, recent advancements in the field of safety and security management suggest that not all undesired events can be prevented or adequately controlled.

4.1.2 The Emphasis on Cyber Resilience

A growing consciousness on this problem has resulted in a major emphasis on the notion of resilience, being seen as the ability to withstand, adapt to and recover from disruption (Aven, 2019). As such, the word "resilience" has increasingly been present in research and popular science literature. This notion has been tailored to various applications, and different definitions exist, stressing one aspect or another with respect to the desired outcome. Thus, resilience became an inner capability to be engineered in systems' peculiar functions. Yet, considering failures strictly related to system physical components remains unsatisfactory although, with the anomalies related to the IT/Operations technology world to be added for completeness. For this reason, the more specific notion of cyber resilience has emerged, pointing at methodologies to assess and quantify resilience of CSTSs (Björck et al., 2015). Among the industrial sectors, the water industry has been pioneering this idea, due to the increased frequency and magnitude of disruptive events affecting the water infrastructure systems. Moreover, water systems are recognized as critical infrastructures, as systems are essential for the maintenance of vital societal functions. As such, resilience studies in the water sector must integrate information from cyberspace, physical processes, and social behaviors through their underlying interdependencies, improving the overall management of risks to optimize operations management and thus minimize detrimental consequences for the environment. Indeed, while cyber resilience remains an essential focus of water systems' research, the emerging impact of climate change introduces an additional layer of complexity (Garcia-Perez et al., 2023). Climate change is already affecting water systems worldwide, causing increased frequency of extreme weather events such as droughts, floods, and heatwaves. These disruptions can significantly affect the physical infrastructure of water systems, while their impact is often compounded by cyber vulnerabilities. For instance, a power outage caused by extreme weather conditions may directly disrupt the operations of a water treatment plant, which is highly digitalized and thus dependent on interconnected sensors and real-time data. A cyberattack on a facility, paired with a natural disaster such as a flood or heatwave, could exacerbate the failure of these systems, potentially leading to a cascading series of failures that affect broader regional water distribution networks, eventually circling back onto an environmental problem, too.

In this context, assessing and engineering the cyber resilience of modern water infrastructure is deemed fundamental. This type of intervention leaves ampler capacity of maneuver, as it means dealing with man-made artifacts (e.g., sensors, actuators, and controllers), ensuring their correct operations despite any disturbances. Ensuring cyber resilience results, in turn, to potential benefits for the environment, ultimately leading to better monitoring and awareness of climate change and its related hazards, too. This strategy inherits some challenges:

- How do the processes that interact with the infrastructure impact stress and recovery of water systems over time? For example, how can a cyber failure at plant level could be amplified – or dampened – by external stressors (e.g., communities' behavior and natural events)?
- How to assess the cyber resilience of water systems during the whole life of the infrastructure? There might be little to no possibility of conducting real experimentation to measure cyber resilience of water infrastructures due to systems' direct impact on community life.
- How can actual data collected from system operations inform design professionals, standards developers, and emergency managers for the purpose of promoting resilience-informed behavior? Adequate data analysis may foster performance-based management of complex distributed water networks.

Besides the definition of these three challenging research paths, this chapter is meant to discuss some perspectives arising from research conducted at Sapienza University of Rome by the Industrial Systems Engineering division in the assessment of cyber resilience in water systems. Through a synthesis of traditional engineering excellence and contemporary cyber resilience strategies, this chapter offers a comprehensive approach to protecting critical water infrastructures. The presented findings are the results of the application of methods in operating scenarios, meant at empowering knowledge management for diverse stakeholders, that is, civil engineers, policymakers, and researchers. Accordingly, this chapter is dedicated to ensuring the sustainability and security of water infrastructures against the backdrop of global climate change in increasingly digitalized settings.

4.2 THE IMPORTANCE OF SYSTEMS-THEORETIC MODELING

Being the water sector identified as a critical infrastructure, water security inherits a wide range of risks. The damage to a critical infrastructure, or generally its disruption, may have a significant negative impact on the security of the society, the well-being of its citizens, and the economy. In turn, risks to be considered are related to: (1) people, (2) natural environment, (3) economy, and (4) interactions between them. This aspect creates an interconnected complexity of the water CSTS, and the consequent difficulty in its tractability for risk management. As such, analyzing the water sector's security must balance its holistic and interconnected nature, with the reductionist approaches to understand processes and elements involved.

Accordingly, the effectiveness of using Systems-Theoretic Accident Model and Processes (STAMP) and System-Theoretic Process Analysis for Security (STPA-Sec) has been explored (Leveson, 2012). STPA-Sec is a hazard analysis technique based on an extended model for threats against Information technology (IT) systems under attack (Young & Leveson, 2013). It consists of an early concept analysis to identify inadequate safety controls in system design, aiming to assist security and safety management in the definition of the requirements and the possible countermeasures against cyberattacks. The method can be effectively used to analyze complex system interactions between their components, or throughout organizational levels. The system approach permits the understanding of the whole and interrelationships, not just specific functioning of system parts. Anyhow, it remains an explorative approach that mainly enables the development of problems related to water systems, not giving precise guidance on how to solve them. In a cyber resilience perspective, STPA-Sec permits to somehow identify the system's performance of interest.

As such, relying on the theoretical aspects proposed by STPA-Sec, a novel integrated approach for cyber resilience analysis had been developed, namely, the System Theoretic Process Analysis for Security with Simulation (STPA-Sec/S). STPA-Sec/S incorporates STAMP/STPA-Sec into cyber resilience quantification instruments based on simulation (Simone et al., 2023b). STPA-Sec/S is motivated by a few basic principles:

- The STAMP analysis enables the development of a process model based on systems theory. Through STAMP, it is possible to highlight all system elements and their relationships in terms of feedback they exchange and control actions one imposes on another;
- STPA-Sec highlights the unsafe exchanges (i.e., unsafe control actions (UCAs)). UCAs may derive from: (1) anomalies related to the controller and (2) anomalies related to the controlled process. The first case comprehends all those cases in which the controller generates a wrong control action due to a modification of its control algorithm; the second case comprehends all those cases in which the wrong control action derives from wrong feedback or a modification of the controlled process;
- The impact of the UCAs on the CSTS under analysis can be quantified through cyber resilience-oriented simulations.

As such, each system's element involved in unsafe control loops, requires a digitalized counterpart (i.e., system entities and their state variables, controlled processes, sensors and feedback, controllers and related control algorithms, and actuators and control actions). Figure 4.1 summarizes the methodological foundation of STPA-Sec/S: dark grey sliders of methods permit, once performed, to obtain the results highlighted in light grey.

While STAMP and traditional STPA provide qualitative results, STPA-Sec/S is capable of quantitatively assessing the water system cyber resilience, eventually supporting the management of the water infrastructure with objective insights. Moreover, the systemic perspective is ensured, permitting to keep a wider view on the CSTS by means of the consequences of its failures and the source of such failures. Indeed, taking a systemic perspective addresses the dual challenges of cyberattacks and climate change in water systems allowing for a more holistic understanding of how these factors may interact, compounding the risks to water infrastructures.

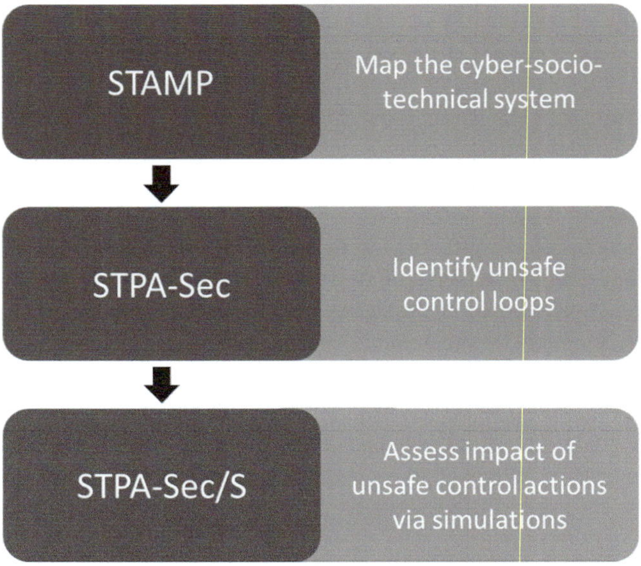

FIGURE 4.1 Methodological foundations of STPA-Sec/S.

4.3 THE IMPORTANCE OF A STRONG DATA-CENTRIC CULTURE

Even if systemic approaches can offer a complete perspective on the water CSTS, studying its cyber resilience quantitatively demands data regarding its critical performance. To measure such data (i.e., how the system will behave during a cyberattack?), real consequences of the disruptions should be experienced by the system itself. However, no one would ever damage a real water infrastructure voluntarily just for measuring its cyber resilience. A solution may be represented by the development of testbeds for experiments, for example, the Secure water treatment testbed at Singapore University of Technology and Design (Goh et al., 2017). Indeed, if not statistically significant data from previous experiences exist, it arises the need for a fail-safe environment in which to produce data.

Simulations are long-lasting, valuable tools in the assessment of water systems response to operational actions and control strategies before applying them to the real water network. Knowing the physical laws governing water system processes, it is possible to, for example, understand how closing or opening a valve would result in the water distribution in a specific area. However, critical infrastructures' resilience is nothing but the result of intertwined physical and human processes. The performance shape presented in Figure 4.2 is a trivialization of real-world cases: the way in which the performance degrades or restores would hardly be identifiable in such a linear and deterministic way (Eisenberg et al., 2025).

This demands for stochastic simulations to be used as a basis for robust methodologies to assess the resilience of water systems under different hazards, including cyber-related ones. By definition, a stochastic simulation converts an unknown probability (in this case, the value of CSTS cyber resilience) into a conditional probability

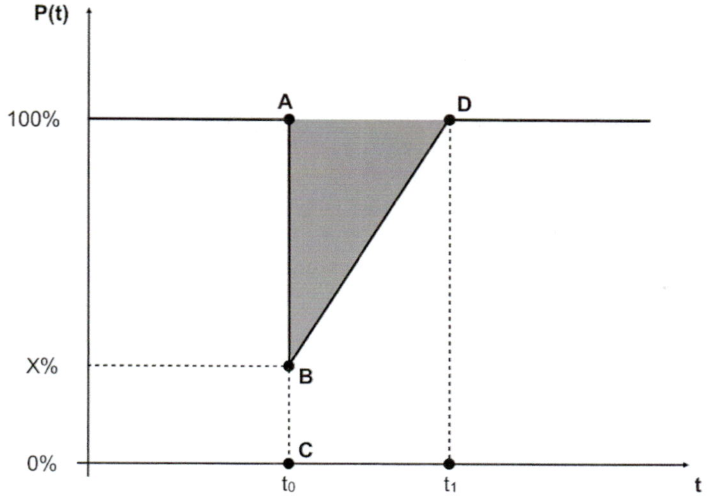

FIGURE 4.2 Plot of a generic system performance over time.

estimated from the result of a set of observations, or inputs. These latter could be either deterministic or stochastic themselves, allowing the existence of uncertainties in the model's parameter and simulated processes. As it will be easy to describe the deterministic output of a partially closed valve, it would be also possible to draw from a probability distribution function the time moment in which a cyberattack occurs, or the population's water consumption throughout the distribution network. As such, simulations represent a promising solution for the development of cyber resilience assessment methodologies since:

- The solution will not require a detailed modeling of every process related to the water system's cyber resilience before, during, and after the disruptive event occurs. Such modeling would have been a major barrier in the development of a viable and practical tool for cyber resilience assessment. To achieve valuable outcomes, not all the complex interactions and feedback loops that exist need to be considered, as well as the interdependencies with other infrastructure systems. The level of details, ranging from low-fidelity to high-fidelity, depends on the resolution and on the scope of the analysis;
- Some of the processes involved in the system loss of performance (and/or recovery) from the disruptive event may compose a complex problem. Considering them the result of a stochastic process permits to manage associated complexity without significant loss of generality;
- Stochastic simulation does not require the definition of all possible what-if scenarios. Once identified the parameters of interest – through system view – failure modes are randomly generated, eliminating any possible analysis bias toward the failure scenarios that are easier to imagine. Randomly

defined failure scenarios can provide analytic results independently of the scenario probability of occurrence. Anyhow, this latter information could be embedded in the simulation scenario through probability functions.

Based on these principles, a well-defined simulation environment combines stochastic with deterministic performance analyses to reduce uncertainty associated with the cyber resilience calculation. Although failure scenarios are randomly stressed, the hydraulics of water flows for each scenario can be safely assumed to be deterministic. The simulation methodology comprehends the following steps (Simone & Patriarca, 2022):

- **Step 1**: Problem identification: to identify a problem statement in terms of: (1) the water system to be investigated and its nominal functioning; (2) a successful cyberattack hypothesis (or a set of them); (3) the cyberattack(s) potential consequences in a systemic perspective;
- **Step 2**: System description: to choose proper system's boundaries (i.e., a testbed to define the simulation model), and to retrieve cyber-physical properties of each system element and their respective relationships. This analysis should span from pure tangible components to communication and control aspects;
- **Step 3**: Simulation model design: to define dedicated mathematical or stochastic expressions for capturing the structure and behaviors identified in Step 2, and to design their virtual counterpart in a modeling environment. At this stage, the model validation is also required to obtain a representative nominal scenario;
- **Step 4**: Definition of the performance: to define metrics for the problem at hand. Metrics shall be grounded in the chosen definition of cyber resilience, and on the key variables of the model designed in Step 3;
- **Step 5**: Modeling faults and effects: to model the effects of successful cyberattack(s) on the model. Depending on its parameters, each cyberattack may generate an individual disrupted performance curve to be compared with the nominal performance defined in Step 3;
- **Step 6**: Resilience assessment: to compute the performance (cf. Step 4) for both nominal and disrupted scenario(s) and then use the obtained results to calculate quantitative cyber resilience metrics.

Iterating this methodology, tools can be developed, such as the one coded in MATLAB/Simulink, and then integrated with software for hydraulic analysis, that is, EPANET (Patriarca et al., 2022). The integrated MATLAB–EPANET simulation tool enables the cyber resilience assessment of the considered water utility under a multiple disruptive event.

The simulation model must be capable of reproducing human actions involved in critical operations, too. Current research activities at Sapienza University of Rome are focusing on the development of integrated simulation models, capable of assessing system cyber resilience and also acknowledging dependencies from human actions (Simone et al., 2024; Simone et al., 2025). As such, the digital model of the

CPS is coupled with a digital twin for human operators through testbeds involving human-hardware-in-the-loop (HHIL) technologies (e.g., motion capture and augmented/mixed reality). Accordingly, the integrated model is meant to capture the way humans and machines interact. This aspect is particularly interesting for systems exposed to cyber vulnerabilities, since human operators share very flat learning curves due to low practice and few – or even no – past experience against challenging events such as cyberattacks (Kadivar, 2014).

Mastering integrated human–hardware–digital simulation design represents the ultimate goal to assess system's cyber resilience, as it permits large possibilities for evaluating – or validating – water systems behaviors in spite of cyberattacks occurrence. Simulations may feed decision support tools for management decision-makers, meant to improve system preparedness and responsiveness, and to guide the prioritization of critical investments in the infrastructure. On this path, human-in-the-loop integration allows to bridge other dimensions of resilience, that is, climate and societal, when simulating, assessing, and optimizing human interventions (Tamasiga et al., 2024). Human-in-the-loop approaches provide a deeper understanding of how societal behaviors and decision-making processes influence the resilience of infrastructure in the face of climate change: by incorporating human feedback, these simulations can better reflect real-world responses to crises, ensuring that strategies for climate risk management are not only technologically sound but also socially feasible and effective. Furthermore, engaging individuals and communities in the process helps to create more adaptive and sustainable solutions, as it fosters collaboration between technology developers, policymakers, and local stakeholders. In doing so, human-in-the-loop frameworks are essential for creating resilient systems that can withstand both cyber threats and the growing challenges posed by climate change.

4.4 THE IMPORTANCE OF SMART INDICATORS

Data-driven methods are essential tools to measure water systems cyber resilience. This latter focuses on system's performance: the resilience perspective does not aim to prevent cyberattacks by reinforcing the system cybersecurity but at preserving the required behavior. Thus, if simulations help in the collection of functional data, their volume can be overwhelming. Essentially, it becomes fundamental to select the useful ones among them (i.e., the ones related to the performance of interest), and to manipulate them properly (i.e., presenting them in a useful manner). Romans acknowledged the importance of measures, too, as they defined standardized basic units for length (i.e., *pes*), and weight (i.e., *libra*), but, nowadays, we are interested in finer measurements. In general, system measures can be of two types (Badawy et al., 2016):

- **Results Indicators (RI)**: are retrospective, as they indicate how the system has behaved. Such indicators could be used to assess historical cyber resilience based on past data;
- **Performance Indicators (PI)**: have a prospective nature, focusing on what it is necessary to do to maintain or even increase performance over time.

In the literature, the notions of key results indicators (KRIs) and key performance indicators (KPIs) exist, pointing at the subset of particularly critical RIs and PIs. KRIs give a measure of what the system has achieved by means of its critical success factors, while KPIs represent the core performance to achieve success. KPIs are the measures to be used to assess and ensure cyber resilience. To be effective, these cyber resilience indicators should tend to be as SMART as possible (Selvik et al., 2021):

- **Specific**: the cyber resilience indicator should be precise and clear in what it expresses;
- **Measurable**: the cyber resilience indicator should be quantified and comparable to other data. Also, the indicator must be collectable and available;
- **Achievable**: the cyber resilience indicator should be based on a realistic objective, and it must describe that objective properly;
- **Relevant**: the cyber resilience indicator should be aligned with the objective;
- **Time-Based**: the cyber resilience indicator should be dynamic and able to describe an appropriate time.

Properly defined and understandable indicators may support the water sector's decision-makers in gaining confidence that the right aspects are managed and tracked. Once defined, the cyber resilience indicators should be available to decision-makers at all organizational levels. Business Intelligence (BI) tools could be beneficial in this sense, as they combine business analytics, data mining, and data visualization to enable data-driven decisions. An example is reported in Figure 4.3 where the performance of the water system over time is analyzed through the definition of two KPIs: the first depicting the cyber resilience of the network by means of nodes supplied with not sufficient pressure (i.e., Quantity KPI), the second one depicting the cyber resilience of the network by means of nodes supplied with not quality-compliant water (i.e., Quality KPI). Numerical indicators in Figure 4.3 show actual values of the two-performance metrics, also giving information about the number of people affected by the disservice at the moment. The two measures are also reported at a more granular level showing the performance of each node by means of the two defined metrics. As such, each node is represented by the percentage of noncompliant checks with respect to each time a control on quality/quantity is made. Please note that the data presented in Figure 4.3 are for illustrative purposes only, and they have been adjusted to enhance readability. The values in Figure 4.3 result from a series of simulations conducted on a model of a water distribution network. Specifically, the water production was simulated using a MATLAB model of a treatment plant, while the resulting data were then input into a distribution network comprising 28 nodes, modeled in EPANET (Patriarca et al., 2022). Despite these simulated values were used to demonstrate the dashboard's functionalities, the BI solution is ultimately designed to operate with real-world data from an actual infrastructure. Nevertheless, data from a digital twin – such as the one described above – can also be integrated, enhancing the analysis through the evaluation of various "what-if" scenarios and projecting possible future system states.

In this context, to fully reflect the complexity of modern water infrastructure, cyber resilience BI dashboards can also incorporate climate-related KPIs. Augmenting

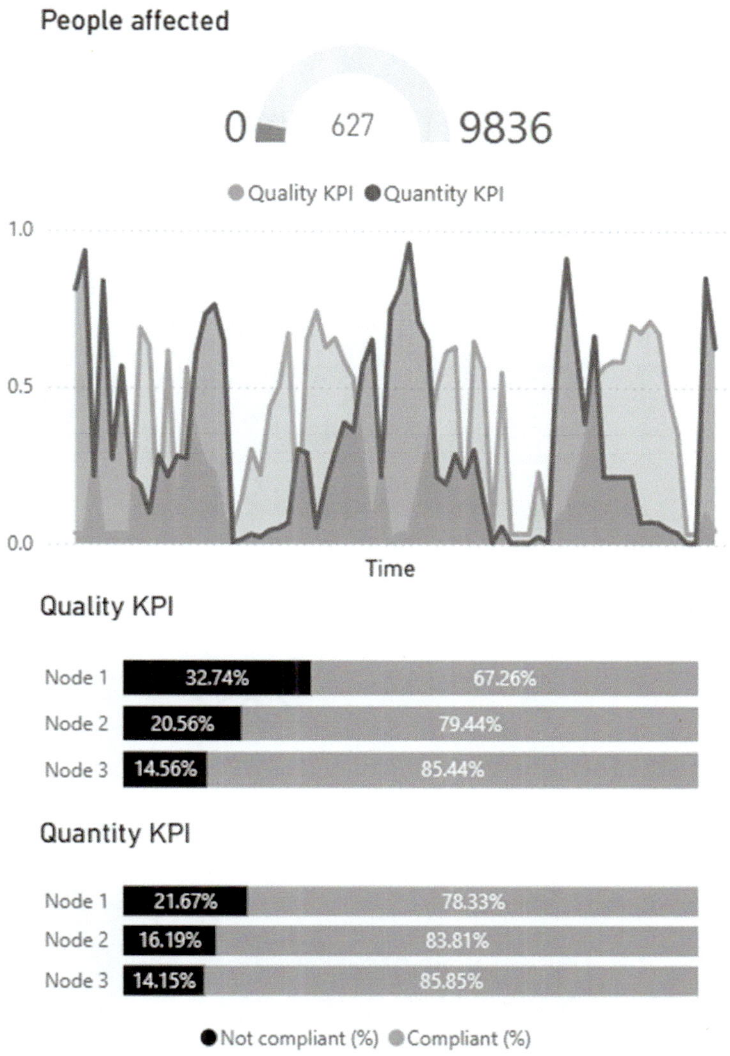

FIGURE 4.3 Sample dashboard for cyber resilience KPIs.

the cyber resilience metrics shown in Figure 4.3, with the climate-related KPIs in Figure 4.4, would offer a complete view of both sudden attacks and chronic stress factors for the water infrastructure. One type of KPI could be the drought exposure KPI, marking nodes' monthly supply drops. The visual shows the number of junctions (in percentage) failing to meet demand at least once in a month. While the values are simulated, this trend shows comparable values to the 2025 Lebanese drought that left Lake Qaraoun with only 45 mm^3 of water versus its long-term 350 mm^3 average (Reuters, 2025). Complementing this view, a thermal stress KPI flags nodes where distributed water remains above 22°C for more than six hours at least once in a day, as a probable effect of climate change. The threshold is chosen as higher

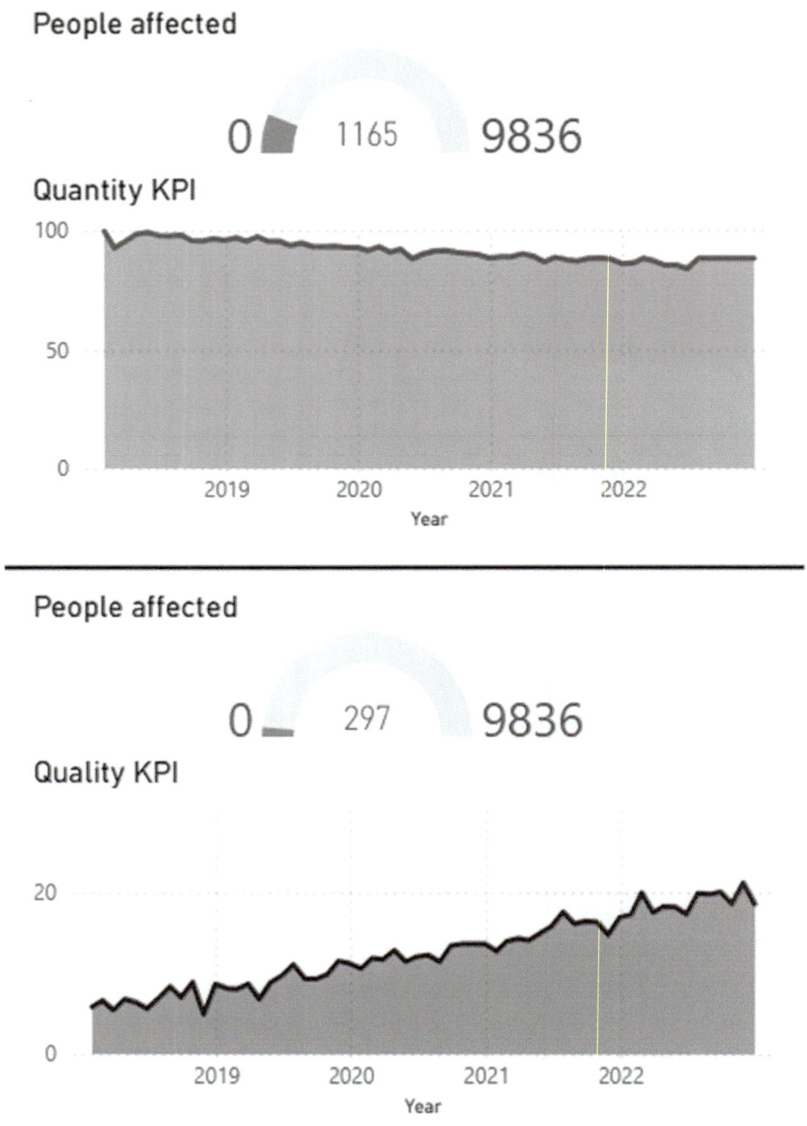

FIGURE 4.4 Sample dashboard for climate change KPIs.

temperatures accelerate *Legionella* growth and expose the users to taste, odor, and microbiological risks (World Health Organization, 2022).

But how can such dashboards presenting quantitative data in the form of KPIs practically contribute to resilience and cyber resilience management?

Displayed alongside the cyber KPIs, the climate indicators let decision-makers compare instantaneous attack-driven outages with latent, weather-driven degradations in a single dashboard, prompting integrated actions such as pressure management, seasonal blending, and reservoir-cover installation. Leveraging the exemplary formulation by Hollnagel (2009), a resilient system is characterized by four essential abilities: (1) monitor incoming critical situations; (2) anticipate the occurrence of future events; (3) respond to challenges when they verify; and (4) learn from past events and occurrences. This type of dashboard inherently proposes a monitoring framework, with the potential to extend its usage to anticipation via predictive techniques. While the responding capacity is not directly proposed by the definition of the KPIs, it is clear that such a granular approach supports responding actions. A learning cycle is instantiated every time.

In this regard, BI visualizations help in monitoring how the system performance is – or is not – being affected in near real time, providing a clear and direct understanding of the system state (Nakhal Akel et al., 2021). Similarly, they foster the resilience ability to anticipate, that is, knowing what to expect from changing operating conditions as extended visibility on the system performance in time is enabled. Such visibility on system performance also permits a correct response to both regular and irregular changes, helping in the adjustment of system functioning. In addition, as historical data are stored, it is possible to know what has previously happened, learning from past experiences, and preparing effective response actions for the future.

4.5 CONCLUSION

Water utilities provide a critical service to the community, and keeping it safe is fundamental. However, to optimize systems operations, a large amount of valuable data is hosted, making the water sector vulnerable to cyberattacks. Instead of strictly focusing on defensive cybersecurity strategies, this chapter suggests adopting a cyber resilience perspective. The focus shifts on ensuring high-level performance besides the cyber disruption that may occur. Understanding the relationship between climate change and the resilience of CPSs in the water sector requires a shift from a linear, siloed view of system failure to a more interconnected interpretation. Climate events can trigger cascading failures in water systems that involve both physical infrastructure breakdowns and failures of digital controls or communication channels. The integration of climate indicators into a cyber resilience evaluation framework is more than a technical necessity, it is a strategic imperative. As climate change continues to stress physical infrastructures and societies, the ability to monitor, anticipate, respond to, and learn from compound threats defines the resilience ability of the water systems of the future. Cyber resilience strategies must, therefore, extend beyond the purely cyber dimension to account for the dynamic interaction among the cyber components, the physical environment, and the societal behavior. To achieve this purpose, in this chapter we stressed:

- the need to model water infrastructures as a system-of-systems. Assuming a systemic view should help highlight critical hotspots. The proposed STPA-Sec/S methodology represents a solution. Yet, in addition to developing frameworks for protecting against cyberattacks, resilience studies in the water sector should also prioritize the system's ability to adapt to and recover from the physical impacts of climate change. This involves the integration of predictive climate models within CSTSs' models that enable proactive management of water infrastructure, ensuring that systems are not only protected against cyberattacks but are also able to function effectively under the increased stress caused by extreme weather events;

- the need for simulation-based quantification. Stochastic simulations are a useful tool in this sense, but they find it difficult to model human factors. HHIL simulations, instead, seem capable of solving this issue. This strategy should be pointed at generating not just another layer of data (big data), but rich, useful descriptions (thick data);
- the need for SMART indicators. Large simulation and real-world operations exercises generate large volume of data, which requires exploitation strategies. BI tools represent an excellent solution in this sense, to be paired with more advanced logics, such as knowledge graphs (Simone et al., 2023a), giving – almost – limitless possibilities for data manipulation.

Overall, a cyber resilience-oriented culture should be adopted and preserved to keep high-level performance. Even history proves that in the Roman Empire, the good performance and – a kind of – resilience of water infrastructures was ensured by the Romans' continuous maintenance and protection. As such, the "modern" design technologies of the Roman water infrastructure were meant to favor maintenance practices, permitting an excellent planning to preserve system's vital requirements. An example was the underground sections of the aqueducts made accessible to the numerous maintenance personnel, or the possibility engineers had to temporarily divert water flows to perform major repairs in critical areas. This thesis is strengthened by the fact that after the fall of the Roman Empire, the aqueducts fell into disuse through lack of organized maintenance and this abandonment had tremendous impact on the city with a drastic population decrease. The story has a happy ending: some aqueducts are still in use nowadays, even if more than 2,000 years have passed. They are still visible, becoming iconic structures of the Eternal City – and in other places of the world, and still providing modern water engineers with interesting insights. This observation calls for systematic and organized cyber resilience planning and execution in the water industry to follow the lead of Roman traditions and share their value worldwide.

One final question that might shake our assumptions: does digitalization truly offer a lifeline in the face of our climate crisis, or is it quietly exacerbating the very crisis it seeks to address?

Here, we assumed digitalization to be capable of championing efficiency and adaptability in response to the increasing unpredictability of climate risks, emphasizing the need to secure digital systems so they, in turn, can help mitigate both climate change causes and effects. However, the rapid growth of data centers, cloud computing, and digital infrastructure significantly contributes to energy consumption and carbon emissions, too, raising important questions about the overall impact of digitalization on climate change.

It remains open the assessment of whether the growth of digital infrastructure is unintentionally worsening climate challenges by increasing demand for resources and energy. As we push for smarter, more connected systems, are we, analytically, politically, and ethically, overlooking the hidden toll on our planet?

NOTE

1 https://climate.ec.europa.eu/climate-change/causes-climate-change_en

REFERENCES

Aven, T. (2019). The call for a shift from risk to resilience: What does it mean? *Risk Analysis*, 39(6), 1196–1203. https://doi.org/10.1111/risa.13247

Badawy, M., El-Aziz, A. A. A., Idress, A. M., Hefny, H., & Hossam, S. (2016). A survey on exploring key performance indicators. *Future Computing and Informatics Journal*, 1(1–2), 47–52. https://doi.org/10.1016/j.fcij.2016.04.001

Björck, F., Henkel, M., Stirna, J., & Zdravkovic, J. (2015). Cyber resilience – Fundamentals for a definition. *Advances in Intelligent Systems and Computing*, 353, 311–316. https://doi.org/10.1007/978-3-319-16486-1_31

Eisenberg, D. A., Seager, T. P., & Alderson, D. L. (2025). The rebound curve is a poor model of resilience. *PNAS Nexus*, pgaf052. https://doi.org/10.1093/pnasnexus/pgaf052

Garcia-Perez, A., Sallos, M. P., & Tiwasing, P. (2023). Dimensions of cybersecurity performance and crisis response in critical infrastructure organisations: An intellectual capital perspective. *Journal of Intellectual Capital*, 24(2), 465–486. https://doi.org/10.1108/JIC-06-2021-0166

Goh, J., Adepu, S., Junejo, K. N., & Mathur, A. (2017). A dataset to support research in the design of secure water treatment systems. *Lecture Notes in Computer Science* (Including Subseries Lecture Notes in Artificial Intelligence and Lecture Notes in Bioinformatics), 10242 LNCS, 88–99. https://doi.org/10.1007/978-3-319-71368-7_8

Hollnagel, E. (2009). The four cornerstones of resilience engineering. In *Resilience Engineering Perspectives, Volume 2 – Preparation and Restoration* edited by Hollnagel, E., Nemeth, C.P. CRC Press, London. https://doi.org/10.1201/9781315244389

Humayed, A., Lin, J., Li, F., & Luo, B. (2017). Cyber-physical systems security – A survey. *IEEE Internet Things Journal*, 4(6), 1802–1831. https://doi.org/10.1109/JIOT.2017.2703172

Kadivar, M. (2014). Cyber-attack attributes. *Technology Innovation Management Review*, 4(11), 22–27. https://doi.org/10.22215/timreview/846

Leveson, N. (2012). *Engineering a Safer World: Systems Thinking Applied to Safety*. The MIT Press, Cambridge.

Nakhal Akel, A. J., Patriarca, R., Di Gravio, G., Antonioni, G., & Paltrinieri, N. (2021). Investigating occupational and operational industrial safety data through Business Intelligence and Machine Learning. *Journal of Loss Prevention in the Process Industries*, 73, 104608. https://doi.org/10.1016/j.jlp.2021.104608

Patriarca, R., Falegnami, A., Costantino, F., Di Gravio, G., De Nicola, A., & Villani, M. L. (2021). WAx: An integrated conceptual framework for the analysis of cyber-socio-technical systems. *Safety Science*, 136(105142). https://doi.org/10.1016/j.ssci.2020.105142

Patriarca, R., Simone, F., & Di Gravio, G. (2022). Modelling cyber resilience in a water treatment and distribution system. *Reliability Engineering & System Safety*, 226, 108653. https://doi.org /10.1016/j.ress.2022.108653

Rajkumar, R., Lee, I., Sha, L., & Stankovic, J. (2010). Cyber-physical systems: The next computing revolution. *Proceedings – Design Automation Conference*, 731–736. https://doi.org/10.1145/1837274.1837461

Reuters (2025), Lebanon's worst drought on record drains largest reservoir. https://www.reuters.com/sustainability/boards-policy-regulation/lebanons-worst-drought-record-drains-largest-reservoir-2025-07-15/ (accessed on July 23, 2025).

Selvik, J. T., Bansal, S., & Abrahamsen, E. B. (2021). On the use of criteria based on the SMART acronym to assess quality of performance indicators for safety management in process industries. *Journal of Loss Prevention in the Process Industries*, 70, 104392. https://doi.org/10.1016/j.jlp.2021.104392

Simone, F., Ansaldi, S. M., Agnello, P., Di Gravio, G., & Patriarca, R. (2023a). Knowledge in graphs: Investigating the completeness of industrial near miss reports. *Safety Science*, 168, 106305. https://doi.org/10.1016/j.ssci.2023.106305

Simone, F., Montaruli, A., Patriarca, R. (2025). AI-aided STPA-Sec for managing cyber security in industrial plants – an ontological approach. *Proceedings of 2025 IEEE International Conference on Cyber Security and Resilience* (submitted)

Simone, F., Nakhal Akel, A. J., Gravio, G. Di, & Patriarca, R. (2023b). Thinking in systems, sifting through simulations: A way ahead for cyber resilience assessment. *IEEE Access*, 11, 11430–11450. https://doi.org/10.1109/ACCESS.2023.3241552

Simone, F., Nakhal Akel, A. J., Lombardi, M., Di Gravio, G., Patriarca, R. (2024). Human-Hardware in the Loop (HHIL) STAMP-based simulations to model cyber-physical complexity in experimental high-risk plants. *P Proceedings of the 11th European STAMP Workshop and Conference ESWC "Advancing Safety in a Complex World",* Alexandroupolis, Greece - October 2-4, 2024.

Simone, F., & Patriarca, R. (2022). A simulation-driven cyber resilience assessment for water treatment plants. *Proceedings of the 32nd European Safety and Reliability Conference*, 2836–2843. https://doi.org/10.3850/978-981-18-5183-4_S23-02-289-cd

Tamasiga, P., Mogomotsi, P. K., Onyeaka, H., & Mogomotsi, G. E. J. (2024). Amplifying climate resilience: The impact of social protection, social cohesion, and social capital on public support for climate change action. *Sustainable Environment*, 10(1). https://doi.org/10.1080/27658511.2024.2361568

World Health Organization (WHO) (2022). Legionellosis. https://www.who.int/news-room/fact-sheets/detail/legionellosis (accessed on July 23, 2025).

Young, W., & Leveson, N. (2013). Systems thinking for safety and security. *ACM International Conference Proceeding Series*, 1–8. https://doi.org/10.1145/2523649.2530277

5 Unpacking the 'BLACK Box' of Enabling Infrastructure Resilience to Climate Change

Time to Address the 'Soft' Elements of Organisations for Ensuring Better Improvements

Henry J. Liu
School of Design and the Built Environment,
The University of Canberra, Bruce, ACT, Australia

5.1 INTRODUCTION

Worldwide, having resilient infrastructure in place to withstand climate change has been a priority of many governments. A case in point is the transportation sector, where disrupted services occurred frequently after flooding, downpours and heatwaves. For instance, the 2022–2023 New South Wales (NSW) floods in Australia resulted in severe damage to transportation, for example, AU$29m in financial losses, as well as traffic chaos like road closures and disruptions to train services. Moreover, the ex-tropical cyclone Alfred had brought heavy wind and rain, causing system-wide technical issues with the G:Link, the tram network of Gold Coast. Consequently, temporary closures and service disruptions were recorded. Australia is not alone, as others such as the United Kingdom (UK), the US and China, to name a few, have confronted intensive environmental impacts, for example, the 2010–2011 snowstorm across Europe, 2019 downpours within England, UK, 2022 Hurricane Ian in the US and the 2021 flood in China's Henan Province, all of which had hindered the local infrastructure assets.

Resilience in the infrastructure context describes how the assets positively adapt and respond to changing conditions (Bosher and Dainty, 2011; Hughes and Healy, 2014). Central to its epistemology are four inquiries: *resilience of what*, *resilience to*

DOI: 10.1201/9781003567332-5

what, *resilience for whom* and *how to be resilient* (Vale, 2014; Chmutina et al., 2016). These '*what*(s)' (i.e., technical abilities and external impacts respectively), '*whom*' (i.e., key stakeholders) and '*how*' (i.e., process and organisational management) have shaped an epistemological framework that involves asset adaptability, vulnerability and recoverability along with those organisational activities leading to resilience (Liu et al., 2019). In this stance, enabling infrastructure resilience is a sophisticated and systemic process, which integrates engineering and technical issues and the managerial elements of the organisations engaged with infrastructure asset delivery (Desouza and Flanery, 2013, Love et al., 2021).

Organisational capabilities are pivotal in determining the robustness of an organisation's managerial elements (Teece et al., 1997). They are interdependently as assets and also are viewed as '*an outcome of investment in staffing, training, compensation, communication and other human resource areas*' (Smallwood and Ulrich, 2004, p. 119). Put simply, an organisation's capability refers to the capacity and resources (inputs) necessary for enabling business functions (e.g., planning, leadership, human resources, systems and procedures, innovation and network coordination) to navigate and achieve a strategic goal (Grant, 1991; Teece, 2007; Inan and Bititci, 2015). Notwithstanding their importance, there is 'no magic list' of the capabilities to ensure infrastructure resilience. Lamenting this critical concern, an understanding of how the capabilities interact with one another in shaping a resilient asset to support governments' resilience policymaking and strategy development is urgently needed (Liu et al., 2021b). This perspective is acknowledged by the UK's Cabinet Office (2017), which states that lacking knowledge of organisational capabilities has contributed to its inability to assess risk and uncertainty.

Despite the UK Government cognisant of inadequate knowledge above (Cabinet Office, 2017), there remains limited research in this field. In recognition of this 'void', this chapter exploits the author's previous studies, e.g., Liu et al. (2019; 2021b), aiming to interpret the systemic interdependency of the organisational capabilities in determining infrastructure resilience over the asset lifecycle, with a focus on the government, head contractor and asset operator. Accordingly, the remainder of this chapter begins with a review of the infrastructure resilience literature. Then, the methodology is presented to address the line of this chapter's inquiry. Next, analysis, implication discussion and future research are presented sequentially before concluding this chapter.

5.2 RESEARCH ON INFRASTRUCTURE RESILIENCE

Considerable attentions have been placed to resilience for decades owing to increasingly changing environments (e.g., economic, social and natural) (Wan et al., 2018; Li et al., 2022). For instance, detailed reviews of resilience emerged between 2011 and 2016 (c.f., Cox et al., 2011; Reggiani, 2013; Reggiani et al., 2015; Chmutin et al., 2016), while the recent ones relied on the systematic approach (i.e., systematic review) such as Wan et al. (2018), Wang et al. (2023) and Gu and Liu (2025), which were conducted in the context of economic infrastructure (e.g., power system and transport).

5.2.1 Defining Infrastructure Resilience

Interestingly, there is no universally accepted definition of resilience albeit wealth of research. This is a particular case of the infrastructure-related studies (Liu et al., 2019). Thus, the normative literature abounds with attempts to define infrastructure resilience (c.f., Solomon et al., 2007; Berche et al., 2009; McBain et al., 2010; Reggiani, 2013; Becker et al., 2015) (see Table 5.1).

As noted from Table 5.1, there is a need for shifting the fundamental paradigm of resilience in the context of infrastructure to encompass not only the adaptability, vulnerability and recoverability of assets *per se* but also the predictability of the engaged organisations (e.g., government, head contractor, etc.) to risks, uncertainties and hazards juxtaposed with the lessons learned from their impacts, for facilitating future practice (i.e., asset design, construction, operations and maintenance) (Hallet, 2013; US Environmental Protection Agency, 2015; Liu et al., 2019). This paradigm shift endorses the necessity of research into the nexus between

TABLE 5.1
Definitions of Infrastructure Resilience

Publications	Asset Abilities	Resilience to What	Detailed Notes
Solomon et al. (2007)	Adapt	Disturbance, stress and change	Retaining the same basic structure and function
McDaniels et al. (2008)	Maintain, alternate and return	Stress and emergencies	Robustness, redundancy and rapidity
Berche et al. (2009)	Withstand and recover	Disturbances	Acceptable degradation and timely recovery
Mansouri et al. (2010)	Recover	Risks of service disruptions	Post-impact recovery in an efficient timeframe
McBain et al. (2010)	Continue, recover and return	Unusual events	Continuing essential services with normal operations
Omer et al. (2012)	Resist and recover	Natural hazards	Adapting to shocks and efficiently recovering to an acceptable level
Hallet (2013)	Prevent, withstand, recover and learn	Weather hazards	Timely predicting and then preventing hazards with a post-impact recovery in an efficient way
Reggiani (2013)	Absorb	External shocks	No catastrophic changes
Becker et al. (2015)	Absorb and retain	Natural hazards	Remaining basic function and structure
US Environmental Protection Agency (2015)	Anticipate, prepare for, respond and recover	Hazards	Minimum damage to socio-economic well-being

Source: Liu et al. (2019).

organisational capability and resilience, as the 'predictability' and 'learning' are viewed as the 'abilities' of an organisations in managing and utilising resources to foresee and assess risks and then achieve the predetermined strategy (Liu et al., 2021a). In addition, it has also placed a sound underpinning for those studies aimed at advancing resilient performance measurement, for example, Hughes and Healy (2014) and Liu et al. (2019; 2021b).

5.2.2 Resilient Performance and Organisational Capability

Measuring resilience is a prerequisite step of resilience-enabling process (UK Cabinet Office, 2016; Liu et al., 2019). Nonetheless, the prevailing research tends to be limited to mathematical metrics or indexes (c.f., Ip and Wang, 2011; Adams et al. 2012; Miller-Hooks et al. 2012; Omer et al., 2012; Adjetey-Bahun et al., 2016; Dobie and Schneider, 2017; Johansen et al., 2017). For instance, emerging from Cox et al. (2011), Baroud et al., (2014), Minaie and Moon (2017) and Wang et al. (2023) were those operational, quantitative metrics that are useful for understanding the resilience status of passenger transportation systems, bridges and inland waterway networks, respectively. Continuing with this theme, Li et al. (2022) quantified the resilience of emergency response networks for supporting infrastructure operations during the occurrence of emergent events. Despite the aforementioned works, a comprehensive framework robust in guiding the identification of resilient measures or metrics was lacking prior to the study by Hughes and Healy (2014), which addresses the organisational perspective (e.g., leadership, communication, decision-making, etc.) other than engineering and technical aspects in resilient measurement.

Liu et al. (2019) then leveraged Hughes and Healy's (2014) conceptual principle as a departure point to conceptualise a novel, life-cycle resilient performance measurement system (PMS) for transport infrastructure, which comprises: (1) asset's technical ability; (2) government's inputs and capability; (3) continuing improvement for asset vulnerability; and (4) contractor's inputs and capability. They also visualise such elements as governance and policy, learning and technologies of asset delivery and management juxtaposed with end-user/local resident expectations and resilience objective, in addition to the four key perspectives above. Noteworthy, Liu et al's (2019) paper is the first study dedicated to theoretically highlighting the vital role of organisational capabilities in the performance measurement of infrastructure resilience. To empirically test a component of their PMS, Liu et al. (2021b) adopted the social network analysis (SNA) grounded in the outcomes of those studies presented in Table 5.2, for a purpose of identifying the influences of organisational capabilities on transport resilience, with a particular to government.

Indeed, a wealth of studies has thrived in understanding organisational capabilities and their impacts on business performance and functions of the organisations since the 1990s (c.f., Collis, 1994; Teece et al., 1997; Winter, 2003; Ambrosini et al., 2009; Raffoni et al., 2018; Khalil and Belitski, 2020). Nevertheless, a detailed examination of the literature reveals a paucity of research that explores the interactions between the capabilities of the organisations and infrastructure resilience. Although Liu et al. (2021b) engendered solid contributions to fill the abovementioned void, they eschewed to examine resilience through the lens of dynamic thinking, which is

TABLE 5.2
Key Studies on Organisational Capability and Infrastructure Resilience

Capability	Types of Infrastructure	Publications
Management of supply chain and stakeholders	Entire infrastructure system	Stewart et al. (2009)
Governance	Communication systems	Carmeli and Markman (2011)
Skilled workforce	Entire infrastructure system	Santos et al. (2014)
Decision-making	Transport networks	Giezen et al. (2015)
Performance measurement	Transport networks	Loo and Leung (2017)
Management of information	Transport networks	Blake et al. (2019)
Planning and resource management	Ports	Shaw et al. (2019)

Source: Liu et al. (2021b).

(1) critical for studying organisational capability under a changing environment in accordance with Teece et al.'s (1997) *Dynamic Capability Theory* and (2) also this chapter aims to address so that headway can be made.

5.3 METHODOLOGY

The study presented in this chapter seeks to interpret how the organisational capabilities of key organisations (e.g., government, head contractor and asset operator) systemically engender resilient infrastructure. To this end, cumulative case studies followed by the system dynamics (SD) are adopted to address the germane inquiry (i.e., Figure 5.1). The rationale for doing so relates to the robustness of the cumulative case-study approach in enabling a greater generalisation via aggregate information (Yin, 2013; Liu et al., 2021b). Rather than a reliance on separate cases or pure literature review, studying a collection of cases concurrently can solicit and ensure a deeper insight into the targeted issue from a real-world perspective. This 'benefit' is particularly important for this chapter, as the proposed research aim is in line with a concern on those core organisations engaging in the delivery of resilient infrastructure assets.

The causal loop under the SD, as indicated by Figure 5.1, is conceptually developed to supplement the case studies for addressing one of the critical concerns of this chapter, i.e., provision of a dynamic insight. SD has been widely applied to visualise and understand dynamic relationships within a system (Sing et al., 2019). It is an approach to comprehending complex systems by considering the interactions and interrelationships between their components (Jackson, 2019). SD also examines a given system as a whole, analysing how the elements interact with each other to produce specific behaviours and outcomes. The notion of the 'feedback (causal) loop' is important for the SD analysis and is a self-reinforcing or self-correcting process that influences a system's behaviour (Jackson, 2019). As such, examining feedback loops is essential for analysing system behaviour and identifying potential leverage points for intervention.

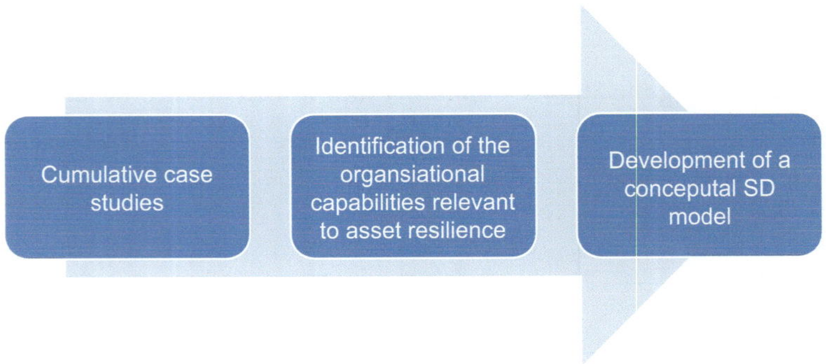

FIGURE 5.1 Methodological design for the study presented in this chapter.

TABLE 5.3
The Selected Infrastructure Networks/Assets for Cumulative Case Studies

Networks/Systems/Assets	Data Sources
WestConnex, Sydney, Australia	Audit Office of New South Wales (2014; 2021; 2023), City of Sydney (2017), Legislative Council (2018) and Transport for NSW (2023)
M25 Motorway, UK	Department for Transport (2014) and New Civil Engineer (2024)
Motorway Network (e.g., M1, M4, M5, M18, M40, M50 and M54), UK	Department for Transport (2014)
A390 (road), Cornwall, UK	Department for Transport (2014) and Cornwall Council (2019)
London Gatwick Airport, UK	Department for Transport (2014), McMillan (2014) and BBC (2019)
A303 (trunk road), Deptford, UK	Department for Transport (2014) and UK Parliament (2014)
Wokingham Borough Road System, UK	Department for Transport (2014) and Cabinet Office (2017)
Dawlish Railway lines, UK	Department for Transport (2014) and NetworkRail (2019)
London Underground, UK	Greater London Authority (2019)

Source: Adapted from Liu et al. (2021b).

As clarified above, this chapter leverages the author's published study, i.e., Liu et al. (2021b), in which a total of seven (7) cases were chosen from the UK for an SNA that draws on the grey literature. In this instance, these cases are picked up for a further digging along with an addition of the M25 motorway, UK and Sydney's WestConnex, Australia, to enable a greater generalisability. Table 5.3 presents the relevant details. Essentially, all the infrastructure assets selected for this research were subjected to severe impacts of extreme weather hazards and/or will confront damages due to detected vulnerability. For instance, the Rozelle Interchange and Iron Cove Link of the WestConnex are more vulnerable to storm surge and flooding like the 2022–2023 NSW floods. In addition, '*the electrical switchgear serving the North Terminal of London Gatwick Airport was inundated by the flooding, which caused a cloudburst in December 2013*' (Liu et al., 2021b, p. 4).

Thematic analysis is performed in the first instance for the chosen cases above in line with the methodological design depicted by Figure 5.1. It aims to code and filter those core organisational capabilities that can determine the resilience of the observed assets from the perspectives of the governments, public authorities (asset operators) and contractors.

5.4 DATA ANALYSIS AND DISCUSSION

NVivo 12 is used for analysing the contents of the documents shown in Table 5.2 and identifying the key themes through the software package's '*Queries*' under the '*Search*' function. To avoid 'double digging' and duplicating those elements that have been well interpreted by the author in Liu et al.'s (2021b) SNA, the coding of the new-added cases including the M25 Motorway and WestConnex is the focus of this chapter and is also triangulated with the coding process and finally filtered capabilities (e.g., leadership, reflective learning, organisational culture, supporting mechanism from top management and continuous improvement) of Liu et al. (2021b), which are visualised by Figure 5.2, to ensure consistency and reliability and enrich findings (Love et al., 2002; Liu et al., 2018).

Having benefited from the thematic analysis, the organisational capabilities in relation to the selected cases are mapped, filtered and illustrated by Figure 5.3. As noted, the identified capabilities are less than those originally listed by Liu et al. (2021b) before the SNA but are more specific. For instance, the leadership- and top-management-support-related capabilities raised in the author's past study are

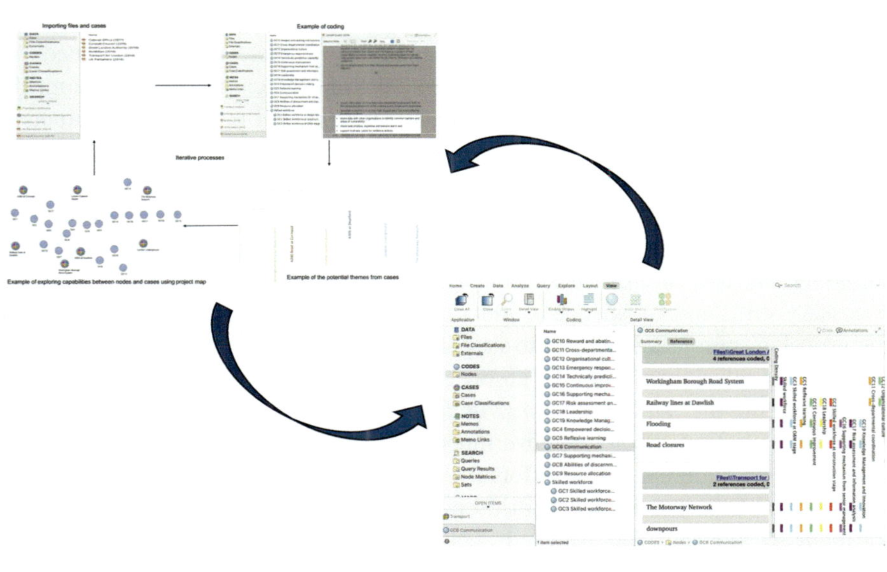

FIGURE 5.2 Coding process and coding of the author's previously published study. (Adapted from Liu et al., 2021b.)

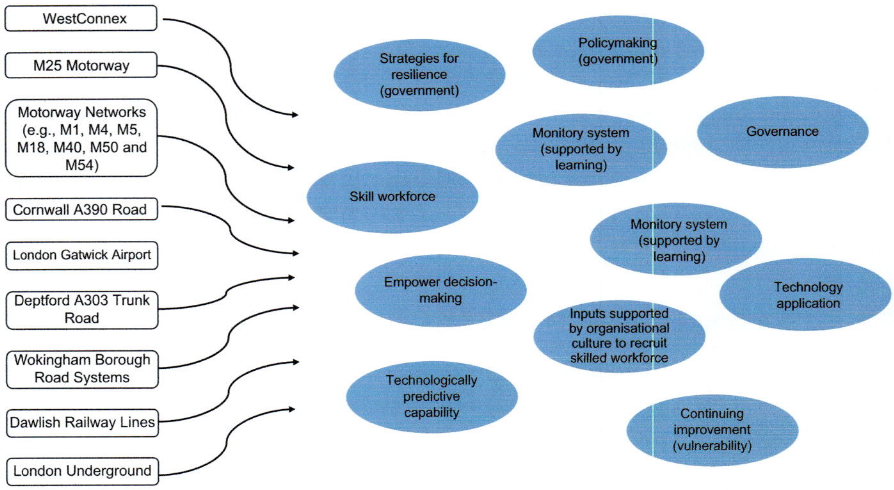

FIGURE 5.3 Mapped organisational capabilities leading to infrastructure resilience.

elaborated to the 'empower decision-making', while the 'reflective learning' and 'continuous improvement' are respective to the 'monitoring system' and 'asset vulnerability'. Apart from these, the coding of the cumulative case studies also narrowed the 'organisational culture', which is vague in Liu et al. (2021b), down to the: (1) 'skilled workforce' (with knowledge of enabling and managing resilience) across the engaged organisations (e.g., government, contractor and asset operator) and (2) 'strategies' (for resilience). Noteworthy, this finding complies with the conventional wisdom that human resources are critical in nurturing organisational culture.

Anchored on Figure 5.3, a conceptual SD model is developed in line with the life-cycle perspective of infrastructure asset to satisfy end-user's expectations (i.e., resilience for whom) (Figure 5.4). As indicated below, there are multiple feedback loops that illustrate how the capabilities coded from the case studies dynamically interact with each other to influence the resilient level, as the inputs and outputs of the organisations.

Conceptually, the government, the client of an infrastructure project, is expected to enact effective resilience strategies, which necessitate robust predictive (adaptive) thinking about future risks and hazards that must be addressed in the '*Initiation & Planning*' stage of an infrastructure project. Consequently, the workforce with the necessary skills and knowledge is undoubtedly needed to practise the strategies and then develop appropriate policies to support leveraging the inputs required to manage the strategic adjustments along with the changing environment. A similar feedback loop regarding the '*Design & Construction*' is also noted in Figure 5.4, where the outputs (i.e., policies) from the government would encourage the head contractor to form the right team (skilled workforce) and deploy a system on the basis of the lessons learned from the client's prioritised concern(s) to design, construct and monitor the asset to ensure the expected 'technical ability'. As a result, the contractor's activities above will shift their organisational culture to stay focused on resilience. Obviously, the 'design and construction' is not the end of the whole 'story' of resilience

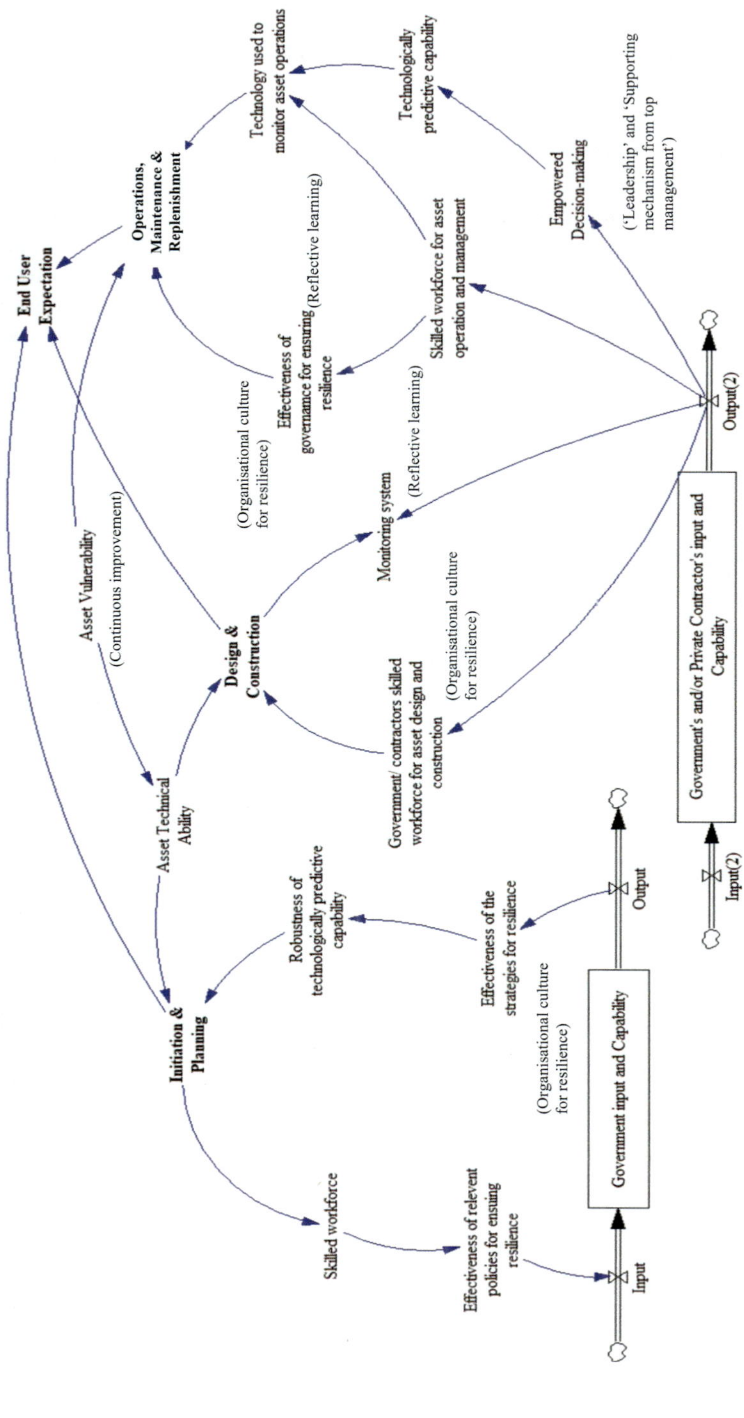

FIGURE 5.4 Conceptual SD model of the organisational capabilities determining resilience.

Note: Detailed '+' or '–' relationships between the factors above will be determined in future research.

enablement, while the asset's '*Operations, Maintenance & Replenishment*' are more important as this stage directly relates to services provided to the end users and lasts much longer than the 'Initiation & Planning' and 'Design & Construction'. As noted in Figure 5.4, there is a loop embedded with such capabilities as the 'technology' (for asset operations and monitoring), 'governance' and 'skilled workforce', which are the breakdowns of the 'reflective learning' and 'organisational culture' raised by the author's past research (i.e., Liu et al., 2021b). The dynamic relationships between these three capabilities in the 'Operations, Maintenance & Replenishment' stage are straightforward. In short, the skills and knowledge of the appointed workforce can determine the utilisation of technologies and also the effectiveness of governance, thereby affecting the activities taken to operate, maintain and replenish the assets.

Importantly, the conceptual SD visualises the significance of the 'empowered decision-making', which requires support from the top management of the government (i.e., if the asset is operated by the public authority) or the head contractor (i.e., if the asset is operated by the private entity under the contractual arrangement of public-private partnerships). This point of view further unearths Liu et al.'s (2021b) identification of the relationship between leadership and resilience, as empowered decision-making is closely linked to democratic leadership. More specifically, the team members will respond to the emergent events like disasters and crises in a faster and more efficient manner to enable less losses if they are empowered or encouraged to make decisions by themselves rather than being mandated to run through a long process to report and wait for the feedback from the 'top' (Huntsman et al., 2021). Put simply, the empowerment will allow the team members to predict hazards/crisis more quickly and then recommend the state of the art technologies or techniques for risk mitigation, to the best of their knowledge, rather than compromising their decision to consciously avoid 'punishment' (i.e., if the decision is unsuccessful), hence improving technologically predictive capability (as highlighted by Figure 5.4).

5.5 IMPLICATIONS FOR FUTURE RESEARCH AND PRACTICE

The developed SD in this chapter is a conceptual model grounded in the narratives and assumptions that were identified and implied from the cumulative case studies based on the grey literature of transport infrastructure. With the limitations of the data, the detailed '+' and '–' relationships embedded into the feedback loops[1] are pending, hence leaving a 'space' for future research to address. As such, more in-depth case studies of different types of assets (i.e., social infrastructure) via interviews and/or observations as well as survey are needed to finalise the conceptual SD for running a scenario simulation to enrich the discussion on practice. Despite the shortcoming above, the model has generated novel insights into the dynamic relationships that ensure the government, contractor and asset operator can make informed decisions and identify better actions taken to have the right strategy and policy in place, recruit a skilled team (workforce), implement appropriate monitoring system and adopt innovative technologies, for enabling or uplifting asset resilience to climate or environmental change. In other words, this chapter has advanced the knowledge of taxonomy in terms of the management of organisational capability and infrastructure resilience.

Additionally, future practice and research should place more attention on the decision-making-related work by addressing the questions such as 'How to enhance empowerment to improve resilience?' or 'How to improve existing decision-making approach to enable higher resilience?' Noteworthy, past research has underscored the relationship between decision-making and asset's resilient level like Zhao et al. (2022; 2023), whose DDCM (dynamic discrete choice model) and value-for-money-oriented framework comprise the resilience-related factors. In essence, the vulnerable aspects of many infrastructure assets resulting in low resilience are caused by some bad decisions, for example, Melbourne's West Gate Tunnel, Australia, and Houston Highways I-45, USA. Referring back to the UK's M25 Motorway selected for this chapter's case studies above, an external engineering association has long advised the National Highways that there have been too much sediment entering the run-off ponds of the motorway situated to collect surface water that exits the carriageway via drainage systems, leading to a very high risk of flood and road closures if extreme weather events like downpour occur. Nonetheless, the National Highways, a government-owned entity in charge of operating, maintaining and improving motorways and major A roads in England, fully ignored this external advice and decided to take no action because the scheduled maintenance is not due (New Civil Engineer, 2024). This interesting example not only reveals that a poor decision made by the asset operator would lower the resilient level of infrastructure but also reinforces a finding above that the predictability to risk/hazards/uncertainties, as an organisational capability, is essential for ensuring resilience.

5.6 CONCLUSIONS

Infrastructure resilience has been a worldwide concern because of the intensive impacts that climate change has on the functionality of a certain asset *per se* and/or the germane entire network. Although substantial attempts towards the enablement and enhancement of resilience are being documented in the normative literature, there remains an absence of critical research aimed at unpacking how the dynamic interdependency of the engaged organisations' capabilities leads to resilient assets. The upshot, in this instance, has been an inability of key stakeholder organisations to enact, engender and practise promising plan, policy and process to continuously improve resilience. Acknowledging this gap, the study presented in this chapter has initiated cumulative cases studies of nine (9) transport infrastructure assets. Therefore, thematic analysis was conducted, filtering out a series of organisational capabilities from the sides of the observed assets' clients (governments), contractors and operators. Then, a conceptual SD model was identified taking advantage of the case-study findings to facilitate a further discussion on those implications useful for future research and practice. To recap, the contribution of this chapter is three-fold, *viz.:* (1) advancement of knowledge through a conceptual SD; (2) novel insights into practice; and (3) implications addressing the valuable areas of future research. However, research limitations are imperative, and these are attributed to the 'nature' of the data on transport infrastructure. Therefore, future studies are encouraged in a wider context to conduct a simulation for the purpose of enabling a greater generalisability.

NOTE

1 While a positive "+" feedback loop amplifies change, a negative "–" loop stabilises or counteracts it (Sterman, 2000).

REFERENCES

Adams, T.M., Bekkem, K.R. and Toledo-Durán, E.J. (2012). Freight resilience measures, *Journal of Transportation Engineering*, 138(11), 1403–1409, https://doi.org/10.1061/(ASCE)TE.1943-5436.0000415

Adjetey-Bahun, K., Birregah, B., Châtelet, E. and Planchet, J. (2016). A model to quantify the resilience of mass railway transportation systems, *Reliability Engineering & System Safety*, 153, 1–14, https://doi.org/10.1016/j.ress.2016.03.015

Ambrosini, V., Bowman, C. and Collier, N. (2009). Dynamic capabilities: an exploration of how firms review their resource base, *British Journal of Management*, 20, 9–24, https://doi.org/10.1111/j.1467-8551.2008.00610.x

Audit Office of New South Wales (2014). *WestConnex: Assurance to the government*, New South Wales Auditor-General Report: Performance Audit, Sydney, Australia.

Audit Office of New South Wales (2021). *WestConnex: Changes since 2014*, New South Wales Auditor-General Report: Performance Audit, Sydney, Australia.

Audit Office of New South Wales (2023). *Transport 2023*, New South Wales Auditor-General Report: Performance Audit, Sydney, Australia.

Baroud, H., Barker, K., Ramirez-Marquez, J.E. and Rocco, C.M. (2014). Importance measures for inland waterway network resilience, *Transportation Research Part E: Logistics and Transportation Review*, 62, 55–67, https://doi.org/10.1016/j.tre.2013.11.010

BBC (2019). Heathrow and Gatwick passengers hit with more delays. Available at: https://www.bbc.com/news/uk-49122644 (accessed on 16th June 2025).

Becker, A.H., Matson, P., Fischer, M., and Mastrandrea, M.D. (2015). Towards seaport resilience for climate change adaptation: Stakeholder perceptions of hurricane impacts in Gulfport (MS) and Providence (RI). *Progress in Planning*, 99, 1–49, https://doi.org/10.1016/j.progress.2013.11.002

Berche, B., von Ferber, C., Holovatch, T. and Holovatch, Y. (2009). Resilience of public transport networks against attacks, *European Physical Journal B*, 71(1), 125–137, https://doi.org/10.1140/epjb/e2009-00291-3

Blake, D.M., Stevenson, J., Wotherspoon, L., Ivory, V. and Trotter, M. (2019). The role of data and information exchanges in transport system disaster recovery: A New Zealand case study, *International Journal of Disaster Risk Reduction*, 39, 101–124, https://doi.org/10.1016/j.ijdrr.2019.101124

Bosher, L. and Dainty, A. (2011). Disaster risk reduction and 'built-in' resilience: Towards overarching principles for construction practice, *Disasters*, 35(1), 1–18, https://doi.org/10.1111/j.1467-7717.2010.01189.x

Cabinet Office (2016). Steps for increasing community resilience, London, United Kingdom.

Cabinet Office (2017). Public summary of sector security and resilience plan, London, United Kingdom.

Carmeli, A. and Markman, G.D. (2011). Capture, governance, and resilience: Strategy implications from the history of Rome, *Strategic Management Journal*, 32, 322–341, https://doi.org/10.1002/smj.880

Chmutina, K., Lizarralde, G., Dainty, A. and Bosher, L. (2016). Unpacking resilience policy discourse, *Cities*, 58, 70–79, https://doi.org/10.1016/j.cities.2016.05.017

City of Sydney (2017). *The WestConnex challenge*, Public Summary Report, Sydney, Australia.

Collis, D.J. (1994). Research note: How valuable are organisational capabilities? *Strategic Management Journal*, 15(2), 143–152, https://doi.org/10.1002/smj.4250150910.

Cornwall Council (2019). Preliminary risk assessment: Annex 5, Cornwall, United Kingdom.

Cox, A., Prager, F. and Rose, A. (2011). Transportation security and the role of resilience: A foundation for operational metrics, *Transport Policy*, 18(2), 307–317, https://doi.org/10.1016/j.tranpol.2010.09.004.

Department for Transport (2014). Transport resilience review: A review of the resilience of the transport network to extreme weather events, CM 8874, London, United Kingdom.

Desouza, K.C. and Flanery, T. (2013). Designing, planning, and managing resilient cities: A conceptual framework, *Cities*, 35, 89–99, https://doi.org/10.1016/j.cities.2013.06.003

Dobie, S. and Schneider, J. (2017). Development of a process improvement matrix to measure community-level sustainability and resilience: a collaborative case study of Rochester, NY, USA, *2017 IEEE International Symposium on Technologies for Homeland Security (HST)*, 25–26 April 2017, IEEE, Waltham, MA, USA.

Giezen, M., Salet, W. and Bertolini, L. (2015). Adding value to the decision-making process of mega projects: Fostering strategic ambiguity, redundancy, and resilience, *Transport Policy*, 44, 160–178, https://doi.org/10.1016/j.tranpol.2015.08.006

Grant, R.M. (1991). The resource-based theory of competitive advantages: Implications for strategy formulation, *California Management Review*, 33(3), 114–135, https://doi.org/10.2307/41166664

Great London Authority (2019). Floods shut tube stations for 137 hours. Available at: https://www.london.gov.uk/press-releases/assembly/caroline-russell/floods-shut-tube-stations-for-137-hours (accessed 16th June 2025).

Gu, B. and Liu, J. (2025). A systematic review of resilience in the maritime transport, *International Journal of Logistics Research and Applications*, 28(3), 257–278, https://doi.org/10.1080/13675567.2023.2165051

Hallet, S. (2013). *Community resilience to extreme weather – the CREW Project*, Final Report. 110.

Hughes, J.F. and Healy, K. (2014). *Measuring the resilience of transport infrastructure*. NZTAR Report 546, 82pp.

Huntsman, D., Greer, A., Murphy, H. and Haynes, S. (2021). Enhancing adaptive performance in emergency response: Empowerment practices and the mediating role of tempo balance, *Safety Science*, 134, 105060, https://doi.org/10.1016/j.ssci.2020.105060

Inan, GG and Bititci, U.S. (2015). Understanding organisational capabilities theories in the context of micro enterprises: A research agenda, *Procedia – Social and Behavioral Science*, 210, 310–319, https://doi.org/10.1016/j.sbspro.2015.11.371

Ip, W.H. and Wang, D. (2011). Resilience and friability of transportation networks: Evaluation, analysis and optimization, *IEEE Systems Journal*, 5(2), 189–198, https://doi.org/10.1109/JSYST.2010.2096670.

Jackson, M.C. (2019). *Critical Systems Thinking and the Management of Complexity*, John Wiley & Sons, Hoboken, NJ

Johansen, C., Horney, J. and Tien, I. (2017). Metrics for evaluating and improving community resilience, *Journal of Infrastructure Systems*, 23(2), https://doi.org/10.1061/(ASCE)IS.1943-555X.0000329

Khalil, S. and Belitski, M. (2020). Dynamic capabilities for firm performance under the information technology governance framework, *European Business Review*, 32(2), 129–157.

Legislative Council (2018). The impact of the WestConnex Project, Sydney, New South Wales, Australia.

Li, L., Ding, Y., Yuan, J., Ji, W., Zhao, J. and Shen, L. (2022). Quantifying the resilience of emergency response network to infrastructure interruptions through an enhanced metanework-based framework, *ASCE's Journal of Management in Engineering*, 38(5), 04022047, https://doi.org/10.1061/(ASCE)ME.1943-5479.0001080

Liu, H.J., Love, P.E.D., Ma, L. and Sing, M.C.P. (2021a). Predicting production-output performance within a complex business environment: from singular to multi-dimensional observations in evaluation, *International Journal of Production Research*, 59(24), 7390–7399, https://doi.org/10.1080/00207543.2020.1841316

Liu, H.J., Love, P.E.D., Sing, M.C.P., Niu, B. and Zhao, J. (2019). Conceptual framework of life-cycle performance measurement: Ensuring the resilience of transport infrastructure assets, *Transportation Research Part D: Transport and Environment*, 77, 615–626, https://doi.org/10.1016/j.trd.2019.10.002

Liu, H.J., Love, P.E.D., Smith, J., Sing, M.C.P. and Matthews, J. (2018). Evaluation of public-private partnerships: A life cycle performance prism for ensuring value for money, *Environment and Planning C: Politics and Space*, 36(6), 1133–1153, https://doi.org/10.1177/2399654417750879

Liu, H.J., Love, P.E.D., Zhao, J., Lemckert, C. and Muldoon-Smith, K. (2021b). Transport infrastructure asset resilience: Managing government capabilities, *Transportation Research Part D: Transport and Environment*, 100, 103072, https://doi.org/10.1016/j.trd.2021.103072

Loo, B.P.Y. and Leung, K.Y.K. (2017). Transport resilience: The occupy central movement in Hong Kong from another perspective, *Transportation Research Part A: Policy and Practice*, 100–115, https://doi.org/10.1016/j.tra.2017.09.003

Love, P.E.D., Holt, G.D. and Li, H. (2002). Triangulation in construction management research, *Engineering, Construction and Architectural Management*, 9(4), 294–303, https://doi.org/10.1108/eb021224

Love, P.E.D., Ika, L.A., Matthews, J., Li, X. and Fang, W. (2021). A procurement policy-making pathway to future-proof large-scale transport infrastructure assets, *Research in Transportation Economics*, 90, 101069, https://doi.org/10.1016/j.retrec.2021.101069

Mansouri, M., Nilchiani, R. and Mostashari, A. (2010). A policy-making framework for resilient port infrastructure systems. *Marine Policy*, 34(6), 1125–1134, https://doi.org/10.1016/j.marpol.2010.03.012

McBain W., Wilkes, D. and Retter, M. (2010). Flood resilience and resistance for critical infrastructure, CIRIA, London, United Kingdom.

McDaniels T., Chang, S., Cole, D., Mikawoz, J. and Longstaff, H. (2008). Fostering resilience to extreme events within infrastructure systems: Characterizing decision contexts for mitigation and adaptation, *Global Environmental Change*, 18(2), 310–318, https://doi.org/10.1016/j.gloenvcha.2008.03.001

McMillan, D. (2014). *Disruption at Gatwick Airport: Christmas Eve 2013*, Report to the Board of Gatwick Airport Limited, United Kingdom.

Miller-Hooks, E., Zhang, X. and Faturechi, R. (2012). Measuring and maximizing resilience of freight transportation networks, *Computers & Operations Research*, 39(7), 1633–1643, https://doi.org/10.1016/j.cor.2011.09.017

Minaie, E. and Moon, F. (2017). A practical and simplified approach for quantifying bridge resilience, *Journal of Infrastructure Systems*, 23(4), 04017016, https://doi.org/10.1061/(ASCE)IS.1943-555X.0000374

NetworkRail (2019). Five years since we reopened Dawlish. Available at: https://www.networkrail.co.uk/stories/five-years-since-we-reopened-dawlish/ (accessed on 16th June 2025).

New Civil Engineer (2024). Expert ‘massively disputes’ National Highways claim that M25 runoff ponds aren’t overdue maintenance. Available at: https://www.newcivilengineer.com/latest/expert-massively-disputes-national-highways-claim-that-m25-runoff-ponds-arent-overdue-maintenance-09-09-2024/ (accessed on 16th June 2025).

Omer, M., Mostashari, A., Nilchiani, R. and Mansouri, M. (2012). A framework for assessing resiliency of maritime transportation systems, *Maritime Policy & Management*, 39(7), 685–703, https://doi.org/10.1080/03088839.2012.689878

Raffoni, A., Visani, F., Bartolini, M. and Silvi, R. (2018). Business performance analytics: exploring the potential for performance management systems, *Production Planning and Control*, 29(1), 51–67, https://doi.org/10.1080/09537287.2017.1381887

Reggiani, A. (2013). Network resilience for transport security: Some methodological considerations, *Transport Policy*, 28, 63–68, https://doi.org/10.1016/j.tranpol.2012.09.007

Reggiani, A., Nijkamp, P. and Lanzi, D. (2015). Transport resilient and vulnerability: The role of connectivity, *Transportation Research Part A: Policy and Practice*, 81, 4–15, https://doi.org/10.1016/j.tra.2014.12.012

Santos, J.R., Herrera, L.C. and Yu, K.D.S. (2014). State of the art in risk analysis of workforce criticality influencing disaster preparedness for interdependent systems, *Risk Analysis*, 34(6), 1056–1068, https://doi.org/10.1111/risa.12183

Shaw, D.R., Achuthan, K., Sharma, A. and Grainger, A. (2019). Resilience orchestration and resilience facilitation: How government can orchestrate the whole UK ports market with limited resources – the case of UK port resilience, *Government Information Quarterly*, 36, 252–263, https://doi.org/10.1016/j.giq.2018.12.003

Sing, M.C.P., Love, P.E.D. and Liu, H.J. (2019). Rehabilitation of existing building Stock: A system dynamics model to support policy development, *Cities*, 87, 142–152, https://doi.org/10.1016/j.cities.2018.09.018

Smallwood, N. and Ulrich, D. (2004). Capitalising on capabilities. *Harvard Business Review*, 82(6), 119–127.

Solomon, S, Qin, D., Manning, M., Chen, Z., Marquis, M., Averyt, K.B., Tignor, M. and Miller, H.L. (2007). Climate change 2007: the physical science basis, *Contribution of Working Group I to the Fourth Assessment Report of the Intergovernmental Panel on Climate Change*, Cambridge University Press, Cambridge, UK.

Sterman, J.D. (2000). *Business Dynamics—Systems Thinking and Modeling for a Complex World*, McGraw Hill, Boston, MA.

Stewart, G.T., Kolluru, R. and Smith, M. (2009). Leveraging public-private partnerships to improve community resilience in times of disaster, *International Journal of Physical Distribution and Logistics Management*, 39(5), 343–364, https://doi.org/10.1108/09600030910973724

Teece, D.J. (2007). Explicating dynamic capabilities: the nature and micro foundations of (sustainable) enterprise performance, *Strategic Management Journal*, 28, 1319–1350, https://doi.org/10.1002/smj.640

Teece, D.J., Pisano, G. and Shuen, A. (1997). Dynamic capabilities and strategic management, *Strategic Management Journal*, 18(7), 509–533, https://doi.org/10.1002/(SICI)1097-0266(199708)18:7<509::AID-SMJ882>3.0.CO;2-Z

Transport for NSW (2023). WestConnex M8 road network performance review plan, NSW Government, Australia.

US Environment Protection Agency (EPA) (2015). Glossary of climate change terms. Available at: https://19january2017snapshot.epa.gov/climatechange/glossary-climate-change-terms_html (accessed on 16th June 2025).

UK Parliament (2014). A303: Debated on Tuesday 4 March 2014. Available at: https://hansard.parliament.uk/Commons/2014-03-04/debates/14030456000001/A303 (accessed on 16th June 2025).

Vale, L. (2014). The politics of resilient cities: Whose resilience and whose city? *Building Research and Information*, 42(2), 191–201, https://doi.org/10.1080/09613218.2014.850602

Wan, C., Yang, Z., Zhang, D., Yan, X. and Fan, S. (2018). Resilience in transportation systems: a systematic review and future directions, *Transport Review*, 38(4), 479–498, https://doi.org/10.1080/01441647.2017.1383532

Wang, J., Liao, F., Wu, J., Sun, H., Wang, W. and Gao, Z. (2023). Measurement of functional resilience of transport network: The case of the Beijing subway network, *Transport Policy*, 140, 54–67, https://doi.org/10.1016/j.tranpol.2023.06.016

Winter, S. (2003). Understanding dynamic capabilities, *Strategic Management Journal*, 24, 991–995, https://doi.org/10.1002/smj.318

Yin, R.K. (2013). *Case Study Research: Design and Methods*, Sage Publications, Thousand Oaks, USA.

Zhao, J., Liu, H.J., Love, P.E.D., Greenwood, D.J. and Sing, M.C.P. (2022). Public-private partnerships: A dynamic discrete choice model for road projects, *Socio-Economic Planning Sciences*, 82, 101227, https://doi.org/10.1016/j.seps.2022.101227

Zhao, J., Liu, H.J., Love, P.E.D., Greenwood, D. and Sing, M.C.P. (2023). Value for money assessment for public-private partnerships: Characteristics, research directions, and policy implications, *Developments in the Built Environment*, 16, 100246, https://doi.org/10.1016/j.dibe.2023.100246

6 Adapting Infrastructure to a Changing Climate with Extreme Event Attribution Assessments

Yating Zhang[a] *and Zachary M. Labe*[b]
[a]College of Computer, Mathematical, and Natural Sciences, University of Maryland, College Park, Maryland
[b]Geophysical Fluid Dynamics Laboratory, Office of Oceanic and Atmospheric Research, National Oceanic and Atmospheric Administration, Princeton, New Jersey

6.1 INTRODUCTION

As global temperatures rise, the need to prepare for and respond to extreme weather-driven events increases. Extreme event attribution (EEA) is an emerging science encompassing a collection of techniques that quantify the influence of human-induced changes in the global climate system on the probability and characteristics of extreme events. Compared to long-term climate projections, EEA offers the potential for greater reliability by adhering to strict protocols for event definition, data collection, model selection, statistical characterization, results validation, and results communication (Philip et al., 2020). Event definition involves specifying key components such as climate variables, impact areas, and analysis time frames. Moreover, EEA can be extended to quantify the social, economic, ecological, and other impacts arising from the influence of climate change on extreme weather, raising awareness of climate risks and informing strategies for mitigating future extremes (Perkins-Kirkpatrick et al., 2024; Philip et al., 2020; van Oldenborgh et al., 2021).

In the context of infrastructure adaptation, EEA has the potential to support multiple areas of research and decision-making (Zhang et al., 2024) (Figure 6.1). First, EEA helps assess the changing likelihood and intensity of a weather event, particularly low-frequency, high-impact events. When integrated with vulnerability assessments, EEA can improve the understanding of how population and infrastructure vulnerabilities are changing. Building codes and standards—regulations that govern how buildings should be designed and constructed—are typically developed based on the identified vulnerabilities and are regularly updated with new scientific knowledge, technological advancements, and lessons learned from past events. For instance, after catastrophic events, investigations are conducted to determine the causes of building failures. These findings are then used to inform revisions to

DOI: 10.1201/9781003567332-6

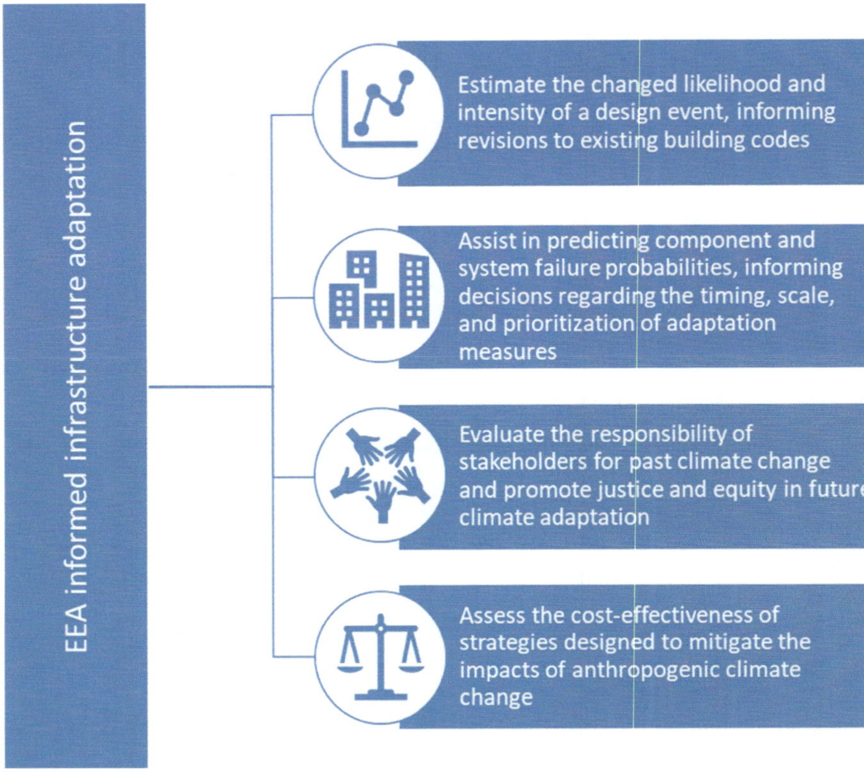

FIGURE 6.1 A schematic illustration of how extreme event attribution (EEA) assessments can be incorporated into adaptation planning.

existing building codes, aiming to prevent similar failures in the future (Vaughan and Turner, 2013). This process integrates the expertise of stakeholders, professionals, and local communities, fostering solutions that align with both community needs and engineering best practices. However, it is important to recognize that this approach is inherently localized and site specific, as vulnerabilities and risks vary significantly across different regions. To enable broader application, a substantial number of case studies across various geographic and climatic contexts should be collected. Systematic analysis should be performed on these case studies to identify overarching patterns and develop generalized frameworks that guide regional and national adaptation strategies (Carter et al., 2007; Jones and Boer, 2004).

Second, EEA helps project the changing probability distribution of a specific type of event, allowing for the estimation of changing component and system failure probabilities. These insights can inform decision-makers about the urgency of mitigating climate change and reducing climate risks at their source. On the other hand, a forward-looking approach to infrastructure adaptation, increasingly advocated in recent years, involves the progressive enhancement of infrastructure based on the most reliable, state-of-the-art future climate model projections (ASCE, 2018). This approach integrates climate projections into impact models to assess the future probability of component and system failures (Ghosn and Ellingwood, 2023; Zhang and Ayyub, 2020). The results inform decisions regarding the timing, scale, and prioritization of adaptation measures, which are dependent on acceptable risk and adaptive capacity. The forward-looking approach requires a greater effort to interpret uncertainties related to climate scenarios, model responses and biases, downscaling methods, and climate variability in an engineering context. Moreover, non-climate factors should be incorporated into this forward-looking approach, such as policy and socioeconomic conditions, to ensure that infrastructure improvements are not only scientifically sound but also socially equitable, economically feasible, and aligned with existing political and governance contexts (Carter et al., 2007; Jones and Boer, 2004).

Third, EEA helps assess the responsibility of stakeholders for past climate change and promote justice and equity in future climate adaptation (Thompson and Otto, 2015; Jézéquel et al., 2019). By establishing the causality between greenhouse gas emissions and extreme events, EEA can facilitate strategic collaborations among stakeholders to address both past and future impacts of climate change (Huggel et al., 2016; Otto et al., 2017). Additionally, EEA can be integrated with exposure and vulnerability assessments to attribute the disproportionate impacts of extreme events to climate change (Smiley et al., 2022; Newman and Noy 2023). Identifying such inequalities is vital for developing equitable and targeted interventions that address the specific needs and challenges faced by vulnerable groups or areas (Smiley et al., 2022; Newman and Noy 2023).

Fourth, attribution results can be used to assess the cost-effectiveness of strategies designed to mitigate the impacts of anthropogenic climate change (Zhang et al., 2024), as discussed in the next section. The next section also describes the methodology for integrating EEA into the probabilistic risk model for evaluation and optimization of adaptation strategies. The section following presents case studies to illustrate this methodology. The subsequent section provides an overview of recent advances in climate science and their implications for engineering practice. The final section summarizes and concludes this chapter. Overall, this chapter explores the methods, opportunities, and challenges of integrating EEA into engineering adaptation for extreme climate events.

6.2 METHODOLOGY

6.2.1 EEA Approach

Two primary methods are used for EEA: the probabilistic approach (or risk-based approach) and the storyline approach (or process-based approach). The probabilistic approach uses a statistical model to estimate the likelihood of an observed event occurring under the current climate and a plausible counterfactual climate (e.g., without anthropogenic Earth-system change). This estimation can be derived from statistical trend models or targeted climate model outputs. The difference in these likelihoods quantifies the influence of forced climate change (Stott et al., 2004; Stone and Allen, 2005), often expressed using the fraction of attributable risk (FAR):

$$\text{FAR} = 1 - \frac{p_0}{p_1} = 1 - \frac{1}{\text{PR}} \tag{6.1}$$

where p_0 is the probability that the event occurs in a counterfactual world without climate change. For example, the counterfactual world can be estimated by considering the preindustrial mean climate conditions (1850–1900). p_1 is the probability that the event occurs in the modern-day real world. FAR describes the contribution of human-induced change to the changed likelihood of an event, such as the proportion of occurrences in the actual world that would not have occurred in the counterfactual world. PR is the probability ratio, indicating that the event is PR times more likely to occur in the current climate than in a preindustrial climate.

It should be noted that the probabilistic approach involves various sources of uncertainties, including differences in definitions and indices for extreme events, inherent climate model biases used to reproduce the events, bias correction methods applied to raw climate model outputs, lack of high-quality observations, and unconditional (full) or conditional probability employed to depict FAR (Jézéquel et al., 2019; Hauser et al., 2017). This chapter focuses on the probabilistic approach as it is straightforward and well-studied. In contrast, the storyline approach is more often used to estimate changes in the intensity of events than in the frequency (Shepherd et al., 2018; Lloyd and Oreskes, 2018). It can isolate the influence of specific physical aspects of climate change, which is pivotal for establishing the causal chain between climate change and associated damage and loss. A description for the storyline approach can be found in Trenberth et al. (2015), Shepherd (2016), or Sillmann et al. (2021).

6.2.1.1 Confidence Level of Extreme Event Attribution

One of the earliest assessment reports to evaluate our confidence in the ability to detect and attribute extreme events was a 2016 report by the National Academies of Sciences, Engineering, and Medicine (NASEM, 2016). This report outlined that confidence is highest for the anthropogenic effect on temperature extremes through long-term warming, and it showed that there is moderate confidence in the attribution of drought and heavy precipitation extremes. On the other hand, confidence in the detection and attribution of smaller-scale phenomena, like severe thunderstorms and (extra-)tropical cyclones, remains low. More recent studies have reinforced these findings (e.g., Kretschmer et al., 2024; Shepherd, 2016; Stott et al., 2016; Otto, 2017; Bellprat et al., 2019; Kirchmeier-Young et al., 2019; Naveau et al., 2020; van Oldenborgh et al., 2021; Clarke et al., 2022; Perkins-Kirkpatrick et al., 2022; Oldenborgh et al., 2022; Clarke et al., 2023; Otto, 2023; Jézéquel et al., 2024).

In addition, several agencies and organizations have developed systems for the rapid attribution of extreme events and apply a wide variety of different statistical and modeling methods (Philip et al., 2020; Barsugli et al., 2025; Gilford et al., 2022; Faranda et al., 2024). Some groups, like from the National Oceanic and Atmospheric Administration (NOAA), primarily detect and attribute the higher-confidence extremes of temperature and hydrologic events (Schreck et al., 2024), while others focus on a broader range of extremes like by World Weather Attribution (https://www.worldweatherattribution.org/). However, there remain substantial uncertainties in understanding the dynamical response to external forcing—in other words, addressing the role of forced changes in the atmospheric circulation for EEA, especially at local and regional scales, remains a challenge (Shepherd, 2014; Shaw et al., 2024a,b).

6.2.2 Probabilistic Risk Model

Climate risk reflects a combination of climate-related hazards, exposure, vulnerability, and human responses to climate change (IPCC, 2021). In system engineering, annual climate risk (R) is calculated as follows (Stewart and Deng, 2015; ASCE, 2018):

$$R = \sum_{C_c}\sum_{C_s}\sum_{F}\sum_{L} P(C_c)P(C_s \mid C_c)P(F \mid C_s)P(L \mid F)L \tag{6.2}$$

where

- R denotes the absolute climate risk; that is, the expected value considering various climate-related factors.
- C_c denotes climate scenarios.
- C_s denotes climate stressors, such as more frequent and intense precipitation events, heat waves, coastal flooding, and wildfires.
- F denotes system failure. It can refer to a single or a complex infrastructure system.
- L comprises direct and indirect economic losses due to physical damage of components and loss of services. L is a function of hazard levels.
- $P(C_c)$ is the probability that a climate change occurs (e.g., a global warming level (GWL) of 1.5°C, 2°C, or 3°C). This is discussed in Section 6.4.2.
- $P(C_s \mid C_c)$ is the probability that a stressor intensifies when the climate changes.
- $P(F \mid C_s)$ is the probability of system failure when the stressor intensifies.
- $P(L \mid F)$ is the probability of loss when the system fails physically or functionally.

In the case that loss L is directly proportional to the magnitude of an extreme event (or inversely proportional to the exceedance probability of an extreme event), the added risk due to climate change (ΔR) can be estimated as follows (Newman and Noy, 2023; Smiley et al., 2022; Frame et al., 2020b):

$$\Delta R = \text{FAR} \times R \tag{6.3}$$

However, the loss L is often not linearly proportional to the magnitude of extreme events (Frame et al., 2020a). For an extreme type under a given climate scenario, where C_c is fixed and thus $P(C_c)$ equals one, Equation 3 can be rewritten as follows (Zhang et al., 2024):

$$R = \sum_{F}\sum_{L} \text{PR}_E \cdot P(x \geq E)P(F \mid x \geq E)P(L \mid F)L \tag{6.4}$$

where

- E denotes the hazard level, such as the 90th percentile of the probability distribution in the historical climate condition.

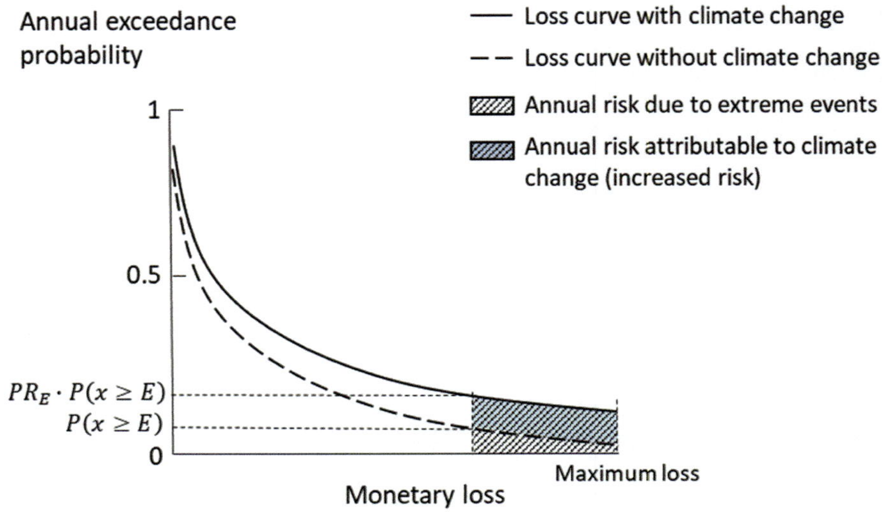

FIGURE 6.2 Loss curves with and without climate change (Zhang et al., 2024). $P(x \geq E)$ is the probability that the extreme event exceeded the level E in a historical reference climate. PR_E is the probability ratio that adjusts the exceedance probability to the given mean climate condition. The maximum loss here refers to the total property loss.

- PR_E is the probability ratio that adjusts the probability of experiencing an event more extreme than E from the baseline to the given climate scenario. PR_E is estimated through the attribution analysis via Eq. (6.1).
- $P(x \geq E)$ is the probability that the extreme exceeded the level E in the historical climate condition.
- $P(F \mid x \geq E)$ is the probability of system failure when the extreme exceeds the level E, which can be estimated through numerical or statistical modeling, including but not limited to simple extrapolation (NASEM, 2020).

Figure 6.2 shows a schematic of loss curves with and without climate change, highlighting the additional risk due to human-induced climate change (see Zhang et al., 2024). The shaded area represents the annual risk incurred by extreme events in a changing climate. The shaded area with a white background depicts the annual risk without climate change. The shaded area with a blue background describes the increased annual risk due to anthropogenic change.

6.2.2.1 Cost-Effectiveness of Adaptation Strategies

Cost-effectiveness is a key decision metric for stakeholders and policymakers. An adaptation strategy is considered cost-effective when its total benefits exceed its total costs over the planning horizon. As outlined in Zhang et al. (2024), the annual risk after the implementation of an adaptation strategy can be estimated as follows:

$$R_{\text{adapt}} = \sum_F \sum_L \mathrm{PR}_E \cdot P(x \geq E) P_\alpha(F \mid x \geq E) P_\alpha(L \mid F) L_\alpha \tag{6.5}$$

where

- $P_\alpha(F \mid x \geq E)$ is the probability of system failure after adaptation.
- $P_\alpha(L \mid F)$ is the probability of loss after adaptation.
- L_α is the loss function after adaptation.

The benefit of adaptation (B_{adapt}) can be estimated as follows (Zhang et al., 2024; Stewart and Deng, 2015):

$$B_{\text{adapt}} = \sum_{t=t_0}^{T} R_{\text{adapt}} (1+r)^{-(t-t_0)} + B_c \tag{6.6}$$

where,

- T is the planning horizon (a future time period), over which benefits and costs are evaluated. Planning horizon is typically aligned with the expected lifespan of a building or structure (e.g., 50 years for ordinary buildings, 75 years for bridges).
- r is the annual discount rate, used to adjust future cash flows back to the present value.
- t_0 is the year when adaptation strategy is implemented.
- B_c is the co-benefit of adaptation, such as reduced losses to moderate events and other hazard types.

Equation (6.5) typically uses a static estimate for PR_E based on historical data. If PR_E is expected to increase over time, Eq. (6.6) will be a conservative estimate of the benefit. For example, the likelihood of exceeding a given temperature threshold is likely to increase over time, in which case PR_E will also increase over the lifetime of the structure being designed. However, if PR_E is expected to decrease over time, which is the case for cold temperature extremes, Eq. (6.6) will lead to an inflated estimate of future benefits. It is straightforward although to allow PR_E to vary with t in Eqs. (6.4) and (6.5) to avoid this concern (outlined in Zhang et al., 2024).

6.3 REVIEW OF CASE STUDIES

Several studies have used attribution results to evaluate the contribution of anthropogenic climate change to the economic impacts of extreme events. For example, Newman and Noy (2023) selected 185 attribution studies covering extreme events worldwide and combined their FAR estimates with estimated socioeconomic costs. Their findings indicate that human-induced change may be responsible for 53% of annual losses from extreme events, with loss of human life comprising 63% of the climate change-attributable losses. The majority of climate change–related damages were associated with storms, such as hurricanes, while heat waves accounted for 16% of the losses. Flooding and drought each contributed 10% of total losses, and wildfires were responsible for 2%.

Similarly, Frame et al. (2020b) analyzed the economic costs of 14 drought and flood events in New Zealand from 2007 to 2017. By integrating estimated economic costs with FAR results, they attributed approximately $140 million (2017 NZ dollars) losses from 12 extreme rainfalls (or flood events) and $300 million losses from 2 drought events to human-induced climate change. This study relied on insurance data to estimate economic losses and did not account for long-term costs, potentially leading to a substantial underestimation of climate change's contribution to these losses.

Three key studies focused on hurricanes. Frame et al. (2020a) analyzed the economic impact of the 2017 Hurricane Harvey in Harris County, Texas. Their results show that climate change–driven rainfall may have contributed to one-third of the total costs. With total damages from Harvey estimated at approximately $90 billion (2016 US dollars), $67 billion could be attributed to human-induced climate change, with a likely lower bound of $30 billion (Frame et al. 2020a).

Smiley et al. (2022) assessed climate change–attributed flood depths and damages during Hurricane Harvey. Using detailed land-parcel and census tract socioeconomic data, they found that 30%–50% of flooded properties would not have been inundated without climate change. Underrepresented communities, particularly low-income neighborhoods outside FEMA's 100-year floodplain, were disproportionately affected.

Lastly, Strauss et al. (2021) examined the impact of the 2012 Hurricane Sandy in New York, New Jersey, and Connecticut. Using building data at the census block level and depth–damage curves (describing flood damage that would occur at specific water depths per asset) for 33 building types, they found that approximately $8.1 billion (2012 US dollars) in damages can be attributed to anthropogenic sea-level rise. The increased flood extent affected an estimated 71,000 additional people (9.2% of the total population in these states) and 36,000 additional housing units (8.8% of the total housing stock).

Beyond these striking figures, Newman and Noy (2023) argued that commonly cited economic costs of climate change derived from integrated assessment models (IAMs) may be significantly underestimated. However, the IAM approach and the EEA-based approach are not directly comparable due to different economic data and model assumptions used. Additionally, existing estimates are often incomplete and likely understate the full economic costs of climate change, particularly the long-term economic effects (Frame et al., 2020b). These include shifts in economic activity, such as changes in employment and production across industries, following a damaging event (Strauss et al., 2021).

However, there are still many challenges associated with attributing economic losses to climate change. One major issue is the uneven geographical coverage of attribution studies and the lack of research on certain extreme weather events (Newman and Noy, 2023). Data availability poses another challenge, as researchers frequently rely on insured damages while excluding uninsured losses and broader economic impacts of flood events (Frame et al., 2020b). Additionally, studies on droughts typically estimate indirect economic losses using models, as there is no insurance coverage for drought-related damage to infrastructure (Frame et al., 2020b). Another challenge lies in the nonlinear relationship between loss and the magnitude of extreme events (Perkins-Kirkpatrick et al., 2022), which may include trigger points, as infrastructure is designed to withstand a certain level of extreme events (Zhang et al., 2024). Equation (6.4) captures this nonlinear relationship and should be validated in future studies.

6.4 DISCUSSION

Next, we discuss recent advances in climate science and their implications for EEA (Figure 6.3). The first notable advancement is decadal climate modeling, which provides a greater focus on near-term predictions for climate variability and change compared to standard centennial-scale climate models, such as via the Coupled Model Intercomparison Project (CMIP). Decadal prediction is the modeling component of EEA for some extreme events, particularly for those that can persist over years to decades (e.g., drought). The second advancement is the wider adoption of GWLs as a new way to communicate and frame climate uncertainties. Climate scenarios, such as the Representative Concentration Pathways (RCPs) and Shared Socioeconomic Pathways (SSPs), have long been used to describe potential future climates. However, these scenarios can be complex, restrictive, and challenging for engineers and policymakers to interpret, often resulting in the selection of a single scenario for developing climate plans (Burgos et al. 2024). In contrast, GWLs have been used in recent EEA studies to project change in extreme events due to human influence. The third is the expansion of regional climate services (encompassing various governmental agencies, nongovernmental organizations, and private-sector consulting entities), designed to provide real-time, trustworthy, and actionable climate information. This presents an opportunity to integrate short-term climate projections with EEA to support adaptation decisions.

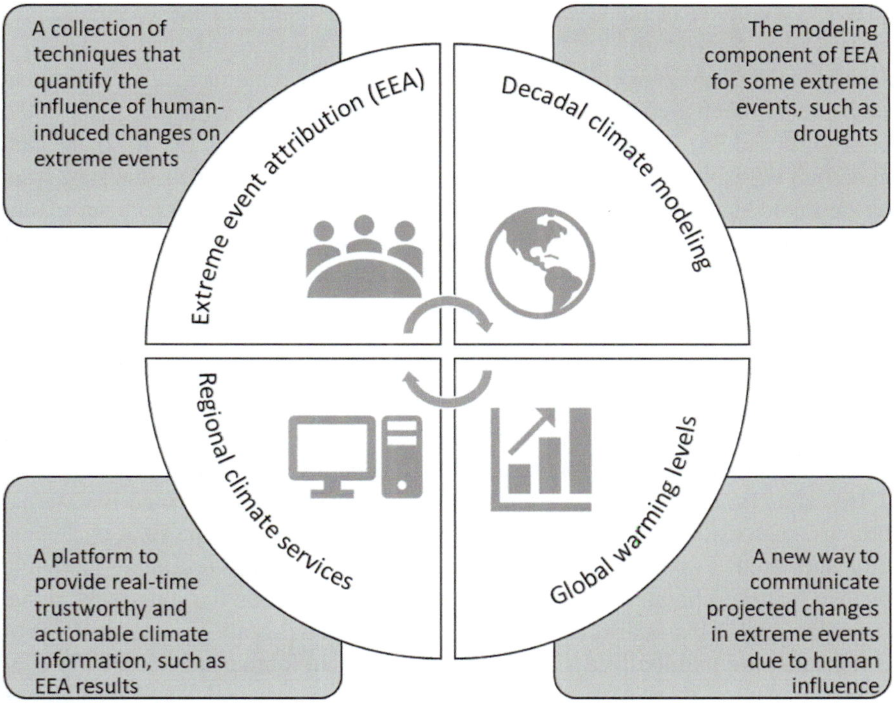

FIGURE 6.3 Summary of recent advances in climate science and their implications for EEA.

6.4.1 Decadal Climate Modeling

A rapidly growing desire from stakeholders is accessible and usable climate information within the multiyear to decadal time horizon (Towler et al., 2018; Done et al., 2021; Dunstone et al., 2022). This window, typically up to 30 years, bridges the gap between seasonal climate prediction and long-term climate change projections, and it is also key for near-term adaptation planning and decision-making. For example, a water manager in the western United States may be interested in climate information related to the risk of a multiyear extreme drought for preparing local water supplies and allocating resources to related engineering infrastructure (Done et al., 2021). Or, as another example, the agricultural sector may look for near-term predictions on the probability of sudden shifts in the expected temperature distribution which could result in a multiyear amplification of excessive heat stress and evapotranspiration changes for crops and irrigation (Solaraju-Murali et al., 2019).

New climate services are even being developed to skillfully predict the prevalence of hurricane activity and translate these forecasts for relevant stakeholders like the insurance industry (Caron et al., 2014; Lockwood et al., 2023). To help meet the increasing demand for this type of information, the World Meteorological Organization established the Lead Centre for Annual-to-Decadal Climate Predictions in 2017 to help coordinate experiments across major climate modeling center institutions (Hermanson et al., 2022). However, significant challenges still remain for capturing and communicating high-impact weather and climate extremes at longer time scales beyond the subseasonal range (Merryfield et al., 2020).

While a number of modeling centers now regularly produce these decadal predictions (e.g., Boer et al., 2016; Yeager et al., 2018; Sospedra-Alfonso et al., 2021; Yang et al., 2021; Yeager et al., 2022), this area of study remains relatively new and experimental compared to the broader field of climate science (Smith et al., 2007; Keenlyside and Ba, 2010). Similar to seasonal forecasts (Vitart et al., 2012), decadal prediction is based on global climate models or Earth system models that start from the observed state of the climate system. This is otherwise known as a set of initial conditions and usually includes real-world information about the current state of the atmosphere, ocean, land, and sea ice (Meehl et al., 2021). Incorporating observational climate information into the decadal prediction systems typically involves algorithms of data assimilation, which are crucial for aligning the start of the model runs with the real-world phases of internal climate variability and related persistence existing from climate inertia (Zhou et al., 2024), such as from patterns of sea surface temperature anomalies and deeper ocean heat content. In general, this is a significant source of prediction skill for forecasts within this multiyear to decadal window.

Decadal climate modeling systems also include information related to external forcings, such as the evolution of greenhouse gases, changes in the solar cycle, and influences from explosive volcanic eruptions, and thus, decadal climate modeling involves natural climate variability superimposed on top of the long-term global warming trend. This is distinct from standard centennial-scale climate model projections that are uninitialized and rely only on external forcings and boundary conditions, including those associated with the Intergovernmental Panel on Climate Change (IPCC) and CMIP (Durack et al., 2025).

Despite the rapid growth of this field, challenges persist in both addressing technical issues and effectively communicating findings to society. In particular, improvements are needed related to simulating the physics of extreme weather and climate phenomena, and this involves reducing related model biases and forcing sensitivities (Meehl et al., 2013). Increasing the capability to capture this fine spatial or vertical information through higher-resolution models is currently one active area of research geared toward improving decadal prediction (Meehl et al., 2021). Work is also ongoing to assess the potential for artificial intelligence methods in improving forecast skill and identifying related forecasts of opportunity (Toms et al., 2021; Gordon and Barnes, 2022; Labe and Barnes, 2022). Lastly, it is important to note that the formation of decadal climate services is just beginning. Making sure the information that is communicated for these outlooks and probabilistic information is usable needs to remain a priority when working together across the diverse network of user interests and needs (Dunstone et al., 2022; O'Kane et al., 2023).

6.4.2 Global Warming Levels

At longer time horizons, such as after a decade, errors grow and any potential prediction skill gained from memory related to the initial state of the Earth system begins to decline; eventually, this leads to a boundary value problem that is more commonly associated with multi-decadal to century-time scale climate model projections (i.e., changes are constrained in response to external forcing from greenhouse gas levels) (Collins and Allen, 2002; Meehl et al., 2021). Extending these decadal predictions seamlessly toward multi-decadal time scales remains an important but difficult task (Delworth et al., 2020; Befort et al., 2022). This is especially an important question for services around climate projections (Findlater et al., 2021; Sillmann et al., 2024; Mariotti et al., 2024), which have their own unique contributions of uncertainty and overall societal utility.

With greater lead time, new uncertainties begin to grow related to specific climate scenario pathways (Hawkins and Sutton, 2009). In other words, this uncertainty is associated with the future level of radiative forcing that is primarily driven by greenhouse gas emissions. For climate model projections, this is typically structured by prescribing the model with a range of different climate scenario designs. For some of the recent CMIP generations, these scenarios were referred to as SSPs (O'Neill et al., 2016), RCPs (van Vuuren et al., 2011), or the Special Report on Emissions Scenarios (SRES; Nakicenovic et al., 2000). The most recent SSP scenarios specifically look into possible future realizations of changes in greenhouse gases, population, and other socioeconomic factors, which ultimately lead to a set amount of radiative forcing imbalance by the end of the 21st century.

One method that has been newly proposed for building codes is the use of "global warming levels" (GWLs) (Ayyub et al., 2023; Barsugli et al., 2025), rather than having to select one or two climate scenarios as in previous efforts incorporating climate information into building practices. Another way of thinking about this convention is an "impacts-by-degree" standardization (Ayyub et al., 2023), which aims to reframe having to consider scenario-specific uncertainties (see Cross-Chapter Box 11.1 in IPCC, 2023). Here, a GWL refers to the change in global mean surface air temperature compared to a given reference period, which for policymaking is often relative to the preindustrial era (1850–1900) as defined by the IPCC (e.g., IPCC, 2018, 2021). For example, one could scale the local changes in heat wave extremes at different GWLs (Dosio et al., 2018), and then infrastructure standards could be designed to account for projected changes in heat extremes at these different levels of future warming.

A key advantage of this approach is its adaptability to future changes in climate modeling efforts and emission scenario designs (Barsugli et al., 2025). As noted in the recent American Society of Civil Engineers (ASCE) workshop report, the GWL approach helps engineers to develop building code solutions to accommodate different levels of warming (Barsugli et al., 2025). Scaling hazards by GWL can also account for either more likely or more extreme outlier outcomes, including for EEA. However, this approach has some limitations. It does not eliminate uncertainties in future climate projections, especially for slow-responding variables like coastal sea-level rise or potential nonlinear interactions in the Earth system that cannot be easily scaled by GWL. Additionally, different climate models can produce different regional responses even at the same level of global mean warming.

Given that land temperatures are warming faster than global mean temperatures (Manabe and Stouffer, 1980; Duan et al., 2020), any additional increases to GWL will be associated with amplified regional warming and subsequent societal impacts. According to the Fifth National Climate Assessment (NCA5), the average United States temperature is expected to rise approximately 1.4°C for every 1°C increase to the GWL (Marvel et al., 2023). Similarly, with any increase to the GWL, there is a resulting amplification of extremes, like for heat and precipitation. Projections of future GWLs are highly dependent on both future greenhouse gas emissions and the sensitivity of Earth system feedbacks to these emissions (Marvel et al., 2023). Recent estimates from NCA5 indicate that the average timing to reach the policy-relevant threshold of a GWL of 2°C is about 2053 for a more probable middle-to-high emission scenario (SSP2-4.5) (Peters and Hausfather, 2020; Hausfather and Peters, 2020; Srikrishnan et al., 2022).

There is a growing consensus that current policies are aligned with approximately 2.7°C of warming by the end of the 21st century (Hausfather, 2025). While significant uncertainties remain in this projection, recent estimates indicate that staying below 1.5°C is almost impossible (Betts et al., 2023; Forster et al., 2023, 2024). As a result, the ASCE code committee has adopted a 3°C GWL scenario for 2080–2100 as the basis for developing a new chapter that addresses future conditions for flood, ice, snow, rain, and wind hazards (Barsugli et al., 2025). Further research is needed to support the shifts in engineering practices to accommodate various GWLs. This involves building codes and standards, which establish minimum requirements for the design, construction, maintenance, repair, and renovation of structures.

6.4.3 Climate Services

There is a growing interest in developing regional climate services, where climate information and data are tailored and translated to meet the needs of stakeholders and local communities through coproduction. Jakob et al. (2023) proposed the concept of operational climate modeling, with the goal to provide actionable near-real-time climate information for decision-making. Van den Hurk et al. (2018) noted that several key dilemmas (i.e., purpose of development, spatial and temporal scales, and application domain) should be acknowledged and addressed to enhance the usability and effectiveness of climate model data for climate services. Additionally, NOAA and ASCE are collaborating to integrate climate projections into engineering practices across the United States. One example is the proposed updates for ASCE standards and manuals of practice (Ayyub et al., 2023). Buildings and infrastructure are conventionally designed to withstand environmental loads based on historical data, with extreme events modeled as the tails of statistical distributions (e.g., the 90th percentile). However, as climate stresses intensify, leading to greater losses, it becomes essential to reassess and update building codes to reflect future climate conditions and nonstationarity. The latest ASCE–NOAA workshop outlines a methodological framework for civil engineers to incorporate meteorology and climate science into ASCE standards and manuals of practice (Barsugli et al., 2025). This framework highlights three key areas needed for future research: (1) detailed review of methodologies and datasets cited in codes, standards, or other relevant engineering guidance documents to identify key environmental variables; (2) development of projected values as input and appropriate methodologies to account for changes in natural and anthropogenic climate forcing functions; and (3) provision of engineering values for calculations described in relevant engineering guidance documents.

6.5 CONCLUSIONS

Integrating EEA into infrastructure adaptation planning offers numerous benefits. It provides a scientific basis for probabilistic risk analyses, enabling engineers and policymakers to assess the likelihood of future extreme events. This integration also facilitates cost-effectiveness analyses for adaptation measures, ensuring that resources are allocated efficiently to reduce vulnerabilities and mitigate damages. This chapter outlines a methodology for incorporating EEA results into these analytical frameworks, supported by illustrative case studies that demonstrate practical applications across diverse event types.

In addition to the methodology, this chapter identifies key areas for future research, including

- **Enhancing Decadal Climate Modeling**: Expanding the capacity to project climate variations over the next 5–30 years is critical for bridging the gap between near-term adaptation needs and long-term climate resilience.
- **Adapting Engineering Practices**: Preparing for shifts in engineering design standards and practices to accommodate varying global warming scenarios is essential for ensuring infrastructure remains safe and resilient under changing climate conditions.
- **Developing Climate Services**: Creating advanced climate services that integrate EEA capabilities with near-term climate projections will provide actionable insights for stakeholders, from municipal planners to private-sector developers.

In the United States, some local jurisdictions have already begun adopting forward-looking approaches by establishing guidelines for incorporating climate projections into the design and maintenance of municipal facilities (Burgos et al., 2024). These efforts demonstrate the growing recognition of the importance of integrating climate science into practical decision-making and highlight the potential of EEA to shape resilient infrastructure systems in the face of a changing climate. This chapter emphasizes the need for collaborative efforts among researchers, practitioners, and policymakers to harness the full potential of EEA in addressing the challenges of climate adaptation for civil infrastructure systems.

ACKNOWLEDGMENTS

We thank Dr. Kirsten Findell and Dr. Keith Dixon for their insightful comments and suggestions on the NOAA GFDL internal review of this chapter. The statements, findings, conclusions, and recommendations are those of the authors and do not necessarily reflect the views of NOAA, or the U.S. Department of Commerce.

REFERENCES

ASCE (American Society of Civil Engineers) (2018), *Climate-Resilient Infrastructure: Adaptive Design and Risk Management*. Edited by Ayyub, B. M., ASCE manuals and reports on engineering practice No.140, ISBN 9780784415191.

Ayyub, B. A., DeAngelo, B., Walker, D., and Barsugli, J. (2023), *ASCE–NOAA Workshops on Leveraging Earth System Science and Modeling to Inform Civil Engineering Design*. doi:10.25923/e8kn-n884

Barsugli, J. J., D. Walker, M. N. Edwards, and B. M. Ayyub (2025), Climate risk reduction: Hazards and processes for operationalizing climate information into asce standards and manuals of practice, NOAA technical memorandum OAR CPO, pp. 1–102. doi:10.25923/0D8C-DW27

Befort, D. J., L. Brunner, L. F. Borchert, C. H. O'Reilly, J. Mignot, A. P. Ballinger, G. C. Hegerl, J. M. Murphy, and A. Weisheimer (2022), Combination of decadal predictions and climate projections in time: Challenges and potential solutions, *Geophysical Research Letters*, 49. doi:10.1029/2022GL098568

Bellprat, O., V. Guemas, F. Doblas-Reyes, and M. G. Donat (2019), Towards reliable extreme weather and climate event attribution, *Nature Communications*, 10. doi: 10.1038/s41467-019-09729-2

Betts, R. A., S. E. Belcher, L. Hermanson, A. K. Tank, J. A. Lowe, C. D. Jones, C. P. Morice, N. A. Rayner, A. A. Scaife, and P. A. Stott (2023), Approaching 1.5°C: How will we know we've reached this crucial warming mark? doi:10.1038/d41586-023-03775-z

Boer, G. J., D. M. Smith, C. Cassou, F. Doblas-Reyes, G. Danabasoglu, B. Kirtman, Y. Kushnir, M. Kimoto, G. A. Meehl, R. Msadek, W. A. Mueller, K. E. Taylor, F. Zwiers, M. Rixen, Y. Ruprich-Robert, and R. Eade (2016), The decadal climate prediction project (dcpp) contribution to CMIP6, *Geoscientific Model Development*, 9, 3751–3777. doi:10.5194/GMD-9-3751-2016

Burgos, E., K. Irby, A. Williams, M. Fried, and A. Parris (2024), NIST Grant/Contractor Report (GCR) NIST GCR 24-056. doi:10.6028/NIST.GCR.24-056

Caron, L. P., C. G. Jones, and F. Doblas-Reyes (2014), Multi-year prediction skill of atlantic hurricane activity in cmip5 decadal hindcasts, *Climate Dynamics*, 42. doi: 10.1007/s00382-013-1773-1

Carter, T. R., R. N. Jones, X. Lu, S. Bhadwal, C. Conde, L. O. Mearns, B. C. O'Neill, M. D. A. Rounsevell, and M. B. Zurek (2007), New assessment methods and the characterisation of future conditions. In: Parry, M. L., Canziani, O. F., Palutikof, J. P., van der Linden, P. J., Hanson, C. E., editors. *Climate Change 2007: Impacts, Adaptation and Vulnerability. Contribution of Working Group II to the Fourth Assessment Report of the Intergovernmental Panel on Climate Change*. Cambridge, UK: Cambridge University Press, pp. 133–171.

Clarke, B., F. Otto, and R. Jones (2023), When don't we need a new extreme event attribution study? *Climatic Change*, 176. doi:10.1007/s10584-023-03521-4

Clarke, B., F. Otto, R. Stuart-Smith, and L. Harrington (2022), Extreme weather impacts of climate change: An attribution perspective, *Environmental Research: Climate*, 1, 012001. doi:10.1088/2752-5295/AC6E7D

Collins, M., and M. R. Allen (2002), Assessing the relative roles of initial and boundary conditions in interannual to decadal climate predictability, *Journal of Climate*, 15, 3104–3109. doi: 10.1175/1520-0442(2002)015<3104:ATRROI>2.0.CO;2

Delworth, T. L., W. F. Cooke, A. Adcroft, …, and M. Zhao (2020), Spear: The next generation gfdl modeling system for seasonal to multidecadal prediction and projection. *Journal of Advances in Modeling Earth Systems*, 12, e2019MS001895. doi:10.1029/2019MS001895

Done, J. M., R. E. Morss, H. Lazrus, E. Towler, M. R. Tye, M. Ge, T. Das, A. Mun´evar, J. Hewitt, and J. A. Hoeting (2021), Toward usable predictive climate information at decadal timescales, *One Earth*, 4. doi:10.1016/j.oneear.2021.08.013

Dosio, A., L. Mentaschi, E. M. Fischer, and K. Wyser (2018), Extreme heat waves under 1.5°C and 2°C global warming, *Environmental Research Letters*, 13. doi: 10.1088/1748-9326/aab827

Duan, S. Q., K. L Findell, and J. S Wright, (2020), Three regimes of temperature distribution change over dry land, moist land and oceanic surfaces, *Geophysical Research Letters*, 47(24). doi:10.1029/2020GL090997

Dunstone, N., J. Lockwood, B. Solaraju-Mural, K. Reinhardt, E. E. Tsartsali, P. J. Athanasiadis, A. Bellucci, A. Brookshaw, L. P. Caron, F. J. Doblas-Reyes, B. Früh, N. Gonza´lez-Revirieg, S. Gualdi, L. Hermanson, S. Materia, A. Nicodemou, D. Nicolì, K. Pankatz, A. Paxian, A. Scaife, D. Smith, and H. E. Thornton (2022), Towards useful decadal climate services, *Bulletin of the American Meteorological Society*, 103. doi:10.1175/BAMS-D-21-0190.1

Durack, P. J., K. E. Taylor, P. J. Gleckler, G. A. Meehl, B. N. Lawrence, C. Covey, R. J. Stouffer, G. Levavasseur, A. Ben-Nasser, S. Denvil, M. Stockhause, J. M. Gregory, M. Juckes, S. K. Ames, F. Antonio, D. C. Bader, J. P. Dunne, D. Ellis, V. Eyring, S. L. Fiore, S. Joussaume, P. Kershaw, J.-F. Lamarque, M. Lautenschlager, J. Lee, C. F. Mauzey, M. Mizielinski, P. Nassisi, A. Nuzzo, E. O'Rourke, J. Painter, G. L. Potter, S. Rodriguez, and D. N. Williams (2025), The coupled model intercomparison project (cmip): Reviewing project history, evolution, infrastructure and implementation, *EGUsphere*. doi:10.5194/EGUSPHERE-2024-3729

Faranda, D., G. Messori, E. Coppola, T. Alberti, M. Vrac, F. Pons, P. Yiou, M. S. Lu, A. N. Hisi, P. Brockmann, S. Dafis, G. Mengaldo, and R. Vautard (2024), Climameter: Contextualizing extreme weather in a changing climate, *Weather and Climate Dynamics*, 5, 959–983. doi:10.5194/WCD-5-959-2024

Findlater, K., S. Webber, M. Kandlikar, and S. Donner (2021), Climate services promise better decisions but mainly focus on better data, *Nature Climate Change*, 11. doi:10.1038/s41558-021-01125-3

Forster, P. M., C. J. Smith, T. Walsh, …, and P. Zhai (2023), Indicators of global climate change 2022: Annual update of large-scale indicators of the state of the climate system and human influence, *Earth System Science Data*, 15. doi:10.5194/essd-15-2295-2023

Forster, P. M., C. J. Smith, T. Walsh, …, and P. Zhai (2024), Indicators of global climate change 2023: Annual update of key indicators of the state of the climate system and human influence, *Earth System Science Data*, 16, 2625–2658. doi:10.5194/ESSD-16-2625-2024

Frame, D. J., M. F. Wehner, I. Noy, and S. M. Rosier (2020a), The economic costs of Hurricane Harvey attributable to climate change, *Climatic Change*, 160(2), 271–281.

Frame, D. J., S. M. Rosier, I. Noy, L. J. Harrington, T. Carey-Smith, S. N. Sparrow, D. A. Stone, and S. M. Dean (2020b), Climate change attribution and the economic costs of extreme weather events: A study on damages from extreme rainfall and drought. *Climatic Change*, 162, 781–797.

Ghosn, M., and B. R. Ellingwood (2023), Risk-informed design and safety assessment of structures in a changing climate: A review of U.S. practice and a path forward, *Structure and Infrastructure Engineering*, *20*(7–8), 1159–1173. doi:10.1080/15732479.2023.2265334

Gilford, D. M., A. Pershing, B. H. Strauss, K. Haustein, and F. E. Otto (2022), A multimethod framework for global real-time climate attribution, *Advances in Statistical Climatology, Meteorology and Oceanography*, 8. doi:10.5194/ascmo-8-135-2022

Gordon, E. M., and E. A. Barnes (2022), Incorporating uncertainty into a regression neural network enables identification of decadal state-dependent predictability in CESM2, *Geophysical Research Letters*, 49. doi:10.1029/2022GL098635

Hausfather, Z. (2025), An assessment of current policy scenarios over the 21st century and the reduced plausibility of high-emissions pathways, *Dialogues on Climate Change*. doi:10.1177/29768659241304854

Hausfather, Z., and G. P. Peters (2020), Rcp8.5 is a problematic scenario for near-term emissions. doi:10.1073/pnas.2017124117

Hauser, M., Gudmundsson, L., Orth, R., Jézéquel, A., Haustein, K., Vautard, R., Van Oldenborgh, G.J., Wilcox, L. and Seneviratne, S.I. (2017). Methods and model dependency of extreme event attribution: the 2015 European drought. *Earth's Future*, *5*(10), pp.1034–1043.

Hawkins, E., and R. Sutton (2009), The potential to narrow uncertainty in regional climate predictions, *Bulletin of the American Meteorological Society*, 90, 1095–1108. doi:10.1175/2009BAMS2607.1

Hermanson, L., D. Smith, M. Seabrook, …, and A. Kumar (2022), WMO global annual to decadal climate update a prediction for 2021–25, *Bulletin of the American Meteorological Society*, 103. doi:10.1175/BAMS-D-20-0311.1

Huggel, C., I. Wallimann-Helme, D. Stone, and W. Cramer (2016), Reconciling justice and attribution research to advance climate policy. *Nature Climate Change*, 6, 901–908. doi:10.1038/nclimate3104

IPCC (2018), *IPCC Special Report on Global Warming of 1.5°C*. Cambridge, United Kingdom and New York: Cambridge University Press.

IPCC (2023), *Weather and Climate Extreme Events in a Changing Climate*. Cambridge, United Kingdom and New York: Cambridge University Press. doi:10.1017/9781009157896.013

IPCC (Intergovernmental Panel on Climate Change). (2021), *Climate Change 2021: The Physical Science Basis. Contribution of Working Group I to the Sixth Assessment Report of the Intergovernmental Panel on Climate Change*. Cambridge, United Kingdom and New York: Cambridge University Press.

Jakob, C., A. Gettelman, and A. Pitman (2023), The need to operationalize climate modelling. doi:10.1038/s41558-023-01849-4

Jézéquel A., A. Bastos, D. Faranda, J. Kimutai, N. L. Grix, A. M. Wilson, S. Rufat, T. G. Shepherd, R. F. Stuart-Smith, A. F. V. Loon, E. Bevacqua, F. D'Andrea, F. Lehner, E. A. Lloyd, J. Moemken, A. M. Ramos, S. Sippel, and J. Zscheischler (2024), Broadening the scope of anthropogenic influence in extreme event attribution, *Environmental Research: Climate*, 3. doi:10.1088/2752-5295/AD7527

Jézéquel, A., P. Yiou, and J. P. Vanderlinden (2019), Comparing scientists and delegates perspectives on the use of extreme event attribution for loss and damage, *Weather Climate Extremes*, 26, 100231. doi:10.1016/j.wace.2019.100231

Jones, R., and R. Boer (2004), Assessing current climate risks. In: Lim, B., Spanger-Siegfried, E., Burton, I., Malone, E., Huq, S., editors. *Adaptation Policy Frameworks for Climate Change: Developing Strategies, Policies and Measures*. Cambridge, UK: Cambridge University Press, pp. 91–117.

Keenlyside, N. S., and J. Ba (2010), Prospects for decadal climate prediction, *Wiley Interdisciplinary Reviews: Climate Change*, 1. doi:10.1002/wcc.69

Kirchmeier-Young, M. C., H. Wan, X. Zhang, and S. I. Seneviratne (2019), Importance of framing for extreme event attribution: The role of spatial and temporal scales, *Earth's Future*, 7. doi:10.1029/2019EF001253

Kretschmer, M., A. Jézéquel, Z. M. Labe, and D. Touma (2024), A shifting climate: New paradigms and challenges for (early career) scientists in extreme weather research, *Atmospheric Science Letters*, 25, e1268. doi:10.1002/ASL.1268

Labe, Z. M., and E. A. Barnes (2022), Predicting slowdowns in decadal climate warming trends with explainable neural networks, *Geophysical Research Letters*, 49, e2022GL098173. doi:10.1029/2022GL098173

Lloyd, E. A., and N. Oreskes (2018), Climate change attribution: When is it appropriate to accept new methods? *Earth's Future*, 6, 311–325. doi:10.1002/2017EF000665

Lockwood, J. F., N. Dunstone, L. Hermanson, G. R. Saville, A. A. Scaife, D. Smith, and H. E. Thornton (2023), A decadal climate service for insurance: Skillful multiyear predictions of north atlantic hurricane activity and US hurricane damage, *Journal of Applied Meteorology and Climatology*, 62. doi:10.1175/JAMC-D-22-0147.1

Manabe, S., and R. J. Stouffer (1980), Sensitivity of a global climate model to an increase of CO_2 concentration in the atmosphere, *Journal of Geophysical Research*. doi:10.1029/JC085iC10p05529

Mariotti, A., D. C. Bader, S. E. Bauer, G. Danabasoglu, J. Dunne, B. Gross, L. R. Leung, S. Pawson, W. R. Putman, V. Ramaswamy, G. A. Schmidt, V. Tallapragada, A. Mariotti, D. C. Bader, S. E. Bauer, and C. Bader (2024), Envisioning US climate predictions and projections to meet new challenges, *Earth's Future*, 12, e2023EF004187. doi:10.1029/2023EF004187

Marvel, K., W. Su, R. Delgado, S. Aarons, A. Chatterjee, M. E. Garcia, Z. Hausfather, K. Hayhoe, D. A. Hence, E. B. Jewett, A. Robel, D. Singh, A. Tripati, and R. S. Vose (2023), Chapter 2: Climate trends. In: *Fifth National Climate Assessment*.

Meehl, G. A., J. H. Richter, H. Teng, A., …, and S.P. Xie (2021), Initialized earth system prediction from subseasonal to decadal timescales, *Nature Reviews Earth and Environment*, 2, 340–357. doi:10.1038/s43017-021-00155-x

Meehl, G. A., L. Goddard, G. Boer, R., …, and S. Yeager (2013), Decadal climate prediction: An update from the trenches, *Bulletin of the American Meteorological Society*. doi:10.1175/BAMS-D-12-00241.1

Merryfield, W. J., J. Baehr, L. Batt´e, E. J., …, and S. Yeager (2020), Current and emerging developments in subseasonal to decadal prediction, *Bulletin of the American Meteorological Society*, 101, E869–E896. doi:10.1175/BAMS-D-19-0037.1

Nakicenovic, N., J. Alcamo, G. Davis, B., …, and Z. Dadi (2000), *Special Report of Emissions Scenarios (SRES), A Special Report of Working Group III of the Intergovernmental Panel on Climate Change*. ISBN: 92-9169-113-5.

NASEM (National Academies of Sciences, Medicine) (2016), *Attribution of extreme weather events in the context of climate change*. doi: 10.17226/21852

NASEM (National Academies of Sciences, Engineering, and Medicine) (2020), *Incorporating the Costs and Benefits of Adaptation Measures in Preparation for Extreme Weather Events and Climate Change Guidebook*. Washington, DC: The National Academies Press. doi:1017226/25744

Naveau, P., A. Hannart, and A. Ribes (2020), Statistical methods for extreme event attribution in climate science. doi:10.1146/annurev-statistics-031219-041314

Newman, R., and I. Noy (2023), The global costs of extreme weather that are attributable to climate change, *Nature Communications* 14, 6103. doi:10.1038/s41467-023-41888-1

O'Kane, T. J., A. A. Scaife, Y. Kushnir, A. Brookshaw, C. Buontempo, D. Carlin, R. K. Connell, F. Doblas-Reyes, N. Dunstone, K. F¨orster, A. Gra¸ca, A. J. Hobday, V. Kitsios, L. van der Laan, J. Lockwood, W. J. Merryfield, A. Paxian, M. R. Payne, M. C. Reader, G. R. Saville, D. Smith, B. Solaraju-Murali, N. Caltabiano, J. Carman, E. Hawkins, N. Keenlyside, A. Kumar, D. Matei, H. Pohlmann, S. Power, M. Raphael, M. Sparrow, and B. Wu (2023), Recent applications and potential of near-term (interannual to decadal) climate predictions. doi:10.3389/fclim.2023.1121626

O'Neill, B. C., C. Tebaldi, D. P. V. Vuuren, V. Eyring, P. Friedlingstein, G. Hurtt, R. Knutti, E. Kriegler, J. F. Lamarque, J. Lowe, G. A. Meehl, R. Moss, K. Riahi, and B. M. Sanderson (2016), The scenario model intercomparison project (scenariomip) for CMIP6, *Geoscientific Model Development*, 9, 3461–3482. doi:10.5194/GMD-9-3461-2016

Otto, F. E. (2017), Attribution of weather and climate events.

Otto, F. E. (2023), Attribution of extreme events to climate change.

Otto, F. E. L., R. Skeie, J. Fuglestvedt, T. Berntsen, and M. R. Allen (2017), Assigning historic responsibility for extreme weather events, *Nature Climate Change*, 7, 757–759. doi:10.1038/nclimate3419

Perkins-Kirkpatrick, S. E., D. A. Stone, D. M. Mitchell, S. Rosier, A. D. King, Y. T. Lo, J. Pastor-Paz, D. Frame, and M. Wehner (2022), On the attribution of the impacts of extreme weather events to anthropogenic climate change, *Environmental Research Letters*, 17. doi:10.1088/1748–9326/ac44c8

Perkins-Kirkpatrick, S. E., L. V. Alexander, A. D. King, S. F. Kew, S. Y. Philip, C. Barnes, D. Maraun, R. F. Stuart-Smith, A. Jézéquel, E. Bevacqua, and S. Burgess (2024), Frontiers in attributing climate extremes and associated impacts, *Frontiers in Climate*, 6, 1455023. doi:10.3389/fclim.2024.1455023

Peters, G. P., and Z. Hausfather (2020), Emissions – the 'business as usual' story is misleading, *Nature*, 577, 618–620.

Philip, S., S. Kew, G. J. van Oldenborgh, F. Otto, R. Vautard, K. van der Wiel, A. King, F. Lott, J. Arrighi, R. Singh, and M. van Aalst (2020), A protocol for probabilistic extreme event attribution analyses, *Advances in Statistical Climatology, Meteorology and Oceanography*, 6, 177–203. doi:10.5194/ascmo-6-177-2020

Schreck, C. J., D. R. Easterling, J. Barsugli, D. A. Coates, A. Hoell, N. C. Johnson, K. E. Kunkel, Z. M. Labe, J. Uehling, R. S. Vose, and X. Zhang (2024), A rapid response process for evaluating causes of extreme temperature events in the united states: The 2023 texas/louisiana heat wave as a prototype, *Environmental Research: Climate*, 3, 1–16. doi:10.1088/2752-5295/AD8028

Shaw, T. A., J. M. Arblaster, T. Birner, A. H. Butler, D. I. Domeisen, C. I. Garfinkel, H. Garny, K. M. Grise, and A. Y. Karpechko (2024a), Emerging climate change signals in atmospheric circulation, *AGU Advances*, 5, e2024AV001297. doi:10.1029/2024AV001297

Shaw, T. A., P. A. Arias, M. Collins, D. Coumou, A. Diedhiou, C. I. Garfinkel, S. Jain, M. K. Roxy, M. Kretschmer, L. R. Leung, S. Narsey, O. Martius, R. Seager, T. G. Shepherd, A. A. So¨rensson, T. Stephenson, M. Taylor, and L. Wang (2024b), Regional climate change: Consensus, discrepancies, and ways forward, *Frontiers in Climate*, 6, 1391634.

Shepherd, T. G. (2014), Atmospheric circulation as a source of uncertainty in climate change projections. doi:10.1038/NGEO2253

Shepherd, T. G. (2016), A common framework for approaches to extreme event attribution, *Current Climate Change Report*, 2(1), 28–38. doi:10.1007/s40641-016-0033-y

Shepherd, T. G., E. Boyd, R. A. Calel, S. C. Chapman, S. Dessai, I. M. Dima-West, H. J. Fowler, R. James, D. Maraun, O. Martius, and C. A. Senior (2018), Storylines: An alternative approach to representing uncertainty in physical aspects of climate change, *Climate Change*, 151, 555–571. doi:10.1007/s10584-018-2317-9

Sillmann, J., T. G. Shepherd, B. van den Hurk, W. Hazeleger, O. Martius, J. Slingo, and J. Zscheischler (2021), Event-based storylines to address climate risk. *Earth's Future*, 9, e2020EF001783. doi:10.1029/2020EF001783

Sillmann, J., T. H. Raupach, K. L. Findell, M. Donat, L. M. Alves, L. Alexander, L. Borchert, P. B. de Amorim, C. Buontempo, E. M. Fischer, C. L. Franzke, B. Guan, M. Haasnoot, E. Hawkins, D. Jacob, R. Mahon, D. Maraun, M. A. Morrison, B. Poschlod, A. C. Ruane, Shampa, T. Stephenson, N. van der Wel, Z. Wang, X. Zhang, and J. Zupani´c (2024),ˇ Climate extremes and risks: Links between climate science and decision-making, *Frontiers in Climate*, 6, 1499765.

Smiley, K.T., I. Noy, M. F. Wehner, D. Frame, C. C. Sampson, and O. E. Wing (2022), Social inequalities in climate change-attributed impacts of Hurricane Harvey, *Nature Communications*, 13(1), 3418.

Smith, D. M., S. Cusack, A. W. Colman, C. K. Folland, G. R. Harris, and J. M. Murphy (2007), Improved surface temperature prediction for the coming decade from a global climate model, *Science*, 317, 796–799. doi:10.1126/SCIENCE.1139540

Solaraju-Murali, B., L. P. Caron, N. Gonzalez-Reviriego, and F. J. Doblas-Reyes (2019), Multi-year prediction of European summer drought conditions for the agricultural sector, *Environmental Research Letters*, 14. doi:10.1088/1748-9326/ab5043

Sospedra-Alfonso, R., W. J. Merryfield, G. J. Boer, V. V. Kharin, W. S. Lee, C. Seiler, and J. R. Christian (2021), Decadal climate predictions with the canadian earth system model version 5 (CANESM5), *Geoscientific Model Development*, 14. doi:10.5194/gmd-14-6863-2021

Srikrishnan, V., Y. Guan, R. S. Tol, and K. Keller (2022), Probabilistic projections of baseline twenty-first century co2 emissions using a simple calibrated integrated assessment model, *Climatic Change*, 170. doi:10.1007/s10584-021-03279-7

Stewart, M. G., and X. Deng (2015), Climate impact risks and climate adaptation engineering for built infrastructure, *ASCE–ASME Journal of Risk and Uncertainty in Engineeri ng Systems, Part A: Civil Engineering*, 1(1), 04014001. doi:10.1061/AJRUA6.0000809.

Stone, D. A., and M. R. Allen (2005), The end-to-end attribution problem: From emissions to impacts. *Climate Change*, 71, 303–318. doi:10.1007/s10584-005-6778-2

Stott, P. A., D. A. Stone, and M. R. Allen (2004), Human contribution to the European heatwave of 2003, *Nature*, 432, 610–614. doi:10.1038/nature03089

Stott, P. A., N. Christidis, F. E. Otto, Y. Sun, J. P. Vanderlinden, G. J. van Oldenborgh, R. Vautard, H. von Storch, P. Walton, P. Yiou, and F. W. Zwiers (2016), Attribution of extreme weather and climate-related events, *Wiley Interdisciplinary Reviews: Climate Change*, 7. doi:10.1002/wcc.380

Strauss, B. H., P. M. Orton, K. Bittermann, M. K. Buchanan, D. M. Gilford, R. E. Kopp, S. Kulp, C. Massey, H. de Moel, and S. Vinogradov, (2021), Economic damages from Hurricane Sandy attributable to sea level rise caused by anthropogenic climate change, *Nature Communications*, 12, 2720. https://doi.org/10.1038/s41467-021-22838-1

Thompson, A., and F. E. L. Otto (2015), Ethical and normative implications of weather event attribution for policy discussions concerning loss and damage. *Climate Change*, 133, 439–451.

Toms, B. A., E. A. Barnes, and J. W. Hurrell (2021), Assessing decadal predictability in an earth-system model using explainable neural networks, *Geophysical Research Letters*, 48, e2021GL093842. doi:10.1029/2021GL093842

Towler, E., D. Paimazumder, and J. Done (2018), Toward the application of decadal climate predictions, *Journal of Applied Meteorology and Climatology*, 57.

Trenberth, K. E., J. T. Fasullo, and T. G. Shepherd (2015), Attribution of climate extreme events, *Nature Climate Change*, 5, 725–730. doi:10.1038/nclimate2657

Van Den Hurk, B, C. Hewitt, D. Jacob, J. Bessembinder, F. Doblas-Reyes, and R. Döscher (2018), The match between climate services demands and Earth System Models supplies. *Climate Services*, 12, 59–63. doi:10.1016/j.cliser.2018.11.002

van Oldenborgh, G. J. V., K. van der Wiel, S. Kew, S. Philip, F. Otto, R. Vautard, A. King, F. Lott, J. Arrighi, R. Singh, and M. van Aalst (2021), Pathways and pitfalls in extreme event attribution. *Climate Change*, 166, 13. doi:10.1007/s10584-021-03071-7

van Oldenborgh, G. J. V., M. F. Wehner, R. Vautard, F. E. Otto, S. I. Seneviratne, P. A. Stott, G. C. Hegerl, S. Y. Philip, and S. F. Kew (2022), Attributing and projecting heatwaves is hard: We can do better. doi:10.1029/2021EF002271

van Vuuren, D. P., J. Edmonds, M. Kainuma, K. Riahi, A. Thomson, K. Hibbard, G. C. Hurtt, T. Kram, V. Krey, J.-F. Lamarque, T. Masui, M. Meinshausen, N. Nakicenovic, S. J. Smith, and S. K. Rose (2011), The representative concentration pathways: An overview, *Climatic Change*, 109, 5–31. doi:10.1007/s10584-011-0148-z

Vaughan, E., and J. Turner (2013), *The Value and Impact of Building Codes*. Washington, DC: Environmental and Energy Study Institute. https://www.eesi.org/papers/view/the-value-and-impact-of-building-codes

Vitart, F., A. W. Robertson, and D. L. T. Anderson (2012), Subseasonal to seasonal prediction project: Bridging the gap between weather and climate, *Bulletin of the World Meteorological Organization*, 61, 23.

Yang, X., T. L. Delworth, F. Zeng, L. Zhang, W. F. Cooke, M. J. Harrison, A. Rosati, S. Underwood, G. P. Compo, and C. McColl (2021), On the development of GFDL's decadal prediction system: Initialization approaches and retrospective forecast assessment, *Journal of Advances in Modeling Earth Systems*, 13. doi:10.1029/2021MS002529

Yeager, S. G., G. Danabasoglu, N. A. Rosenbloom, W. Strand, S. C. Bates, G. A. Meehl, A. R. Karspeck, K. Lindsay, M. C. Long, H. Teng, and N. S. Lovenduski (2018), Predicting near-term changes in the earth system: A large ensemble of initialized decadal prediction simulations using the community earth system model, *Bulletin of the American Meteorological Society*, 99. doi:10.1175/BAMS-D-17-0098.1

Yeager, S. G., N. Rosenbloom, A. A. Glanville, X. Wu, I. Simpson, H. Li, M. J. Molina, K. Krumhardt, S. Mogen, K. Lindsay, D. Lombardozzi, W. Wieder, W. M. Kim, J. H. Richter, M. Long, G. Danabasoglu, D. Bailey, M. Holland, N. Lovenduski, W. G. Strand, and T. King (2022), The seasonal-to-multiyear large ensemble (smyle) prediction system using the community earth system model version 2, *Geoscientific Model Development*, 15. doi:10.5194/gmd-15-6451-2022

Zhang, Y., and B. M. Ayyub (2020), Electricity system assessment and adaptation to rising temperatures in a changing climate with Washington metro area as a case study, *ASCE Journal of Infrastructure Systems*, 26(2), 04020017

Zhang, Y., B. M. Ayyub, J. F. Fung, and Z. M. Labe (2024), Incorporating extreme event attribution into climate change adaptation for civil infrastructure: Methods, benefits, and research needs, *Resilient Cities and Structures*, 3(1), 103–113.

Zhou, W., J. Li, Z. Yan, Z. Shen, B. Wu, B. Wang, R., Zhang, and Z. Li (2024), Progress and future prospects of decadal prediction and data assimilation: A review. doi:10.1016/j.aosl.2023.100441

7 Hazard and Risk Assessment in Nonstationary Environments for Engineering Applications

Si Han Li[a] *and Jayantha T. Obeysekera*[b]
[a]Rowan Williams Davies & Irwin Inc., Guelph, Ontario, Canada
[b]Institute of Environment, Sea Level Solutions Center, Florida International University, Miami, Florida

7.1 EXTREME HAZARD AND ENGINEERING DESIGN IN STATIONARY CLIMATE

7.1.1 Extreme Value Distribution in a Stationary Climate

Climatic design loads in modern building codes have conventionally used historical observations to quantify extreme hazards, such as wind (e.g., Vickery et al., 2009; Hong et al., 2014; Li et al., 2017), snow (e.g., Hong & Ye, 2014; Bean et al., 2021), and ice (e.g., Robbins & Cortinas, 1996; Konrad, 1998; Jones & Eylander, 2017). Most modern building codes and standards, such as ASCE 7, NBCC 2020, assume climate stationarity for various design loads (e.g., Pintar et al., 2015; Vickery et al., 2010; Hong et al., 2014; Liel et al., 2017) and reliability calculations (Ellingwood et al., 1980; Bartlett et al., 2003a,b; McAllister et al., 2018). The extreme value theory is applied when determining the extreme climatic loads or hazards. Three types of extreme value probability distributions include the Fisher–Tippett Type I (Gumbel), Fisher–Tippett Type II (Frechet), and Fisher–Tippett Type III (Weibull) distributions, which can be combined as a single distribution known as the Generalized Extreme Value (GEV) distribution (Castillo, 1988; Coles, 2001).

The Type I or Gumbel distribution can be expressed by,

$$F(x) = \exp\left(-\exp\left(-\frac{x - u_G}{a_G}\right)\right) \tag{7.1}$$

DOI: 10.1201/9781003567332-7

where $F(x)$ is the cumulative distribution, x is the random variable, u_G and a_G are the location and scale parameters for the Gumbel distribution.

The Type II or Frechet distribution can be expressed by,

$$F(x) = \exp\left(-\left(\frac{x - u_F}{a_F}\right)^{k_F}\right) \tag{7.2}$$

where u_F, a_F, and k_F are the location, scale, and shape parameters for the Frechet distribution.

The Type III or Inverse Weibull distribution can be expressed by,

$$F(x) = \begin{cases} 1 - \exp\left(-\left(\frac{x - u_W}{a_W}\right)^{k_W}\right), & x > u_W \\ 0, & x < u_W \end{cases} \tag{7.3}$$

where u_W, a_W, and k_W are the location, scale, and shape parameters for the Weibull distribution. The 3-parameter Weibull distribution can also be reduced to a 2-parameter Weibull distribution.

$$F(x) = \begin{cases} 1 - \exp\left(-\left(\frac{x}{a_W}\right)^{k_W}\right), & x > 0 \\ 0, & x < 0 \end{cases} \tag{7.4}$$

The three types of extreme value distribution can be combined into the GEV distribution,

$$F(x) = \begin{cases} \exp\left(-\left(1 + k\frac{x - u}{a}\right)^{-1/k}\right), & k \neq 0 \text{ and } k\frac{x - u}{a} > -1 \\ 0, & k > 0 \text{ and } \frac{x - u}{a} \leq -1/k \\ 1, & k < 0 \text{ and } \frac{x - u}{a} \geq 1/|k| \end{cases} \tag{7.5}$$

where u, a, and k are location, scale, and shape parameters for the GEV distribution. When $k = 0$, the GEV distribution becomes the Gumbel distribution.

An example of these three types of extreme value probability distributions is provided in Figure 7.1. The horizontal axis is a reduced variate, while the vertical axis is the random variable, x. The k parameter is the shape parameter as defined in Eq. (7.5).

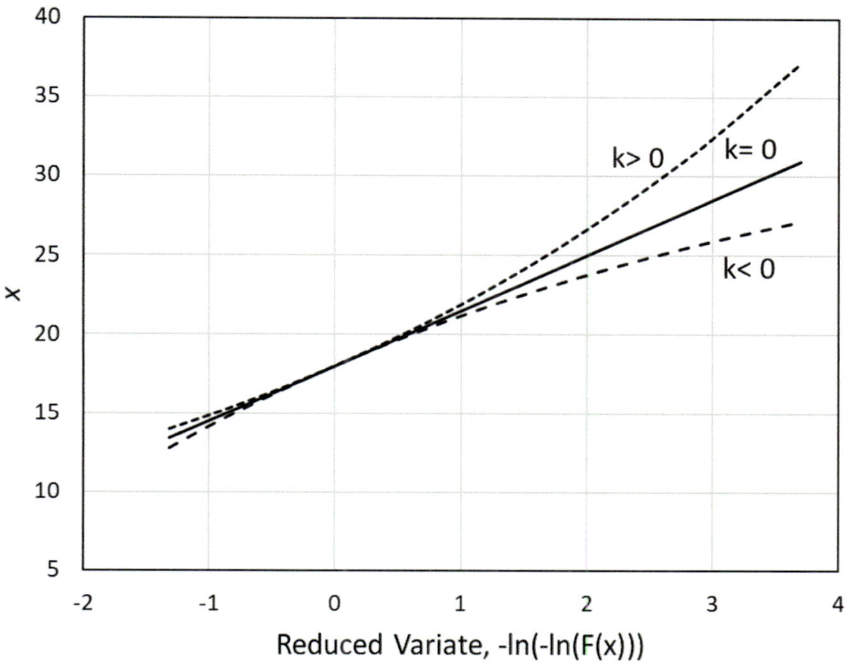

FIGURE 7.1 Extreme value distributions shown in a Gumbel probability paper.

7.1.2 Estimation of Parameters of Extreme Value Distribution in a Stationary Climate

Many techniques can be used to determine the parameters of an extreme value probability distribution (Hong et al., 2013), including the method of moments, method of maximum likelihood (MML), method of L-Moments (Hosking, 1990), ordinary least square (OLS) method (Jordaan, 2005), and generalized least square method (GLS) (Lieblein, 1974; Hong et al., 2013), also known as Best Least Unbiased Estimator (BLUE).

By adapting Eq. (7.1), the ascendingly ordered Gumbel distributed random variable, x_i, can be expressed by,

$$x_i = u + az_i \tag{7.6}$$

where $z_i = \left(-\ln\left(-\ln\left(F\left(x_i\right)\right)\right)\right)$ is the reduced variate, and $F\left(x_i\right)$ is the plotting position.

The Gumbel distribution parameters can be estimated using the above methods (Hong et al., 2013), as listed in Table 7.1.

A widely applied mean plotting position (Blom, 1958; David & Nagaraja, 2003) can be expressed by,

$$\bar{F}\left(x_i\right) = \frac{i}{n+1} \tag{7.7}$$

TABLE 7.1
Methods for Estimating Gumbel Distribution Parameters

Method	Procedures for Estimating Distribution Parameters
MOM	$a = s\sqrt{6}/\pi$, and $u = m - 0.5772 \cdot s\sqrt{6}/\pi$, where s and m are standard deviation and mean of the records, respectively.
MML	Solving $-n - \sum_{i=1}^{n} y_i - \sum_{i=1}^{n} y_i \exp(-y_i) = 0$ and $-n - \sum_{i=1}^{n} \exp(-y_i) = 0$, where $y_i = (x_i - u)/a$
MLM	$a = (2l_1 - l_0)/\ln 2$ and $u = l_0 - 0.5772a$, where $l_i = n^{-1} \sum_{j=1}^{n} \frac{(j-1)\cdots(j-i)}{(n-1)\cdots(n-i)} x_{j:n}$
OLS	$a = \text{cov}(z_i, x_i)/\text{var}(z_i)$, and $u = m - 0.5772a$, where var denotes variance, cov denotes covariance, and m is the average of x.
GLM	$a = \sum_{i=1}^{n} c_{a,i} x_{i:n}$, $u = \sum_{i=1}^{n} c_{u,i} x_{i:n}$, where $c_{a,i}$ and $c_{u,i}$ are BLUE coefficients.

Alternatively, one can use the OLS with the plotting positions derived from the GLS. A formula provided in Hong and Li (2014) for the revised plotting position can be used,

$$\bar{F}(x_i) = \begin{cases} \left(i - 0.37 + 0.232/\sqrt{n}\right) / \left(n + 0.144 + 0.232/\sqrt{n}\right), & i = 1, \ldots, n-1 \\ \exp\left(-\exp(-0.5772)/n\right), & i = n \end{cases} \tag{7.8}$$

The GLM (Hong et al., 2013) performs the best when fitting the Gumbel distribution parameters. The BLUE coefficients up to 100 samples have been provided in Hong et al. (2013).

In practice, if the data is Gumbel distributed, Eq. (7.8) results in mostly the same results as when exact BLUE coefficients and GLS are used (Li et al., 2020).

The GEV distributed variables can be expressed by,

$$x_i = u + \frac{a}{k}\left(\left(-\ln F(x_i)\right)^{-k} - 1\right) \tag{7.9}$$

The methods can be used for fitting the GEV distribution parameters are summarized in Table 7.2.

When samples are not continuously recorded, the issue known as disjunct sampling (DS) (Larsen & Mann, 2006; Li, 2018) can affect the estimated return period values. Analytical corrections for disjunct sampling (C_{DS}) can be applied to the 50-year return period values calculated from disjunct sampled data, given the extreme values follow Gumbel distribution (Li, 2018),

TABLE 7.2
Methods for Estimating GEV Distribution Parameters

Method	Procedures for Estimating Distribution Parameters
MOM	Solving $m = u + a\left(1-\Gamma(1+k)\right)/k$, $s = a(\Gamma(1+2k)-\Gamma^2(1+k))^{1/2}/\left(\text{sign}(k)k\right)$, and $\gamma_3 = \text{sign}(k)\left(-\Gamma(1+3k)+3\Gamma(1+k)\Gamma(1+2k)-2\Gamma^3(1+k)\right)/\left(\Gamma(1+2k)-\Gamma^2(1+k)\right)^{3/2}$, where γ_3 is the sample skewness and $\Gamma(\cdot)$ is the gamma function.
MML	Maximizing $L = -n\ln a - (1-k)\sum_{i=1}^{n} y_i - \sum_{i=1}^{n}\exp(y_i)$, where $y_i = k^{-1}\left(1+k(x_i-u)/a\right)$
MLM	Solving $l_0 = u + a\left(1-\Gamma(1+k)\right)/k$, $2l_1 - l_0 = a\Gamma(1+k)\left(1-2^{-k}\right)/k$, and $(3l_2-l_0)/(2l_1-l_0) = \left(1-3^{-k}\right)/\left(1-2^{-k}\right)$

$$C_{DS} = \begin{cases} 0.042t^{-0.775} + 0.958, & \text{PYD} = 70\% \\ 0.137t^{-0.242} + 0.863, & \text{PYD} = 80\% \\ 0.938t^{-0.039} + 0.062, & \text{PYD} = 90\% \\ -0.554t^{0.075} + 1.554, & \text{PYD} = 100\% \end{cases} \tag{7.10}$$

where t is the sampling time in hours, and PYD denotes Percentage of Years of DS.

Other distributions have also been used in engineering design applications for quantifying the design hazard values, such as lognormal distribution (Essery et al., 1999; Mo et al., 2015), Generalized Pareto distribution (Holmes & Moriarty, 1999; Harris, 2005; Gharib et al., 2017), and penultimate extreme value distribution (Harris, 2009).

The extreme value analysis can be carried out applying the Homogenous Poisson Process. For example, given the annual occurrence of windstorms is λ, and the storm intensity follows a Weibull distribution, $F(x)$, where x is the wind speed, then the probability of the annual maximum values can be expressed by,

$$F_{X\max}(x) = \exp\left(-\lambda\left(1-F_X(x)\right)\right) \tag{7.11}$$

where $F_{X\max}(x)$ is the probability distribution of annual maximum value, $X_{\max}$, and $F_X(x)$ is a Weibull distribution of random variable x.

With any of the above distributions applied, a concept of return period has been widely used to define the design level climatic loads. This concept is implicitly related to the stationary geometric process. The return period is associated with a level directly related to the annual probability of exceeding, p, of this level. The relation $T = 1/p$ is valid for every year in a stationary climate. An important consequence of the memoryless property of the geometric process is that the time to the next occurrence of the load event is invariant. In other words, at any point in time, the expected time to the next occurrence of the event is always equal to the return period ($T = 1/p$).

Besides fitting the historical observed climatic data into a stationary extreme value distribution, advanced numerical models have been employed to simulate hazards, such as tropical cyclones. Tropical cyclone simulations have been employed in engineering applications to quantify extreme wind speeds and derive design level wind speeds for coastal line regions of countries in North Atlantic basin (e.g., Vickery et al., 2000, 2009, 2010; Li & Hong 2014; Hong et al., 2022), western North Pacific regions (e.g., Li & Hong 2015a,b, 2016; Hong et al., 2016), Bay of Bengal (Li & Kumar, 2021 & 2023), western Arabian Sea regions (e.g., Li et al., 2021), and Australian coastal regions (e.g., Harper & Mason 2016). These models consist of similar components mixed with physical modules, such as wind field model and intensity quantification, and probability predictions, such as storm track predictions. A fundamental assumption employed by these models is that the environmental observations or reanalysis used to develop the models are stationary, such that the "black box" developed by employing the historical information does not change with time or other environmental covariates and these developed frameworks can be applied to future environments. The change in future environments used as inputs will not change the "black box," but will affect the outputs.

7.1.3 Basics of Structural Reliability in a Stationary Climate

Modern structural design codes, such as those for buildings and bridges, are based on probabilistic concepts of reliability. The reliability-based framework has provided a rational basis for designing structures that comply with high safety standards. The basic reliability model, the "stress-strength" model, is fundamental to developing reliability-based design codes and standards. This problem, as shown in Figure 7.2, involves comparing two random variables, stress (X) and strength (Y). The probability of failure, P_f, is then defined as the probability of strength being less than the stress:

$$P_f = P[Y \leq X] = \int_0^\infty P[Y \leq x] f_X(x)\,dx = \int_0^1 P[Y \leq x]\,dF_X(x) \tag{7.12}$$

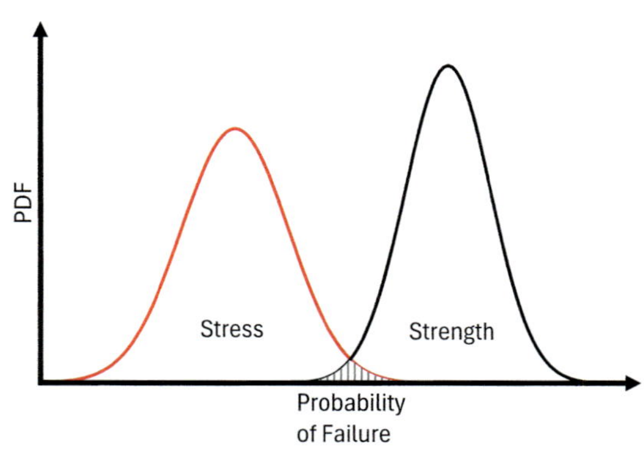

FIGURE 7.2 Illustration of basic reliability problem.

where $f_X(x)$ is the probability density function of any random variable, X and $F_X(x)$ is the cumulative distribution function. The first- and second-order reliability methods, FORM and SORM, emerged as prominent methods for solving practical applications (Zhao & Ono, 1999; Rackwitz, 2001). Simulation-based methods, such as Monte Carlo simulation, are employed to find benchmark solutions for different applications (Ellingwood et al., 1980; Bartlett et al., 2003a,b; Hong et al., 2014; McAllister et al., 2018; Li, 2018).

The design codes adopted the extreme value approach to modeling stochastic load effects. In this approach, the distribution of maximum load, $X_{max}(t)$, over a given time interval (0, t] is derived, and this distribution is subsequently used in the reliability calculation to compute the probability of failure in this time interval.

$$P_f(t) = \int_0^{\infty} F_Y(x) f_{X\max}(x,t)\,dx \tag{7.13}$$

In this manner, the probability of failure over this interval is computed. Annual probability of failure, P_{fa}, can be computed by setting $t=1$ in the above equation. Most structural design codes are based on this conceptual approach.

Using the annual maximum load distribution, the annual probability of structural failure is computed with reference to a designed limit state. Because of stationarity of the load process, the annual probability of failure (P_{fa}) for any annual time interval, (t, $t+1$), remains a constant irrespective of the value of t. This also means that the probability of over an interval, (0, t], is approximately $P_{fa} \times t$ provided that $P_{fa} \ll 1$.

The integration of the annual probability of failure leads to the total (or lifetime) probability of failure, which can be converted into the reliability index, often used in modern building codes and standards to measure the safety of structural design.

7.2 NONSTATIONARITY AND ITS DETECTION FROM HISTORICAL OBSERVATIONS AND CLIMATE PROJECTIONS

Recently, there has been a significant interest in moving from stationarity to nonstationarity (Box 7.1) in risk assessments. Changes in climate and associated extreme weather events are already impacting inland and coastal communities and infrastructure. Notably, the frequency of high-tide flooding in some coastal regions of Florida has increased significantly due to rising sea levels, higher tides, and storm surges. Additionally, there are concerns that extreme rainfall is increasing, leading to more frequent flooding in both inland and coastal areas. Flooding is projected to become more severe in the coming years and decades (Vogel et al., 2011).

Sea level rise (SLR) directly affects the groundwater table, especially along coastal areas, and this impact may extend inland (Befus et al., 2020). Furthermore, increasing storm surges and extreme waves may result from future changes in intensity, frequency, and tracks of tropical and extratropical storms (e.g., Knutson et al., 2019, 2020). In recent years, many locations worldwide have experienced unprecedented rainfall. For instance, rainfall along Hurricane Ian's track in Florida had return periods as rare as 500 years in certain locations. Similar high return period

BOX 7.1 STATIONARITY VERSUS NONSTATIONARITY

We can characterize the frequency, duration, and magnitude of rare or extreme events. Both human and natural systems have mechanisms to address these events using well-established hydrologic methods. For example, it is possible to estimate the magnitude of a 100-year flood (a 1% chance of exceedance in any given year) and design a bridge large enough to accommodate the flood with minimal risk of damage to the bridge or changes in water depth. However, all prediction methods rely on the assumption that the future will resemble the past in statistical terms. This assumption is known as "stationarity" in hydrology.

Specifically, stationarity implies that while there is considerable variability in precipitation and streamflow, this variability fluctuates around a relatively constant average value and has a relatively constant spread. However, changes in land use and climate can result in magnitudes, durations, and frequencies of floods and droughts that fall outside the historical range of observations. This can lead to shifts in the average magnitude of floods or changes in their variability.

Understanding and accounting for such "nonstationarity" in environmental variables is one of the greatest challenges in hydrology today. Accurate future predictions are essential for making informed decisions about flood and drought risks, as well as for designing infrastructure such as roads, bridges, and culverts.

Source: SERC Carleton (2017)

events have been estimated for rainfall associated with other hurricanes, such as Helene and Milton.

Significant research in hydrology and water resources has focused on detecting and attributing changes in hydrological processes at various scales (e.g., Salas et al., 2018; Milly et al., 2008). Historically, key statistical measures like the mean and variance were assumed constant over time—a concept known as stationarity. However, many hydrologists have challenged this assumption, with Milly et al. (2008) declaring, "Stationarity is Dead – Whither Water Management?" This assertion sparked worldwide debate. While some embraced the shift toward accounting for nonstationarity, others responded with opposing views, as reflected in titles such as:

- "Stationarity: Wanted Dead or Alive?" (Lins & Cohn, 2011)
- "Comment on the Announced Death of Stationarity" (Matalas, 2012)
- "Negligent Killing of Scientific Concepts: The Stationary Case" (Koutsoyiannis & Montanari, 2015)
- "Modeling and Mitigating Natural Hazards: Stationarity is Immortal!" (Montanari & Koutsoyiannis, 2014)
- "Stationarity is Undead: Uncertainty Dominates the Distribution of Extremes" (Serinaldi & Kilsby, 2015).

Despite increased attention, there is no universally accepted methodology for performing essential hydrologic analyses, such as Flood Frequency Analysis, under nonstationary conditions. Many agencies have updated flood protection design guidelines to account for nonstationarity (e.g., Stedinger & Griffis, 2011; Prosdocimi et al., 2014). It is worth noting that hydrologic processes with short- and long-term memory (e.g., Koutsoyiannis, 2002) can produce trends and abrupt shifts in hydrologic sequences, which may not necessarily be classified as nonstationary. Similarly, other climatic variables, such as wind speeds and ground snows, could have similar decadal variations or be caused by man-made built-environment changes that produce trends and shifts in the observations, but not necessarily be classified as nonstationary.

The lack of consensus regarding nonstationary approaches stems partly from the uncertainty in detecting, attributing, and modeling/predicting trends. In cases where historical trends are evident, statistically significant, and attributable to factors such as land use changes, climate shifts, or water infrastructure development, it is crucial to update planning and design methodologies accordingly.

7.2.1 Climate Nonstationarity

Until recently, buildings and infrastructure planning and design have assumed "stationarity" in environmental drivers such as flood, wind, snow, rainfall, and sea levels, and other design variables for most modern building codes, standards, and practices. The only exception may be to address SLR but even its treatment tends to be limited to projections, which are often limited to historical trends. Planning for agile, resilient, and quality buildings and infrastructure for the future, depending on the current design criteria and the use of such approaches as safety factors, may require moving from this stationarity assumption to a nonstationary environment characterized by rising sea levels, changing rainfall patterns and floods, stronger and/or more frequent wind storms, increasing ocean waves, increasing frequency and magnitude of tropical storms, changing land use/land cover, and a rising groundwater table especially along the coast and extending to some inland locations. Currently, the "Design Return Period" concept is the standard practice, but recent research on nonstationarity indicates that this concept needs to be revisited since the return period and future risks are dynamic (Salas et al., 2018) as they vary with time due to the impact of global warming level. A new paradigm, which incorporates changing exceedance probabilities (or failure probabilities), risks, and innovative planning criteria based on future projections, is needed.

The recent update for NBCC 2025 has replaced the return period concept with the maximum Annual Probability of Exceedance (APE), equal to a target value, during the design working life for the design of buildings (Li et al., 2025). The 2025 edition of the National Building Code of Canada (NBCC) (Li et al., 2025) offers valuable examples of how model codes can be advanced by comprehensively integrating climate projections into design regulations, defining basic climatic design variables, and assessing structural reliability with climate nonstationarity.

Before embarking on applying nonstationary techniques for planning and designing buildings and infrastructure, a logical process needs to be followed to justify their use. Because many factors (land use change and climate) may contribute to any apparent nonstationarity in data, careful adherence to this process may be needed. Such a process should essentially include the following general steps:

I. Detection of nonstationarity in relevant climatic and non-climatic drivers using historical data or climate projections when available in the vicinity of the project site;
II. Literature review to determine the possible presence of climate change in the project area;
III. Projected change in relevant climate drivers;
IV. Risk and vulnerability assessments for the project features;
V. Develop planning and design options to address impacts caused by nonstationarity;
VI. Lifetime cost and benefit assessments for addressing nonstationarity in future changing climate.

An example of such a process, developed by the US Army Corps of Engineers (USACE) is shown in Figure 7.3.

It should be noted that the prediction of future climatic conditions, including SLR, is inherently associated with significant uncertainties, particularly regarding extreme events. Such uncertainties result from the inadequacies in science and modeling of future projections, and the realization of climate change mitigation is heavily affected by geopolitics. While there are many methods to deal with "deep uncertainties," one approach that is becoming increasingly popular is the scenario testing. When uncertainty estimates of projections are available, they should be used along with the best available trajectories for future conditions for developing adaptation strategies with due consideration to risks of failure. When building and infrastructure planning and design methods include a significant safety factor, it may already cover future conditions reflecting nonstationarity.

7.2.2 Detection of Nonstationarity

Detecting nonstationarity is a key step. Often, natural variability exhibits episodic nonstationary-like characteristics, which may not represent persistent, systematic trends that are statistically significant. This section explored existing tools, particularly those from federal agencies such as the USACE and the US Geological Survey. As examples of application, their approaches are customized for locations in Florida.

7.2.2.1 Methods

Stationarity is assumed when statistical properties do not change over time. A more formal definition of a stationary stochastic process is one where the joint probability distribution of $X_{t1}, X_{t2}, \ldots, X_{tn}$ is the same as the joint distribution of $X_{t1+m}, X_{t2+m}, \ldots, X_{tn+m}$ for all values of m and n (Nason, 2006). A starting point of any time series modeling is verifying whether the underlying time series is stationary (Madsen et al., 2014). A simple plot of the time series may reveal much about possible nonstationarity in the data. Often, the focus is on the temporal variation in the mean and variance.

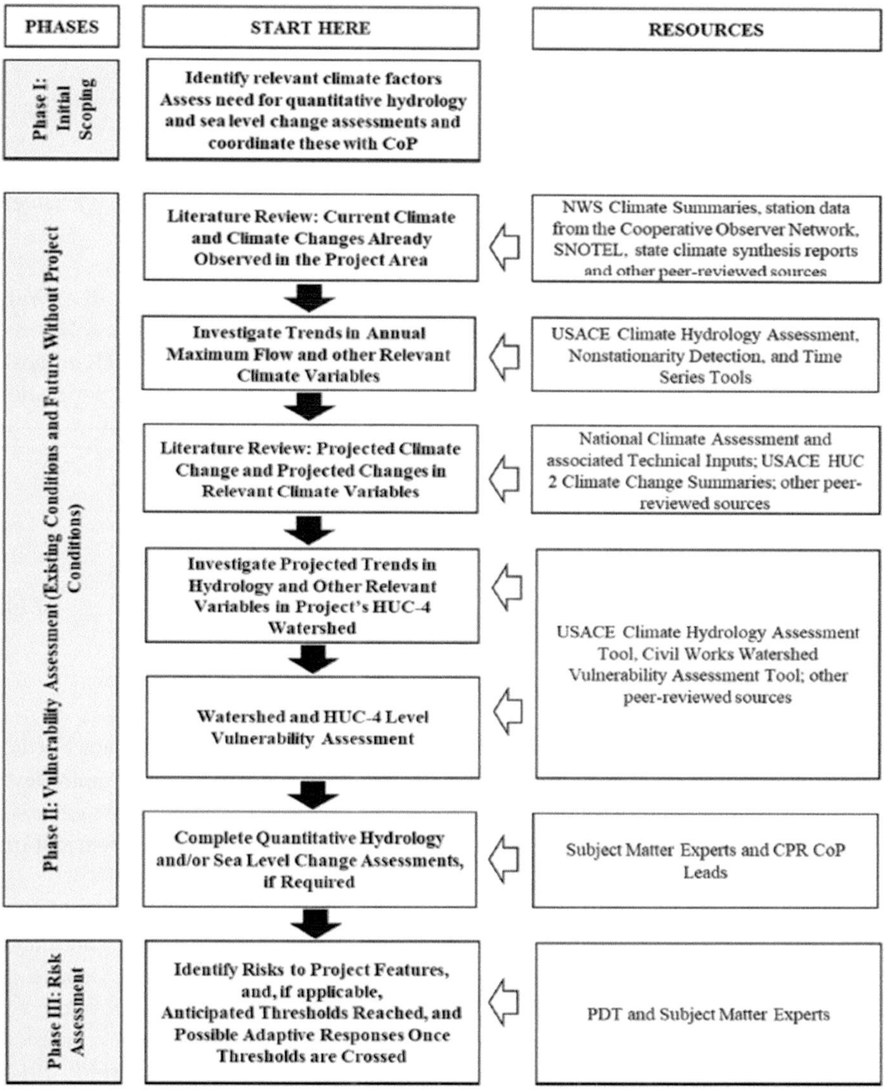

FIGURE 7.3 Flow Chart and Crosswalk Table developed by USACE for incorporating climate change impacts to inland hydrology in civil works (ECB #2018-14).

Another useful technique is to compute the autocorrelation function or spectrum on two different parts of the given time series to investigate some evidence of nonstationarity. However, this generally requires that the sample data is of sufficient length. Often, a realization from a known stationary process may appear as nonstationary, particularly if the process has temporal dependence. For this reason, more formal statistical tests are desirable to determine the choice between stationarity and nonstationarity (Cohn & Lins, 2005).

This section describes commonly used tests for detecting statistically significant changes. For continuous changes, both parametric and nonparametric methods are described.

7.2.2.1.1 Trend Detection by Parametric Methods

The most common parametric method is linear or nonlinear regression. A fundamental assumption in these methods is the normality of the response variables. A popular method for a normality test is described below.

7.2.2.1.1.1 Shapiro–Wilk Normality Test The Shapiro–Wilk test (Razali & Wah, 2011) is used to determine if a sample, X_i, $i = 1, 2, \ldots, n$, is drawn from a Normal Distribution with the population mean μ and variance, σ^2, respectively. Hypothesis testing is used to determine whether the sample was drawn from a normal population (null hypothesis) or a non-normal population.

The Shapiro-Wilk test statistic is given by

$$W = \frac{\left(\sum_{i=1}^{n} a_i x_{(i)}\right)^2}{\sum_{i=1}^{n} (x_i - \bar{x})^2} \tag{7.14}$$

where $x_{(i)}$ is the *i*th order statistic (*i*th-smallest), and $\bar{x}$ is the sample mean. Further details of this method can be found in Shapiro and Wilk (1965). The "shapiro.test" script available in the stats library of R was used to compute W and its p-value associated with the hypothesis testing to determine if data follows a Normal Distribution.

7.2.2.1.1.2 Linear Model for Single Trends Smooth changes with time are referred to as trends. The simplest type of change is a linear trend which may be a continuous increase or decrease over time. In some cases, the continuous trend may depend on other variables, besides time. If the temporal variation of the variable of interest is deemed to be linear, the starting point for modeling such a process is linear regression, and many textbooks and tutorials describe its application (e.g., Draper & Smith, 2014).

The general form of the linear model is,

$$y_t = \beta_0 + \beta_1 X_{t1} + \beta_2 X_{t2} \ldots + \beta_p X_{tp} + \varepsilon_t \tag{7.15}$$

where y is the dependent variable with its subscript t denoting time (1 to n), X values denote the predictors, and ε is the error term.

In matrix form, the linear model may be written as,

$$Y = X\beta + \varepsilon \tag{7.16}$$

where Y is a vector of size $(n \times 1)$ and X is a matrix of size $(n, p + 1)$ consisting of unity as the first column and the remaining columns correspond to each time series of the predictors, X. The size of the parameter vector β is equal to $p+1$. The error term is a vector of size $(n \times 1)$.

Using the least squares methodology, the estimator of the parameter vector β is given by

$$\hat{\beta} = \left(X^{\mathrm{T}}X\right)^{-1} X^{\mathrm{T}}Y \tag{7.17}$$

A property of the least squares estimation approach is that the estimators given by Eq. (7.17) are unbiased. Assuming the error term ε_t has a Gaussian distribution, the estimators of β have a multivariate normal distribution:

$$\hat{\beta} \sim N_{p+1}\left[\beta, \sigma^2\left(X^{\mathrm{T}}X\right)^{-1}\right] \tag{7.18}$$

where N_{p+1} denotes a multivariate normal distribution of dimension $p+1$ and σ^2 is the variance-covariance matrix of the error term. This in turn, can provide confidence intervals for each parameter.

For diagnostic testing purposes, it is often convenient to decompose the variability of the dependent variable, Y, in the form of an Analysis of Variance (ANOVA) table.

Given the sample data, the estimate of σ^2 can be obtained as $S^2 = \mathrm{SSE}/(n-p-1)$ and the standard error of each parameter is estimated as $S_{\hat{\beta}_i} = S\sqrt{C_{ii}}$ where C_{ii} is the corresponding diagonal element of $\left(X^{\mathrm{T}}X\right)^{-1}$. Assuming the standardized values of estimated parameters to follow a Student-t distribution with $n-p-1$ of degrees of freedom, the $100(1-\alpha)$ percent confidence interval for $\hat{\beta}_i$ is given by

$$\left[\hat{\beta}_i - t_{1-\frac{\alpha}{2}}(n-p-1)\,S_{\hat{\beta}_i}, \hat{\beta}_i + t_{1-\frac{\alpha}{2}}(n-p-1)\,S_{\hat{\beta}_i}\right] \tag{7.19}$$

Diagnostic tests for individual parameters estimated from data can be conducted using the Student-t variate, $t = \left(\hat{\beta} - \beta_{i0}\right) / S\sqrt{C_{ii}}$. If the absolute value of t is greater than $t_{1-\frac{\alpha}{2}}(n-p-1)$, then the null hypothesis $\beta_i = \beta_{i0}$ is rejected. For detecting the presence or absence of a trend, often $\beta_{i0} = 0$ is used (i.e., zero slope).

TABLE 7.3

ANOVA Table for Multiple Linear Regression

Source	Sum of Squares	Degrees of Freedom	Mean Square	*F* Ratio
Regression	$\mathrm{SSR} = \hat{\beta}X^{\mathrm{T}}Y - n\bar{Y}^2$	p	$\mathrm{MSR} = \frac{\mathrm{SSR}}{p}$	$F = \frac{\mathrm{MSR}}{\mathrm{MSE}}$
Error	$\mathrm{SSE} = \varepsilon^{\mathrm{T}}\varepsilon$	$n-p-1$	$\mathrm{MSE} = \frac{\mathrm{SSE}}{n-p-1}$	-
Total	$\mathrm{SSTO} = Y^{\mathrm{T}}Y - n\bar{Y}^2$	-	-	-

A simultaneous test for all coefficients, whether they are all equal to zero, may be conducted using the results of the ANOVA table. Since the F ratio (last column in Table 7.3) has an F-distribution with p and $(n-p-1)$ degrees of freedom, the null hypothesis that all coefficients $\beta_i=0$, $i=1, \ldots, p$ is rejected if the computed $F > F_\alpha\left(p, n-p-1\right)$. Another common approach for rejecting the null hypothesis is to demonstrate that $\alpha_c = \text{Prob}\left[F_\alpha\left(p, n-p-1\right) > F\right]$ is greater than α.

7.2.2.1.2 Trend Detection by Nonparametric Methods

The parametric methods require certain distributional assumptions. More specifically, the data should be normally distributed, although special methods may deal with other exceptions such as autocorrelation. However, many climatic time series are skewed and do not follow a normal distribution. Often, transforming the variables through such an approach as taking the log transformation may help satisfy this normality requirement. Thus, alternative methods such as distribution-free nonparametric methods are needed to detect nonstationarity.

7.2.2.1.2.1 Mann–Kendall Test One of the most common, nonparametric tests for detecting trends is the Mann–Kendall test (Mann, 1945; Kendall, 1938). This method improves the OLS regression method because it can handle outliers and influential data. It is also a robust test that can detect nonlinear monotonic trends in non-normally distributed data that may contain missing or censored data.

The test consists of summing the signs of the difference between all pairs of sequential time series values

$$S = \sum_{k=1}^{n-1} \sum_{j=k+1}^{n} \text{sgn}\left(X_j - X_k\right) \tag{7.20}$$

where n is the length of the time series, X_j and X_k are sequential time series values for times t_j and t_k $(j > k)$ and sgn is the sign function defined as

$$\text{sgn}\left(X_j - X_k\right) = \begin{Bmatrix} 1 & \text{if } X_j - X_k > 0 \\ 0 & \text{if } X_j - X_k = 0 \\ -1 & \text{if } X_j - X_k < 0 \end{Bmatrix} \tag{7.21}$$

The null hypothesis, H_0, for the test is that the data, X, are a sample of n independent and identically distributed random variables (i.e., there is no trend and $S=0$). The alternative hypothesis, H_a, states that the distribution of X_j and X_k are not identical for all $k, j \leq n$, and $k \neq j$ (i.e., there is a trend, and $S \neq 0$). The null hypothesis is rejected when S is significantly different from zero, and there is a monotonic trend over time. The test requires the calculation of the standardized test statistic, Z_s, which is assumed to come from a standard normal distribution, $N(0,1)$. The test statistic is given by:

$$Z_s = \begin{cases} \dfrac{s-1}{\sigma_s} & \text{if } S > 0 \\ 0 & \text{if } S = 0 \\ \dfrac{s+1}{\sigma_s} & \text{if } S < 0 \end{cases} \tag{7.22}$$

The standard deviation of S is given by

$$\sigma_s = \sqrt{\frac{1}{18}\left[n(n-1)(2n+5) - \sum_{p=1}^{q} t_p\left(t_p-1\right)\left(2t_p+5\right)\right]} \tag{7.23}$$

where q is the number of tied groups, and t_p is the size of the pth tied group.

7.2.2.1.2.2 Sen–Theil Regression Tests The Sen–Theil trend line (Theil, 1950; Sen, 1968) is a nonparametric alternative to linear regression that can be used in conjunction with the Mann-Kendall trend test. It assumes that residuals are statistically independent and that the relationship between the variables is linear. It can also handle missing and censored values.

The method consists of computing the simple pairwise slope estimate: $S_{kj} = (X_j - X_k)/(t_j - t_k)$ for all possible distinct pairs of measurements, (X_k, X_k) where $t_j > t_k$. For a sample of size n, there will be $N = n\ (n-1)/2$ pairs of slope. The Sen–Theil trend slope can then be computed as the median of all pairwise slopes, $b = \text{median}(S_{kj})$, and the intercept of the trend line is given by $a = \text{median}(X - bt)$. By taking the median pairwise slope instead of the mean, extreme pairwise slopes usually caused by outliers have little impact on the final slope estimate.

7.2.2.1.3 Parametric Versus Nonparametric Methods

Choosing between parametric and nonparametric methods for trend detection depends on the assumptions about the data and the nature of the trend. Parametric approaches, such as OLS regression or generalized least squares, are appropriate when there is a reason to believe the trend follows a specific functional form (e.g., linear or quadratic), the residuals are normally distributed with constant variance, and observations are independent, or autocorrelation is explicitly modeled. These methods are powerful when assumptions hold, allow direct estimation of trend magnitude and rate of change, and can incorporate covariates, but they can produce misleading results if assumptions are violated and are sensitive to outliers. Nonparametric methods, such as the Mann–Kendall test and Theil–Sen slope are preferable when a particular functional form may not be assumed; the data may be non-normally distributed, skewed, or heteroscedastic; or when there are outliers or small sample sizes. In practice, practitioners often start by exploring the data visually and statistically, checking assumptions such as normality, autocorrelation, and homoscedasticity; if these are met, a parametric approach is suitable, whereas if they are violated, a nonparametric approach is safer. In some cases, running both methods and comparing results can provide additional confidence in the detected trend.

7.3 NONSTATIONARY EXTREME VALUE ANALYSIS AND RELIABILITY ANALYSIS

In a changing climate, much evidence calls into question the adequacy of the stationarity assumption for extreme climates, as shown in previous section. The conventional extreme value analyses described in Section 7.1 cannot be appropriately used but need to be adapted. A nonstationary stochastic process implies that any component of the process: the occurrence, intensity, and duration, can be a function of an environmental covariate, for example, global warming level or chronological time. It is unnecessary to model all three components as time variant, which ultimately depend on the climate projections and the type of load process.

The hazards in a nonstationary changing climate have been investigated in many past studies, such as wind (e.g., Chen et al., 2012; Cheng et al., 2014; Jeong & Sushama, 2018; Knutson et al., 2019, 2020; Esmaeili & Barbato, 2021; Li et al., 2017 , 2022; Li, 2023a,b, 2025; Li and Kopp 2023), precipitation (e.g., Semmler & Jacob, 2004, Kyselý & Beranová, 2009; Trenberth, 2011; Costa et al., 2012; Kunkel et al., 2013; O'Gorman, 2015; Tabari & Willems, 2018), snow (e.g., Jeong & Sushma, 2018; Li et al., 2022; Li, 2023a,b, 2025), ice (e.g., Jeong et al., 2018), flood and SLR (Cunderlik & Burn, 2003; Gilroy & McCuen, 2012; Machado et al., 2015; NASEM, 2024). Some of these studies have been used to develop and update building design codes and standards (e.g., Li et al., 2022; Li, 2023a; Li et al., 2025).

In a nonstationary regime, the stationary geometric process model is no longer applicable. Consequently, the concept of return period loses most of its attractive properties and cannot be directly applied to the nonstationary climate. Researchers, especially those in hydrologic engineering, have attempted to define an equivalent return period in the nonstationary climate. Two definitions of the equivalent return period have been proposed (Du et al., 2015, Cooley, 2012). The Waiting Time-Base Concept (Olsen et al., 1998) defines the return period as the expected time of first occurrence of the event. The Frequency-Based Concept (Laurent & Parey, 2007) defines the return period as the duration in which the expected number of occurrences of the event is equal to 1. Both concepts have been analyzed using a discrete nonstationary geometric process in which the APE varies from year to year. For time horizon of n years, the APE takes n different values as $p_1, p_2, \ldots, p_n$ (Cooley, 2012).

However, the proposed return period concept has many limitations. The return period, defined as the mean inter-occurrence time between events, is not identical in a nonstationary process. In case of an increasing trend in the occurrences of events, it is expected that the return period will decrease in the future. Therefore, defining a return period with reference to the first occurrence does not provide complete information about the impact of a nonstationary climate. At any future time, t, the mean time to the occurrence of the next event is not equal to the first occurrence. Therefore, the return period is not a constant measure of safety over the life of the structure. The nonstationary "return period" cannot be determined from a constant

APE, but needs to consider the nonstationary of the APE (Li et al., 2025). During the life of a structure or infrastructure, the design level event is likely to occur multiple times. Therefore, the return period as defined with reference to a single occurrence is not indicative of the reliability of a structure. If occurrences of climatic loads follow the Poisson process, then the frequency-based definition implies that the probability of exceedance over the "return period" is approximately 0.63, a level may be quite high for the safety of any critical infrastructure system (Rootzén & Katz, 2013). There has been an increasing recognition of the fact that the "return period" is not a coherent metric of reliability and safety, especially in a nonstationary climate (Serinaldi, 2015; Read & Vogel, 2015). To quantify the impact of changing climate, Vogel et al. (2011) introduced the concept of the magnification factor for flood level, a ratio of the kth percentile of the flood at the end of design life, $(t + \Delta t)$, to the flood level at the beginning of design life in some year t. If the flood trend is increasing over time, the magnification factor will be greater than 1. Alternatively, Vogel et al. (2011) proposed a recurrence reduction factor with reference to a base year t and the planning horizon of Δt.

There has been a wider acceptance of the probability of exceedance (or probability of failure) over a given design-life period, which is a more direct and accurate measure of safety against environmental hazards in a nonstationary climate. A concept of DLL is introduced (Rootzén & Katz, 2013) as a way to quantify and communicate environmental risk in a changing climate. This approach allows design to consider a specific design life (start- and end-of-life year) as well as the probability of exceeding a hazardous level during this period. Suppose a structure's design lifespans from year t_1 to t_2. The design life level (DLL) is associated with the probability p with which the design level is likely to be exceeded during the interval (t_1, t_2). In essence, the DLL is the pth percentile of the distribution of the maximum load value in the interval (t_1, t_2). In the DLL concept, the APE varies over the lifespan of a structure. Therefore, it is likely that in some year, the risk of failure can reach an unacceptable level within (t_1, t_2). To avoid this situation, Rootzén and Katz (2013) introduced another risk measure, the Minimax design approach. The design level is chosen such that the maximum probability of exceedance in any one year, that is, the maximum annual exceedance probability, during the interval (t_1, t_2) is at the maximum value p.

Several different nonstationary methods proposed in past studies are summarized below.

The expected waiting time (EWT) approach (Bayazit et al., 2001; Salas & Obeysekera, 2014) can be expressed by,

$$E_T = E(k) = 1 + \sum_{k=1}^{\infty} \prod_{i=1}^{k} F_{X_i}(x_d) \tag{7.24}$$

where E_T denotes the EWT, $E()$ denotes the expected value, k is the kth year when annual maximum event is considered, F_{X_i} is the time-varying probabilistic distribution, and x_d is the design level value.

The expected number of events (ENE), being equal to 1 in the time interval $(0, T)$, which can be evaluated by solving the equation below (Du et al., 2015),

$$E\left(N(T)\right)=\sum_{k=1}^{T}\left(1-F_{X_k}\left(x_d\right)\right)=1 \tag{7.25}$$

A more general form can be defined, given the average APE, P_a, as follows,

$$E\left(N(T)\right)=\sum_{k=1}^{T}\left(1-F_{X_k}\left(x_d\right)\right)=T\cdot P_a \tag{7.26}$$

The Minimax approach can be expressed by,

$$P_{a,\text{DL}}(x>x_d)=\max\left(P_{a,t|\,t_1<t<t_2}(x>x_d)\right) \tag{7.27}$$

where $P_{a,\text{ DL}}$ denotes the APE or failure of the Design Level, which sets to be the maximum APE or failure during time interval of t_1 and t_2. When $P_{a,\text{ DL}}$ is taken as the maximum APE for design, it might be also named as the Maximum Allowable APE (Li et al., 2025).

The DLL can be calculated by solving the following equation,

$$G\left(x_d\right)=1-\sum_{i=1}^{N}F_{X_i}\left(x_d\right) \tag{7.28}$$

where $G\left(x_d\right)$ is the probability of exceedance or failure during the nominal design life of N years (e.g., 50 years). Salas et al. (2018) provides tabular comparisons of DLL and other metrics for both stationary and nonstationary conditions.

When transition from stationary climate design to a nonstationary climate condition, to be consistent, one might set the total probability of exceedance or failure as the target (Li, 2023a, 2024), that is, $G\left(x_d\right)=1-\left(1-p_a\right)^N$, where p_a is the APE or failure.

As shown in Li et al. (2022) and Li (2023a), among the methods above, the ENE and EWT generally result in the same results and are mostly the same as the Minimax approach when taken the maximum value during the concerned periods. The DLL uses different risk measures than the Minimax approach. The Minimax approach takes the worst annual risk measure during the concerned periods, while the DLL approach takes the cumulative risk measure during the concerned periods. When the DLL is applied to a constant nominal design life, it can be considered a static approach between year 1 and N. To extend the DLL approach more robustly, dealing with the fact that the actual design life can be longer than the nominal design life, this method can be extended to move with the actual design life and evaluate the risk for any specific period after the nominal design life. This approach can be considered a dynamic DLL approach (Li, 2023a, 2024).

$$G_t\left(x_d\right)=1-\sum_{i=t}^{N+t-1}F_{X_i}\left(x_d\right) \tag{7.29}$$

The above mathematical forms require carrying out nonstationary extreme value analysis when the probability of exceedance is taken as the risk measure; or nonstationary reliability analysis, when the probability of failure is taken as the risk measure.

When carrying out the nonstationary extreme value analysis, the extreme value distribution introduced in Section 7.1 needs to be adapted to be nonstationary. The nonstationarity can be defined as a function of a covariate, which can be either time, or some key climatic variables, such as global warming level. For engineering applications, as the design life becomes important in a nonstationary climate, it seems to be reasonable to use the time as the covariate. The global level warming can be mapped to a specific year when a certain level is reached.

For example, the Gumbel distribution in a stationary climate can be adapted into a nonstationary format,

$$F_t(x) = \exp\left(-\exp\left(-(u_t - x)/a_t\right)\right) \tag{7.30}$$

where u_t and a_t are nonstationary Gumbel distribution parameters.

When a linear trend as a function of time exhibits in future climates, these parameters can be expressed by,

$$\begin{aligned} u_t &= u_0 + u_1 t \\ a_t &= a_0 + a_1 t \end{aligned} \tag{7.31}$$

The GEV distribution can be also adapted into a nonstationary format,

$$F_t(x) = \exp\left(-\left(1 + k_t(u_t - x)/a_t\right)^{-1/k_t}\right) \tag{7.32}$$

When a linear trend as a function of time exhibits in future climates, these parameters can be expressed by,

$$\begin{aligned} u_t &= u_0 + u_1 t \\ a_t &= a_0 + a_1 t \\ k_t &= k_0 + k_1 t \end{aligned} \tag{7.33}$$

While mathematically, a nonlinear nonstationary model can be used in above cases, past studies often employed linear model in developing nonstationary extreme value distributions.

Several methods have been proposed to fit climatic variables into a nonstationary extreme value distribution. These methods include the conventional maximum likelihood method combined with step wise approach (Kharin & Zwiers, 2005), the generalized maximum likelihood method (GMLM) (Martins & Stedinger, 2000), the generalized adaptive model for location, scale, and shape (GAMLSS) (Rigby & Stasinopoulos, 2005), and the Differential Evolution Markov Chain method (DEMC) (Cheng et al., 2014).

The maximum likelihood method with a stepwise approach estimates the constant GEV distribution parameters using the maximum loglikelihood method. The results are used as an initial guess to estimate the distribution parameters (location, scale, and shape parameters) by considering location parameter varying with time. The resulted parameters are used as initial guess to estimate the distribution parameters by considering two distribution parameters varying with time. Finally, the results from this step are used as initial guess to estimate all three distribution parameters as a linear function of time. The loglikelihood calculated from the end of each stepwise estimation can be used to calculate the loglikelihood ratio to determine the preference of the degree of nonstationarity of each parameter.

The GAMLSS model (Rigby & Stasinopoulos, 2005) has been applied in hydrological studies, for example, López and Francés (2013) and Marchado et al. (2015). The GAMLSS model expresses the independent observations at a given time as a cumulative probability distribution function of vector parameters, which can be further expressed as a function of the explanatory variables via a monotonic function.

GMLM (Martins & Stedinger, 2000; El Adlouni et al., 2007) can be used to fit the GEV distribution by constraining the shape parameters into a reasonable range, which is based on prior knowledge on shape parameters from practical experiences.

The differential evolution Markov Chain (DEMC) approach (Cheng et al., 2014) has been used to estimate the nonstationary GEV parameters assuming a linearly varying location parameter. The basic idea of this approach is to use the Bayesian-based Markov Chain Monte Carlo method to obtain the posterior distribution of the GEV parameters. The posterior distribution of the parameters can be evaluated from a large number of realizations using the DE-MC method. The statistics and quantiles of the distribution parameters can then be evaluated.

Most reliability-based modern standards and codes, different design limit states have been used to ensure that a design meets the target reliability requirement. In a typical limit state design procedure, both climatic loads and resistance are specified as adequate percentiles of their distributions. The load and resistance factors are then calibrated with reference to a target reliability. The structural reliability target is typically specified in terms of annual probability of failure or total probability of failure during a nominal service life of the structure. For example, in ASCE 7-22, the reliability index could vary from 2.5 to 4.5 depending on the risk category and the failure mode (ASCE 7-22, Table 1.3). A nominal service life of 50 years of the structure has been assumed in calculation of the reliability index. In essence, the target reliability requirement is specified to ensure that the probability of failure in any given year as well as over the entire service life does not exceed a maximum allowable value.

The above methods focus on hazard level quantification. In a nonstationary climate, the nonstationary reliability analysis is required to evaluate the structural reliability for given climatic design-level loads or given target risk level to quantify the design-level climatic loads.

If a nonstationary load process exhibits an increasing trend, then it is obvious that a structure designed on the basis of stationary conditions would not satisfy the target reliability requirement over its lifetime.

In a stationary climate, the risk measure using annual probability of failure and lifetime cumulative probability of failure, also expressed by the reliability index, is equivalent. This relationship can be expressed by,

$$P_{\text{DL}} = 1 - \left(1 - P_a\right)^N \tag{7.34}$$

where P_{DL} denotes the probability of failure during the design life, P_a is the annual probability of failure.

This form is valid as long as the P_a does not vary with the time during the design life, N. In other words, given the annual probability of failure, one can obtain the total probability of failure during the design life, and vice versa. This is the case why modern codes often do not distinguish between the annual probability of failure and total probability of failure during a nominal design life, that is, reliability index. For example, ASCE 7-22 Table 1.3-1 provides both the reliability index and annual probability of failure for each risk category and failure mode. These values are interchangeable in stationary climate. An annual probability of failure of 3.0×10^{-5} leads to the probability of failure of $1 - (1 - 3.0 \times 10^{-5})^{50} = 0.0015$, corresponding to the reliability index of 2.97 (about 3.0), during a nominal design life of 50 years.

However, in a nonstationary climate, the annual probability of failure varies given a specific design level climatic load. The design life probability of failure needs to be expressed by,

$$P_{\text{DL}} = 1 - \prod_{i=1}^{N} \left(1 - P_{a,i}\right) \tag{7.35}$$

The reliability analysis framework that has been used in a stationary climate can still be used in nonstationary climate conditions, but requires adaptation of the climatic load effect components with nonstationary probabilistic models. The design scheme associated with the risk measure will have to be selected in a nonstationary condition, as abovementioned that the annual risk measure can be different from the lifetime overall risk measure in a nonstationary condition.

Take the structural reliability analysis for wind loads as an example, the load combination when dead load and wind load effects are considered can be expressed by (Li, 2023a),

$$\gamma R_n = \alpha_D D_n + \alpha_w W_n \tag{7.36}$$

where γ is the resistance factor, α_D is the load factor for the nominal dead load, D_n, and α_w is the load factor for the nominal wind load, W_n.

This mathematical form is generic and applicable to most structural reliability analysis involving the dead load and wind load, but the resistance factor and load factors can vary in different codes and standards. For example, NBCC 2020 uses $\alpha_D = 1.25$ and $\alpha_w = 1.4$ for a nominal wind load associated with AEP of 1/50, while the ASCE 7-22 uses $\alpha_D = 1.2$ and $\alpha_w = 1.0$ for a nominal wind load associated with

AEP of 1/700. The updated NBCC 2025, uses $\alpha_w = 1.0$ for a nominal wind load associated with AEP of 1/500 (Li et al., 2025).

There are different ways to derive a state function based on (7.36). Different formats have been presented in some recent studies that investigated nonstationary reliability analysis.

The state function for principal wind loads might be expressed by (Bartlett et al., 2003a,b; Hong & Ye, 2014; Hong et al., 2021; Li, 2023a),

$$g = \frac{1}{\gamma_R} X_R - \frac{1}{1 + I_W R_{W/D}} \left(\frac{X_D}{\alpha_D} + R_{W/D} \frac{Z_W \left(V / v_T \right)^2}{\alpha_W} \right) \tag{7.37}$$

where I_W is the importance factor for wind loads, v_T is the design level wind speed, Z_W is wind load transformation factor, $R_{W/D}$ is the ratio of nominal wind load to nominal dead load.

It might be also expressed by (Ghosn & Ellingwood, 2024),

$$Z = \lambda_R \left[\frac{1.2D_n + 1.0C_{W,n}V_n^2}{0.9} \right] - \lambda_D D_n - \lambda_W \left[C_W \left(\bar{V}_{\max} \right)^2 \right] \tag{7.38}$$

where λ_R and λ_D are nondimensional parameters similar to the X_R and X_D in above state functions. The λ_W is the nondimensional parameter related to wind load effects. This state function follows Ellingwood et al. (1982) and McAllister et al. (2018).

These state functions shown above can be interchangeably transformed from one to the other. The nondimensional parameters can be modeled into different probability distributions in different studies. For example, the nondimensional resistance factor was modeled as a lognormal distribution in all these studies, but the distribution parameters, such as bias and coefficient of variation, can be different. Similarly, the wind load effect was characterized as a transformation factor, modeled as a lognormal distribution, in combination with the squared wind speed ratio (follows a specific extreme value distribution) in Hong et al. (2021) and Li (2023a), but a combined wind load effect (modeled as a Gumbel distribution) in Ghosn and Ellingwood (2024).

There appears to be several methods that have been used in these recent studies in how to carry out the nonstationary reliability analysis.

The first approach is to follow Eq. (7.28) that only considers a single target reliability index in a stationary climate, within the specified design life. In other words, the nonstationary climate was integrated into the state function but only considered the overall effect through targeting the overall reliability index within the nominal design life of 50 years, say between 2025 and 2075. This approach is the DLL static approach (Li, 2023a, 2024) combined with the lifetime probability of failure.

The second approach is to derive an equivalent annual reliability index (Hong et al., 2021).

$$\beta_A = -\Phi^{-1}\left(1 - \left(1 - \Phi\left(-\beta_{50}\right)\right)^{1/50}\right) \tag{7.39}$$

where the β_A is the equivalent annual reliability index, and β_{50} is the target reliability index in a nominal 50 years. The minimum reliability index within a specific period shall meet a target value to derive the corresponding design level load.

The third approach is to calculate the maximum annual probability of failure within a specific period and set its value to be a target value for deriving the design level value. This approach is the Minimax approach combined with the annual probability of failure. This approach follows Eq. (7.27) more closely. The state function (Eq. 7.37) can be rearranged such that the annual probability of failure can be directly calculated without the need to "annualized" the total probability of failure. The cumulative probability of failure through Eq. (7.34) in stationary climate and Eq. (7.35) in nonstationary climate can be achieved.

The Minimax approach and DLL approach introduced above focusing on hazard level were replaced by considering the risk measures. The annual probability of failure was considered in the Minimax approach, while the total probability of failure during a nominal design life was considered for the DLL approach. An extension of the DLL static approach was made to allow the quantified probability of failure during a nominal design life of 50 years to go with the actual year in service. For example, for an original design life between 2030 and 2080 with a nominal design life of 50 years, when a structural reliability analysis was carried out in a stationary climate, the evaluation can be carried out at any given year during this period but still with a nominal lifespan of 50 years considered in the analysis. In other words, the analysis would be carried out between year 2030 and 2080, and also between 2080 and 2130. The results are exactly the same in a stationary climate, as the extreme climates do not change with time, as long as the capacity does not change with time. By neglecting material deterioration, the structural reliability maintains the same at any time with a reference period of 50 years. Similarly, in a nonstationary climate, such an analysis might be considered to be consistent. In such a case, the nonstationary reliability analysis shall be carried out along with the actual year in service with a nominal reference period. It was found in Li (2023a) that the load scaling factor derived based on the DLL dynamic approach can be slightly higher than the Minimax approach, given that the future period beyond the initial nominal design life has been taken into account.

Some graphical illustrations in Figures 7.4 and 7.5 were presented in Li (2024) to demonstrate the difference between the Minimax and DLL approaches and if hazard or risk is used as a measure. In Figure 7.4a, the Minimax design value ensures the maximum annual exceedance probability does not exceed the target value within the design life. When annual probability of failure is considered, as shown in Figure 7.4b, the design value ensures that the maximum annual probability of failure within the design life does not exceed the target value.

Different than the Minimax approach, the DLL approach considers the total probability of exceedance (mathematically expressed by Eq. 7.29), as shown in Figure 7.5a, or the total probability of failure (mathematically expressed by Eq. 7.35) within the design life, as shown in Figure 7.5b. Depending on whether the robustness of the actual service life is considered longer than the initial nominal design life, for example, 50 years, this approach might be used as a single time window (static) or multiple time windows (dynamic), which can vary in length as evaluation requires.

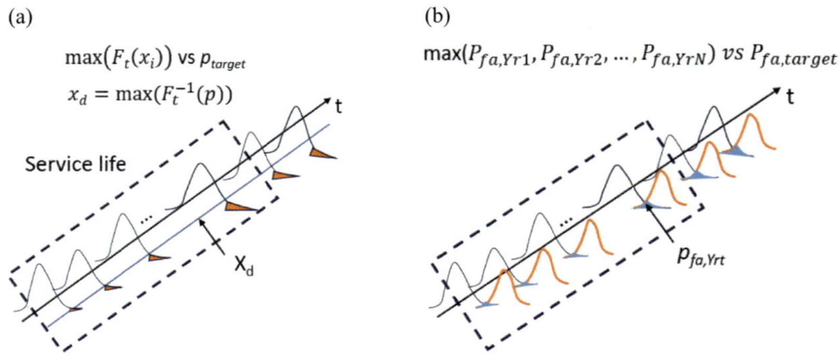

FIGURE 7.4 Minimax approach for (a) deriving the design level value and (b) evaluating annual probability of failure in a nonstationary climate.

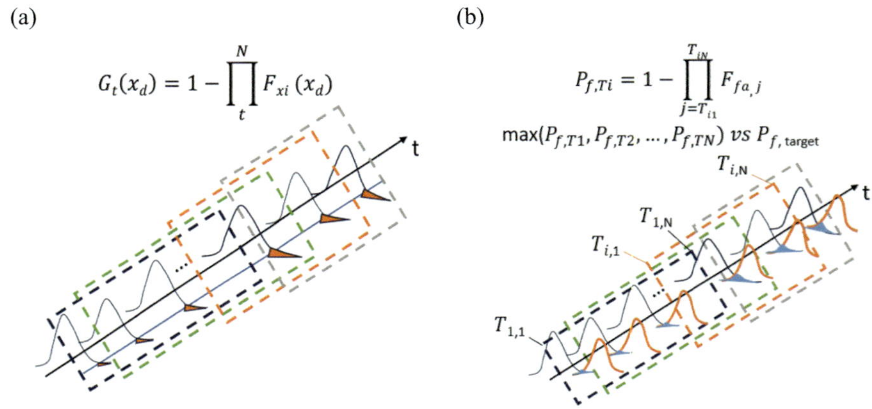

FIGURE 7.5 DLL approach for (a) evaluating the lifetime exceedance probability and (b) evaluating the total probability of failure within a specific period in a nonstationary climate.

The selection of a risk measure in a nonstationary climate requires consideration of the design life explicitly and selection of the risk measure: hazard vs. failure; annual risk vs. nominal lifetime risk. This becomes a primary difference from the design in a stationary climate, in which the annual risk is equivalent to the lifetime risk measure, and both are insensitive to the actual service life of a structure and infrastructure. In a nonstationary climate, the overall risk within a nominal design life averages the annual risk, which is similar to taking the equivalent annual risk. The time when this equivalent annual risk will be reached may be close to the midyear of the nominal design life. In other words, if the climate within such a period is considered, the design aims to prepare the structure close to its mid-lifespan with regard to the annual risk, which might be unconservative. To address this, future climates might be "borrowed" to offset the midyear corresponding to the equivalent annual risk.

A more straightforward approach when considering annual risk is to apply the Minimax approach for risk assessment, which will take the worst annual risk as the design target within the nominal design life. For a linear increasing trend climate, the limit state will be reached close to the end of the design life, while a decreasing trend climate, the limit state is reached at the beginning of the design life, that is, the present climate is the worst-case scenario.

7.4 DEEP UNCERTAINTIES

7.4.1 Engineering Approach for Addressing Deep Uncertainties

Infrastructure systems—including energy, transportation, water, wastewater, and digital networks—are highly vulnerable to environmental hazards due to their complexity and long lifespans. Transforming these systems to address climate change, urbanization, and other socioeconomic shifts is essential but slow, as traditional decision-making approaches tend to be reactive and insufficient. Deep uncertainty arises when the relationships between key drivers of change are difficult to predict or when there is no consensus on how to quantify or value future outcomes. A formal definition of deep uncertainty is a situation in which analysts do not know or cannot agree on: (1) models that relate key forces that shape the future, (2) probability distributions of key variables and parameters in these models, and (3) the value of alternative outcomes (Hallegatte et al., 2012).

This uncertainty complicates long-term planning by making it difficult to assess the impacts of decisions and agree on desirable outcomes. Decision Making under Deep Uncertainty (DMDU) offers an alternative to the conventional "predict-then-act" approach by emphasizing "monitor-and-adapt" strategies. This enables policymakers to prepare for a range of uncertain futures rather than relying on static projections.

Uncertainty, as a concept tied to the limits of knowledge, has been debated since ancient Greece and was formally distinguished from risk by Knight in 1921. Over time, scholars have refined this distinction, emphasizing that decision-making under uncertainty extends beyond unknown future states to include strategic behaviors of decision-makers. Uncertainty is broader than risk and exists on a spectrum, ranging

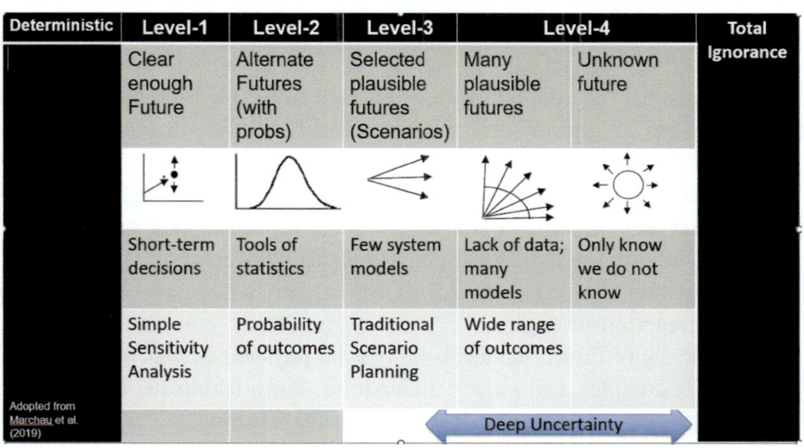

FIGURE 7.6 Level of uncertainties. (Adapted from Marchau et al. 2019).

from complete certainty (which is rarely attainable) to extreme unpredictability. Walker et al. (2003) (also see Marchau et al., 2019) identify four levels of uncertainty based on knowledge about future states, system models, expected outcomes, and stakeholder preferences (Figure 7.6). Level 1 uncertainty involves minor variations in known systems, while Level 2 incorporates probabilistic modeling. Level 3 arises when future scenarios are plausible but lack defined probabilities, requiring scenario analysis. At Level 4, uncertainty becomes unbounded—either due to gaps in knowledge (4a) or the occurrence of unpredictable, high-impact "Black Swan" events (4b), which challenge traditional forecasting and decision-making methods.

Deep uncertainty caused by climate change necessitates a shift from the "predict and act" paradigm to "robust decision making," characterized by continuous learning and dynamic adaptation. The increasing focus on climate change and deep uncertainty has fueled extensive research on DMDU, even leading to a professional society dedicated to this subject (http://www.deepuncertainty.org/).

A clear example of deep uncertainty is the acceleration of SLR, particularly for infrastructure projects with long lifespans and significant societal impact. Current models, especially those related to Antarctic ice sheet melting—have limitations, and future greenhouse gas emissions remain difficult to forecast. As a result, global mean sea level (GMSL) projections for the late 21st century vary widely, necessitating flexible, adaptive strategies rather than traditional cost-benefit analyses. Coastal communities, particularly those in low-lying areas, are already experiencing the physical, social, and economic consequences of rising sea levels. These impacts are often compounded by additional stressors, such as more intense rainfall and rising groundwater levels. Tide gauge data shows that GMSL rose at a rate of 1–2 mm per year until 1990, accelerating to approximately 3 mm per year in recent decades. However, regional and local sea level changes vary due to factors such as ocean density and currents, gravitational shifts caused by melting ice, and vertical land movement from subsidence or past ice mass losses. In many coastal areas, nuisance flooding—also known as high-tide flooding—is already causing economic losses and social disruptions. What is currently a nuisance is likely to become chronic flooding in the near future, increasing the urgency for adaptive strategies. Since adaptation planning and implementation can take decades, decision-making frameworks that incorporate deep uncertainty will be critical in ensuring resilient, long-term solutions.

Making decisions for adaptation to SLR, especially for projects with extended service lives and/or long-term societal effects, is difficult for various reasons.

- Future sea level predictions are uncertain, primarily due to limitations in current ice sheet melting models, particularly for Antarctica, and the inability to accurately predict future emissions. This uncertainty results in wide prediction bounds for late-century GMSL, demanding practitioners adopt robust adaptation strategies.
- The feedback of future sea levels on coastal physical, social, and economic systems is complex and poorly understood, diminishing the usefulness of traditional cost-benefit analysis for adaptation planning.
- Sequential actions in an environment of deep uncertainty cannot be considered independent; choices today will affect future options and societal developments.

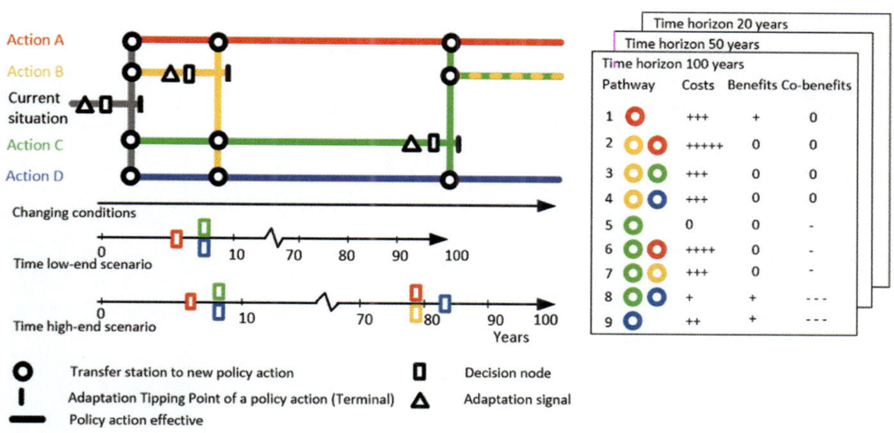

FIGURE 7.7 Example of a pathways map and a scorecard presenting the costs and benefits of the nine alternative pathways presented in the map. (From Haasnoot et al., 2013.)

While many DMDU methods exist, the Dynamic Adaptive Policy Pathways (DAPP) approach is becoming increasingly attractive for project planning and phased implementation. This approach involves a plan with an initial action, a focus on monitoring data, and a sequence of actions over time (pathways) that are contingent on emerging future scenarios. This article provides a concise explanation of the DAPP approach and presents a case study of its application to develop adaptation strategies for future flooding mitigation in a coastal basin in Miami-Dade County, Florida, where climate change has compounding effects.

In scenarios of deep uncertainty, policymakers must make near-term decisions while ensuring that future options for addressing uncertain future conditions are not precluded by other actions and developments. The DAPP approach directly addresses this challenge by incorporating decisions over time, emphasizing dynamic planning for an uncertain future. It integrates adaptive planning, adaptation pathways, and tipping points to identify when an action ceases to fulfill a particular objective.

The overarching DAPP approach operates on the premise that a given policy or action has a design life, after which it may fail due to changing environmental conditions, such as SLR. At this "tipping point," a different policy or action may be necessary to uphold the desired resilience level. As conditions change, alternative policy pathways (sequences of actions) become possible over a project's planning horizon. The best way to illustrate the DAPP approach is through a pathways map (Figure 7.7), which displays various choices and routes under changing conditions. An action might not meet objectives after a period of time (tipping points represented as "transfer stations"), at which point a shift to a different pathway may be needed. Crucially, the decision to transfer must occur before the tipping point, providing policymakers time to plan accordingly.

7.4.2 Case Study

Flood protection planning in highly urbanized coastal basins in Miami-Dade County, Florida, necessitates considering the combined effects of sea level change, rising groundwater levels, and extreme rainfall. The future evolution of these environmental

(a)

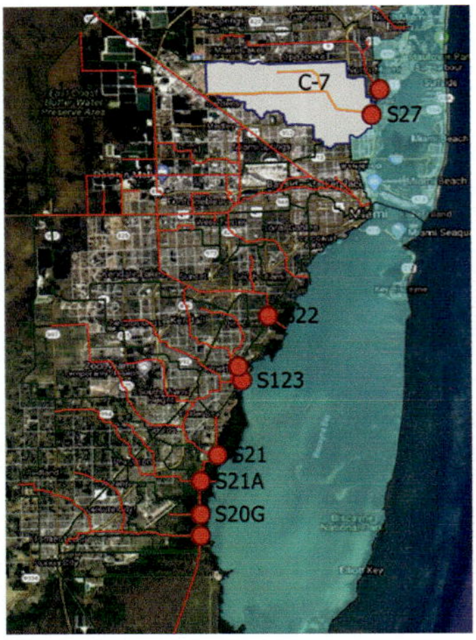

(b)

FIGURE 7.8 (a) Coastal basins in Miami-Dade County, Florida. The study site, C-7 (Little River Basin) is highlighted; (b) S-27 Structure at which drains the Little River Basin to Biscayne Bay.

stresses and shocks is highly uncertain. The DAPP approach was applied to the C-7 basin (Little River Basin) in northeastern Miami-Dade County (Figure 7.8).

The C-7 basin, covering roughly 32 square miles, depends on its primary canal, the C-7 canal, for flood protection, drainage, and groundwater table elevation maintenance to counter saltwater intrusion. The study area, with approximately 254,000 residents, includes single-family and multifamily residential buildings, industrial activities, and public buildings. This basin faces high vulnerability to flooding, with current repetitive property loss data indicating frequent damage from flooding, expected to worsen.

As part of the Central and South Florida Project, the C-7 canal was designed to pass the Standard Project Flood at its outlet structure, S-27, without exceeding the design headwater stage and to maintain a headwater stage that prevents saltwater intrusion. However, the original design, dating back to the early 1960s, did not adequately consider SLR. Since the main C-7 canal drains into Biscayne Bay solely by gravity, the S-27 structure is known to underperform during periods of heavy rainfall and high sea levels. Rising sea levels will further diminish the S-27 discharge capacity due to elevated downstream levels during high tides and coastal storms.

Sea level projections for the South Florida region are broad, and due to the low topography in many watersheds like C-7, the potential impacts of flooding without adaptation are likely to be extensive. The broad projection uncertainty, the varying sensitivity of impacts to SLR, and other uncertainties make this adaptation a classic case of deep uncertainty. Flood depths were analyzed using the XPSWMM model for current sea levels, two sea-level rise scenarios, and four rainfall return periods (5, 10, 25, and 100 years). Because climate projections of extreme rainfall are highly uncertain, possible changes to the design rainfall were not considered, and that may need to be a consideration in future studies. Maximum stage water levels for each sub-basin in the study area were collected for subsequent use by the damage assessment model. Direct flood damages were estimated for each return period using the Delft-Flood Impact Assessment Tool (https://www.deltares.nl/en/software-and-data/products/delft-fiat-flood-impact-assessment-tool). This model uses flood damage functions from the HAZUS model for various building categories and calibrates flood damage estimates based on the impact of Hurricane Leslie in October 2000.

Three alternative adaptation options were evaluated for effectiveness and cost:

- **M1:** Local flood mitigation, consisting of flood walls, additional exfiltration trenches, flap gates, and local pumps.
- **M2:** Regional flood mitigation, involving the installation of forward pumps at the S-27 coastal structure.
- **M3:** Land-use mitigation, comprising improved building codes to raise roads and buildings to a height of 6, 7, or 8 feet, or by elevating buildings to the 100-year flood return period and roads to the 10-year flood return period elevations.

Local and regional flood mitigation measures (M1 and M2) significantly reduced flood risk under moderate SLR, achieving a 30%–50% reduction compared to future risk without any mitigation measures. However, these measures are less effective under the high SLR scenario. Local flood mitigation measures are approximately as effective as regional mitigation measures. While they have a particularly positive effect on several downstream sub-watersheds, they cause increased stage levels and damages upstream, partially offsetting the benefits. Both the local and regional flood mitigation measures considered in this study could not maintain future flood risk at the current level under SLR. Land-use mitigation measures (raising buildings and roads) are highly effective, especially when buildings and roads are elevated to 8 feet. This is the only measure identified that can maintain or reduce current flood risk under future sea-level rise scenarios.

Expected Annual Damage (EAD) was used as a quantitative measure for comparing the adaptation options (Figure 7.9a). The current EAD estimate serves as a threshold, and the effectiveness of future adaptation strategies (M1, M2, and M3) was assessed based on their ability to keep the EAD around the current level. Typically, EAD is used as a standard measure for the planning of flood protection projects under conditions of stationarity. In the case of the DAPP approach, cumulative damage could also be used to determine the tipping point for making decisions regarding the possibility of moving to a different pathway. Figure 7.9b illustrates an example of the adaptation pathways explored for the C-7 basin. The three variations of the M3 mitigation alternatives perform well up to 1.6 feet, or more than 3 feet of SLR if buildings and roads are raised to 7 or 8 feet. It is noted that the EAD will continue to increase under M0 if a project is designed assuming sea levels are not rising.

The pathways analysis suggests that a combination of local or regional flood mitigation measures implemented consecutively can be very effective. This approach reduces near-term risks through flood mitigation measures while mitigating long-term risks by raising roads and buildings. Implementing local and regional pumps in the very near future would help maintain current protection levels while providing time to implement the longer-term solution of raising building and road elevations.

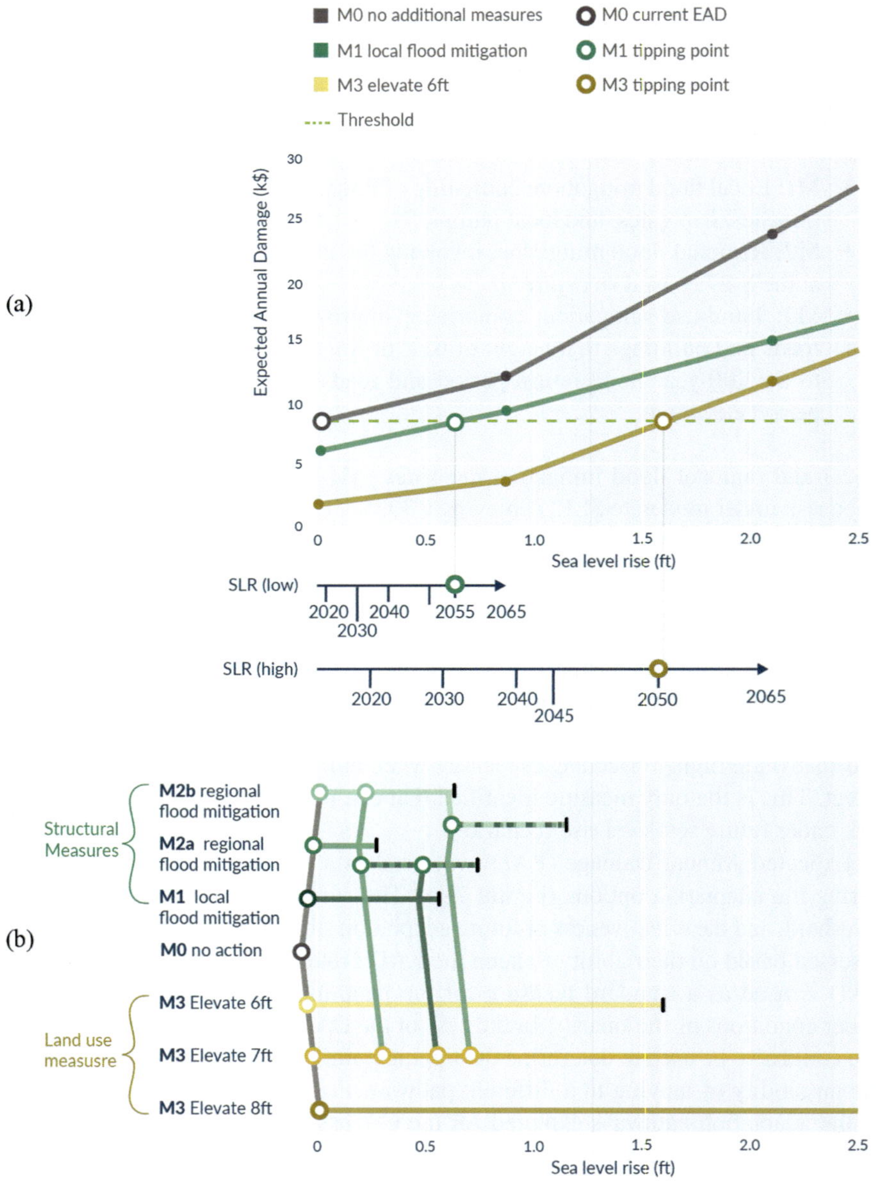

FIGURE 7.9 (a) Expected annual damage (EAD) as a function of sea level rise (SLR) corresponding to M0, M1, and M3; (b) Adaptation Pathways map for the entire basin, based on the simulated expected annual damage for the current sea-level and the two possible future sea level rises.

Given the high uncertainty and the potentially severe consequences of SLR in this area, the pathways analysis highlighted that raising properties, buildings, and transportation infrastructure, including roads, is the most viable long-term option for managing future flood risk and enhancing flood resilience. The DAPP approach application suggests a reasonable set of pathways may include:

- Local and regional pumps in the near term
- Raising properties and infrastructure for the longer term
- Initiating property and infrastructure raising now due to the potentially lengthy implementation time

The deep uncertainties in projections associated with climate change, particularly SLR, require communities to consider new decision-making methods for planning and implementing coastal resilience efforts. The DAPP approach, which integrates adaptive planning, adaptation pathways, and tipping points to address the issue of an action no longer meeting a particular objective, is one such method. As illustrated in this paper's case study of the Miami-Dade region in South Florida, DAPP is an effective tool for developing and phasing adaptation projects under uncertainty.

7.5 DISCUSSIONS

The conventional engineering design and planning under the stationarity assumption appears to be inadequate in light of climate change impacts, causing nonstationary behavior of climatic variables. Methods used for engineering design and planning in stationary conditions can be adapted to address the nonstationary condition. New methods were proposed and adopted in design practices. Nonstationarity in the time series for climatic variables often involves significant uncertainty caused by various factors. While statistical tools like trend tests and change-point analysis may indicate changes, these may actually stem from natural low-frequency climate variability rather than genuine shifts in the climatic or hydrologic regime. Studies, like Serinaldi and Kilsby (2015), caution against relying solely on at-site data, advocating for comprehensive exploratory data analysis and consideration of external non-climatic drivers, such as urbanization. Proper attribution of observed changes requires identifying a clear deterministic mechanism, which is inherently complex. Many studies recommend treating stationarity as the default assumption and suggest evaluating both stationary and nonstationary scenarios probabilistically when nonstationary signals are evident. Case studies, such as the one by Gomes da Silva et al. (2016), even show minimal differences in flood risk estimates between stationary and nonstationary models under certain conditions.

Projecting nonstationarity derived based on historical trend into the future presents further difficulties, especially over a project's design life. For example, if urban development is the primary driver of nonstationarity, the assumption of continued change may be flawed if the area becomes fully developed. Thus, multiple future scenarios may be necessary. Climate change complicates projections even more due to the uncertainty in future extreme climate patterns and significant influence by the global efforts of mitigation, causing the selection of proper global warming level scenarios to be very uncertain. The lack of standardized methods to integrate climate models output into hydrologic design, and transition from conventional design method to forward-looking design method for resilience also add to the sources of uncertainties. Researchers have also pointed out conceptual challenges in applying nonstationary methods, including ambiguities in defining and communicating return period concepts and risk under changing conditions. Traditional interpretations of metrics like return period can be misleading in nonstationary contexts. As a result, there's a growing emphasis on alternative risk metrics and system reliability assessments, such as what has been adopted by the NBCC 2025 edition (Li et al., 2025), although more research is needed to evaluate and compare these approaches and advocate consensus-based approaches for industry-wide engineering applications.

Because there are many factors (land use and climate change) that may contribute to any apparent nonstationarity in data, careful adherence to this process may be needed. Table 7. 4 shows multiple situations relevant to the decision regarding the considerations of nonstationarity.

When applying new nonstationary approaches in design and planning, adequate consistency should be established to ensure that the applied method can result in the same design value should the same historical data be used. One engineering challenge for nonstationarity in a changing condition is the increased design parameters, which are assumed to be constant in the stationary condition, such as service life, annual risk, and lifetime risk. In a nonstationary condition, the annual risk can vary

TABLE 7.4
Decision Matrix for Considering Nonstationarity

Historical Observations / Reanalysis[a]	Future Projections	Considerations
Statistically significant trend absent	Projections do not suggest a systematic trend, or there is insufficient information	Current practice based on stationarity is adequate
Statistically significant trends present	Projections do not demonstrate a systematic trend	Most likely due to when the system is built out or climate change is episodic. The decision to use an extrapolation of historical trends may depend on risks and economic aspects. A Dynamic Adaptive Policy Pathway (DAPP) strategy may be warranted
Statistically significant trend absent	Projections do suggest systematic trend	Apply the new nonstationary techniques unless the current safety factors can account for future changes (e.g., freeboard). Consider DAPP
Statistically significant trends present	Projections do suggest a systematic trend	Apply the new nonstationary techniques unless the current safety factors can account for future changes (e.g., freeboard). Consider DAPP

[a] A careful attribution exercise should be implemented if the trends are statistically significant.

with time, as can the lifetime risk and it can be also significantly affected by the predefined service life of buildings and infrastructures. The future-climate-ready buildings and infrastructure design and planning require holistic consideration of these explicit parameters, cost-benefit analysis, and robust framework, such as DAPP.

REFERENCES

Bartlett, F. M., Hong, H. P., & Zhou, W. (2003a). Load factor calibration for the proposed 2005 edition of the National Building Code of Canada: Companion-action load combinations. *Canadian Journal of Civil Engineering*, 30(2), 440–448.

Bartlett, F. M., Hong, H. P., & Zhou, W. (2003b). Load factor calibration for the proposed 2005 edition of the National Building Code of Canada: Statistics of loads and load effects. *Canadian Journal of Civil Engineering*, 30(2), 429–439.

Bayazit, M., Önöz, B., & Aksoy, H. (2001). Nonparametric streamflow simulation by wavelet or Fourier analysis. *Hydrological Sciences Journal*, 46(4), 623–634.

Bean, B., Maguire, M., Sun, Y., Wagstaff, J., Al-Rubaye, S., Wheeler, J., Jarman, S., & Rogers M. (2021). The 2020 national snow load study. https://doi.org/10.26077/200k-pr86

Befus, K. M., Barnard, P. L., Hoover, D. J., Finzi Hart, J. A., & Voss, C. I. (2020). Increasing threat of coastal groundwater hazards from sea-level rise in California. *Nature Climate Change*, 10(10), 946–952.

Blom, G. (1958). *Statistical Estimates and Transformed Beta-Variables*. John Wiley & Sons, New York.

Castillo, E. (1988). *Extreme Value Theory in Engineering*. Academic Press, New York.

Chen, L., Pryor, S. C., & Li, D. (2012). Assessing the performance of Intergovernmental Panel on Climate Change AR5 climate models in simulating and projecting wind speeds over China. *Journal of Geophysical Research: Atmospheres*, 117(D24). https://agupubs.onlinelibrary.wiley.com/doi/full/10.1029/2012JD017533

Cheng, C. S., Lopes, E., Fu, C., & Huang, Z. (2014). Possible impacts of climate change on wind gusts under downscaled future climate conditions: Updated for Canada. *Journal of Climate*, 27(3), 1255–1270.

Cohn, T. A., & Lins, H. F. (2005). Nature's style: Naturally trendy. *Geophysical Research Letters*, 32(23). https://agupubs.onlinelibrary.wiley.com/doi/full/10.1029/2005GL024476

Coles, S. (2001). *An Introduction to Statistical Modeling of Extreme Values*. Springer, London, UK.

Cooley, D. (2012). Return periods and return levels under climate change. In *Extremes in a Changing Climate: Detection, Analysis and Uncertainty* (pp. 97–114). Springer Netherlands, Dordrecht.

Costa, A. C., Santos, J. A., & Pinto, J. G. (2012). Climate change scenarios for precipitation extremes in Portugal. *Theoretical and Applied Climatology*, 108, 217–234.

Cunderlik, J. M., & Burn, D. H. (2003). Non-stationary pooled flood frequency analysis. *Journal of Hydrology*, 276(1–4), 210–223.

David, H.A. and Nagaraja, H.N. (2003). *Order Statistics*, 3rd Ed., John Wiley & Sons.

Draper, N. R., & Smith, H. (2014). *Applied Regression Analysis* (706 pp), 3rd edition, Wiley, New York.

Du, T., Xiong, L., Xu, C. Y., Gippel, C. J., Guo, S., & Liu, P. (2015). Return period and risk analysis of nonstationary low-flow series under climate change. *Journal of Hydrology*, 527, 234–250.

El Adlouni, S., Ouarda, T. B., Zhang, X., Roy, R., & Bobée, B. (2007). Generalized maximum likelihood estimators for the nonstationary generalized extreme value model. *Water Resources Research*, 43(3). https://agupubs.onlinelibrary.wiley.com/doi/full/10.1029/2005WR004545

Ellingwood, B. R., Galambos, T. V., MacGregor, J. G., & Cornell, C. A. (1980). *Development of a Probability Based Load Criterion for American National Standard A58*. National Bureau of Standards, Washington, DC.

Esmaeili, M., & Barbato, M. (2021). Predictive model for hurricane wind hazard under changing climate conditions. *Natural Hazards Review*, 22(3), 04021011.

Essery, R., Li, L., & Pomeroy, J. (1999). A distributed model of blowing snow over complex terrain. *Hydrological Processes*, 13(14–15), 2423–2438.

Gharib, A., Davies, E. G., Goss, G. G., & Faramarzi, M. (2017). Assessment of the combined effects of threshold selection and parameter estimation of Generalized Pareto Distribution with applications to flood frequency analysis. *Water*, 9(9), 692.

Ghosn, M., & Ellingwood, B. R. (2024). Risk-informed design and safety assessment of structures in a changing climate: A review of US practice and a path forward. *Structure and Infrastructure Engineering*, 20(7–8), 1159–1173.

Gilroy, K. L., & McCuen, R. H. (2012). A nonstationary flood frequency analysis method to adjust for future climate change and urbanization. *Journal of Hydrology*, 414, 40–48.

Gomes da Silva, P., Dalinghaus, C., González, M., Gutiérrez, O., Espejo, A., Abascal, A. J., & Klein, A. H. (2016). Estimating flooding level through the Brazilian coast using reanalysis data. *Journal of Coastal Research*, (75), 1092–1096.

Haasnoot, M., Kwakkel, J. H., Walker, W. E., & Ter Maat, J. (2013) Dynamic adaptive policy pathways: A new method for crafting robust decisions for a deeply uncertain world. *Global Environmental Change*, 23, 485–498.

Hallegatte, S., Shah, A., Lempert, R., Brown, C., & Gill, S. (2012). Investment decision-making under deep uncertainty-application to climate change. Policy research working paper, 6193.

Harper, B. A., & Mason, L. B. (2016, July). A tropical cyclone wind event data set for Australia. In *Proceeding of 18th Australasian Wind Engineering Society Workshop* (pp. 6–8), McLaren Vale, SA, Australia.

Harris, I. (2005). Generalised Pareto methods for wind extremes. Useful tool or mathematical mirage? *Journal of Wind Engineering and Industrial Aerodynamics*, 93(5), 341–360.

Harris, R. I. (2009). XIMIS, a penultimate extreme value method suitable for all types of wind climate. *Journal of Wind Engineering and Industrial Aerodynamics*, 97(5–6), 271–286.

Holmes, J. D., & Moriarty, W. W. (1999). Application of the generalized Pareto distribution to extreme value analysis in wind engineering. *Journal of Wind Engineering and Industrial Aerodynamics*, 83(1–3), 1–10.

Hong, H. P., Mara, T. G., Morris, R., Li, S.H., & Ye, W. (2014). Basis for recommending an update of wind velocity pressures in Canadian design codes. *Canadian Journal of Civil Engineering*, 41(3), 206–221.

Hong, H. P., & Li, S. H. (2014). Plotting positions and approximating first two moments of order statistics for Gumbel distribution: Estimating quantiles of wind speed. *Wind Structure*, 19(4), 371–387.

Hong, H. P., Li, S. H., & Duan, Z. D. (2016). Typhoon wind hazard estimation and mapping for coastal region in mainland China. *Natural Hazards Review*, 17(2), 04016001.

Hong, H. P., Li, S.H., & Mara, T. G. (2013). Performance of the generalized least-squares method for the Gumbel distribution and its application to annual maximum wind speeds. *Journal of Wind Engineering and Industrial Aerodynamics*, 119, 121–132.

Hong, H. P., & Ye, W. (2014). Analysis of extreme ground snow loads for Canada using snow depth records. *Natural Hazards*, 73, 355–371.

Hong, H. P., Tang, Q., Yang, S. C., Cui, X. Z., Cannon, A. J., Lounis, Z., & Irwin, P. (2021). Calibration of the design wind load and snow load considering the historical climate statistics and climate change effects. *Structural Safety*, 93, 102135.

Hong, H. P., Sheng, C., Pozos-Estrada, A., & Gomez, R. (2022). Modeling and estimation of hurricane wind hazard affecting Mexican coastal regions. *Journal of Wind Engineering and Industrial Aerodynamics*, 230, 105199.

Hosking, J. (1990). L-moments: analysis and estimation of distributions using linear combinations of order statistics. *Journal of the Royal Statistical Society. Series B. Statistical Methodology* 52(1), 105–124.

Jeong, D. I., & Sushama, L. (2018). Projected changes to extreme wind and snow environmental loads for buildings and infrastructure across Canada. *Sustainable Cities and Society*, 36, 225–236.

Jeong, D. I., Sushama, L., Vieira, M. J., & Koenig, K. A. (2018). Projected changes to extreme ice loads for overhead transmission lines across Canada. *Sustainable Cities and Society*, 39, 639–649.

Jones, K. F., & Eylander, J. B. (2017). Vertical variation of ice loads from freezing rain. *Cold Regions Science and Technology*, 143(Nov), 126–136.

Jordaan, I.J. (2005), *Decisions under Uncertainty: Probabilistic Analysis for Engineering Decisions*, Cambridge University Press, New York.

Kendall, M. G. (1938). A new measure of rank correlation. *Biometrika*, 30(1–2), 81–93.

Kharin, V. V., & Zwiers, F. W. (2005). Estimating extremes in transient climate change simulations. *Journal of Climate*, 18(8), 1156–1173.

Knutson, T., Camargo, S. J., Chan, J. C., Emanuel, K., Ho, C. H., Kossin, J., ... & Wu, L. (2019). Tropical cyclones and climate change assessment: Part I: Detection and attribution. *Bulletin of the American Meteorological Society*, 100(10), 1987–2007.

Knutson, T., Camargo, S. J., Chan, J. C., Emanuel, K., Ho, C. H., Kossin, J., ... & Wu, L. (2020). Tropical cyclones and climate change assessment: Part II: Projected response to anthropogenic warming. *Bulletin of the American Meteorological Society*, 101(3), E303–E322.

Konrad, C. E. II. (1998). An empirical approach for delineating fine scaled spatial patterns of freezing rain in the Appalachian region of the USA. *Climate Research*, 10(3), 217–227.

Koutsoyiannis, D., & Montanari, A. (2015). Negligent killing of scientific concepts: the stationarity case. *Hydrological Sciences Journal*, 60(7-8), 1174–1183.

Kunkel, K. E., Karl, T. R., Easterling, D. R., Redmond, K., Young, J., Yin, X., & Hennon, P. (2013). Probable maximum precipitation and climate change. *Geophysical Research Letters*, 40(7), 1402–1408.

Kyselý, J., & Beranová, R. (2009). Climate-change effects on extreme precipitation in central Europe: Uncertainties of scenarios based on regional climate models. *Theoretical and Applied Climatology*, 95, 361–374.

Larsen, X. G., Mann, J. (2006). The effects of disjunct sampling and averaging time on maximum mean wind speeds. *Journal of Wind Engineering and Industrial Aerodynamics*, 94 (8), 581–602.

Laurent, C., & Parey, S. (2007). Estimation of 100-year-return-period temperatures in France in a non-stationary climate: Results from observations and IPCC scenarios. *Global and Planetary Change*, 57(1-2), 177–188.

Li, S. H. (2023a). Effect of nonstationary extreme wind speeds and ground snow loads on the structural reliability in a future Canadian changing climate. *Structural Safety*, 101, 102296.

Li, S. H. (2023b). Impact of climate change on wind energy across North America under climate change scenario RCP8.5. 5, *Atmospheric Research*, 288, 106722.

Li, S. H., Dale, J., Irwin, P., Gibbons, M., Dicks, M., Taylor, R., ... & Beaulieu, S. (2025). Updates to the design of buildings for NBCC 2025 for climate adaptation, resilience, and associated initial cost impact. *Journal of Structural Engineering*, 151(7), 04025082.

Li, S. H., & Hong, H. P. (2015a). Observations on a hurricane wind hazard model used to map extreme hurricane wind speed. *Journal of Structural Engineering*, 141(10), 04014238.

Li, S. H., & Hong, H. P. (2015b). Use of historical best track data to estimate typhoon wind hazard at selected sites in China. *Natural Hazards*, 76, 1395–1414.

Li, S. H., & Hong, H. P. (2016). Typhoon wind hazard estimation for China using an empirical track model. *Natural Hazards*, 82, 1009–1029.

Li, S. H. (2018). Effect of disjunct sampling on calibration of design wind speed. *Journal of Wind Engineering and Industrial Aerodynamics*, 183, 283–294.

Li, S. H., & Irwin, P. (2018). Impact of wind speed averaging time on the trend detection in a changing climate. In *CSCE Annual Conference and Annual General Meeting (Fredericton)*, Fredericton, Canada.

Li, S. H., Irwin, P., Kilpatrick, J., Gibbons, M., Sifton, V. (2017). A review of historical extreme wind speeds in a changing climate at some major Canadian cities. In *CSCE Annual Conference and Annual General Meeting (Vancouver)*, Montreal, Canada.

Li, S. H., Sifton, V., Lundgren, J., McClellan, C., & Gibbons, M. (2021). Extreme wind climate of the Arabian Peninsula characterized by using WRF simulation. *Weather and Climate Extremes*, 33, 100330.

Li, S. H., Kilpatrick, J., Browne, M. T., Yakymyk, W., & Refan, M. (2020). Uncertainties in prediction of local peak wind pressures on mid-and high-rise buildings by considering gumbel distributed pressure coefficients. *Journal of Wind Engineering and Industrial Aerodynamics*, 206, 104364.

Li, S. H., Irwin, P., Lounis, Z., Attar, A., Dale, J., Gibbons, M., & Beaulieu, S. (2022). Effects of nonstationarity of extreme wind speeds and ground snow loads in a future Canadian changing climate. *Natural Hazards Review*, 23(4), 04022022.

Li, S. H., & Kopp, G. (2023). Duration of synoptic wind in a changing climate and its engineering implementation. *Structural Safety*, 104, 102349.

Li, S. H., & Kumar, K. S. (2021). Cyclone wind hazard assessments for eastern coastal cities in India using a fast-predictive model. *Journal of Wind Engineering and Industrial Aerodynamics*, 218, 104760.

Li, S. H., & Kumar, S. (2023). Probable maximum tropical cyclone parameters for east and west coast of India. *Natural Hazards*, 118(1), 1–8.

Li, S. H. (2024, June). Incorporation of Climate Nonstationarity into Codes and Beyond. In *Canadian Society of Civil Engineering Annual Conference* (pp. 241–252). Springer Nature Switzerland, Cham.

Li, S. H. (2025). Assessment of the Impact of Climate Change on Thunderstorm Winds. In *Canadian Society of Civil Engineering Annual Conference* (pp. 61–71). Springer, Cham.

Lieblein, J. (1974). Efficient Methods of Extreme-value Methodology. National Bureau of Standards, Washington, ReportNBSIR74-602

Liel, A. B., DeBock, D. J., Harris, J. R., Ellingwood, B. R., & Torrents, J. M. (2017). Reliability-based design snow loads. II: Reliability assessment and mapping procedures. *Journal of Structural Engineering*, 143(7), 04017047.

Lins, H. F., & Cohn, T. A. (2011). Stationarity: wanted dead or alive? 1. *JAWRA Journal of the American Water Resources Association*, 47(3), 475–480.

López, J., & Francés, F. (2013). Non-stationary flood frequency analysis in continental Spanish rivers, using climate and reservoir indices as external covariates. *Hydrology and Earth System Sciences*, 17(8), 3189–3203.

Machado, M. J., Botero, B. A., López, J., Francés, F., Díez-Herrero, A., & Benito, G. (2015). Flood frequency analysis of historical flood data under stationary and non-stationary modelling. *Hydrology and Earth System Sciences*, 19(6), 2561–2576.

Madsen, H., Lawrence, D., Lang, M., Martinkova, M., & Kjeldsen, T. R. (2014). Review of trend analysis and climate change projections of extreme precipitation and floods in Europe. *Journal of Hydrology*, 519, 3634–3650.

Mann, H. B. (1945). Nonparametric tests against trend. *Econometrica: Journal of the Econometric Society*, 13, 245–259. https://www.jstor.org/stable/1907187

Marchau, V. A., Walker, W. E., Bloemen, P. J., & Popper, S. W. (2019). *Decision Making under Deep Uncertainty: From Theory to Practice* (p. 405). Springer Nature.

Martins, E. S., & Stedinger, J. R. (2000). Generalized maximum-likelihood generalized extreme-value quantile estimators for hydrologic data. *Water Resources Research*, 36(3), 737–744.

Matalas, N. C. (2012). Comment on the announced death of stationarity. *Journal of Water Resources Planning and Management*, 138(4), 311–312.

McAllister, T. P., Wang, N., & Ellingwood, B. R. (2018). Risk-informed mean recurrence intervals for updated wind maps in ASCE 7-16. *Journal of Structural Engineering*, 144(5), 06018001.

Mo, H. M., Fan, F., & Hong, H. P. (2015). Snow hazard estimation and mapping for a province in northeast China. *Natural Hazards*, 77, 543–558.

Montanari, A., & Koutsoyiannis, D. (2014). Modeling and mitigating natural hazards: Stationarity is immortal!. *Water resources research*, 50(12), 9748–9756.

Milly, P. C., Betancourt, J., Falkenmark, M., Hirsch, R. M., Kundzewicz, Z. W., Lettenmaier, D. P., & Stouffer, R. J. (2008). Stationarity is dead: Whither water management?. *Science*, 319(5863), 573–574.

Nason, G. P. (2006). Stationary and non-stationary time series. In H. M. Mader, S. G. Coles, C. B. Connor, L. J. Connor (Eds.) *Statistics in Volcanology*. https://pubs.geoscienceworld.org/gsl/books/edited-volume/1732/chapter-abstract/107597496/Stationary-and-non-stationary-time-series?redirectedFrom=fulltext

National Academies of Sciences, Engineering, and Medicine, NASEM. (2024). *Modernizing Probable Maximum Precipitation Estimation*. https://www.nationalacademies.org/projects/DELS-BASCPR-22-04

O'Gorman, P. A. (2015). Precipitation extremes under climate change. *Current Climate Change Reports*, 1, 49–59.

Olsen, J. R., Lambert, J. H., & Haimes, Y. Y. (1998). Risk of extreme events under nonstationary conditions. *Risk Analysis*, 18(4), 497–510.

Pintar, A. L., Simiu, E., Lombardo, F. T., & Levitan, M. L. (2015). Maps of non-hurricane non-tornadic wind speeds with specified mean recurrence intervals for the contiguous United States using a two-dimensional Poisson process extreme value model and local regression. https://www.nist.gov/publications/maps-non-hurricane-non-tornadic-wind-speeds-specified-mean-recurrence-intervals?pub_id=915120

Rackwitz, R. (2001). Reliability analysis—a review and some perspectives. *Structural Safety*, 23(4), 365–395.

Razali, N. M., & Wah, Y. B. (2011). Power comparisons of shapiro-wilk, kolmogorov-smirnov, lilliefors and anderson-darling tests. *Journal of Statistical Modeling and Analytics*, 2(1), 21–33.

Read, L. K., & Vogel, R. M. (2015). Reliability, return periods, and risk under nonstationarity. *Water Resources Research*, 51(8), 6381–6398.

Rigby, R. A., & Stasinopoulos, D. M. (2005). Generalized additive models for location, scale and shape. *Journal of the Royal Statistical Society Series C: Applied Statistics*, 54(3), 507–554.

Robbins, C. C., & Cortinas Jr, J. V. (1996). A climatology of freezing rain in the contiguous United States: Preliminary results. In *Proceedings of 15th AMS Conference on Weather Analysis and Forecasting*, Norfolk, VA.

Rootzén, H., & Katz, R. W. (2013). Design life level: Quantifying risk in a changing climate. *Water Resources Research*, 49(9), 5964–5972.

Salas, J., & Obeysekera, J. (2014). Revisiting the concepts of return period and risk for nonstationary hydrologic extreme events. *Journal of Hydrologic Engineering*, 19(3), 554–568.

Salas, J. D., Obeysekera, J., & Vogel, R. (2018). Techniques for assessing water infrastructure for nonstationary extreme events: A review. *Hydrologic Science Journal*, 63(3), 325–352.

SERC Carleton. (2017). Non-stationary hydrology. https://serc.carleton.edu/integrate/teaching_materials/water_science_society/student_materials/904

Semmler, T., & Jacob, D. (2004). Modeling extreme precipitation events—a climate change simulation for Europe. *Global and Planetary Change*, 44(1–4), 119–127.

Sen, P. K. (1968). Estimates of the regression coefficient based on Kendall's tau. *Journal of the American Statistical Association*, 63(324), 1379–1389.

Serinaldi, F. (2015). Dismissing return periods! *Stochastic Environmental Research and Risk Assessment*, 29(4), 1179–1189.

Serinaldi, F., & Kilsby, C. G. (2015). Stationarity is undead: Uncertainty dominates the distribution of extremes. *Advances in Water Resources*, 77, 17–36.

Shapiro, S. S., & Wilk, M. B. (1965). An analysis of variance test for normality (complete samples). *Biometrika*, 52(3–4), 591–611.

Tabari, H., & Willems, P. (2018). Seasonally varying footprint of climate change on precipitation in the Middle East. *Scientific Reports*, 8(1), 4435.

Theil, H. (1950). A rank-invariant method of linear and polynomial regression analysis. *Indagationes Mathematicae*, 12(85), 173.

Trenberth, K. E. (2011). Changes in precipitation with climate change. *Climate Research*, *47*(1–2), 123–138.

Vickery, P. J., Masters, F. J., Powell, M. D., & Wadhera, D. (2009). Hurricane hazard modeling: The past, present, and future. *Journal of Wind Engineering and Industrial Aerodynamics*, 97(7–8), 392–405.

Vickery, P. J., Skerlj, P. F., & Twisdale, L. A. (2000). Simulation of hurricane risk in the US using empirical track model. *Journal of Structural Engineering*, 126(10), 1222–1237.

Vickery, P. J., Wadhera, D., Galsworthy, J., Peterka, J. A., Irwin, P. A., & Griffis, L. A. (2010). Ultimate wind load design gust wind speeds in the United States for use in ASCE-7. *Journal of Structural Engineering*, 136(5), 613–625.

Vogel, R. M., Yaindl, C., & Walter, M. (2011). Nonstationarity: Flood magnification and recurrence reduction factors in the United States 1. *JAWRA Journal of the American Water Resources Association*, 47(3), 464–474.

Walker, W. E., Harremoës, P., Rotmans, J., van der Sluijs, J. P., van Asselt, M. B. A., Janssen, P., & Krayer von Krauss, M. P. (2003). Defining uncertainty: A conceptual basis for uncertainty management in model-based decision support. *Integrated Assessment*, 4(1), 5–17.

Zhao, Y. G., & Ono, T. (1999). A general procedure for first/second-order reliabilitymethod (FORM/SORM). *Structural Safety*, 21(2), 95–112.

8 Climate Impact Analytics for the US Freight Rail Network

Extreme Temperatures as a Case Study

Nathan Chan and Bilal M. Ayyub
Department of Civil and Environmental Engineering,
Center for Technology and Systems Management (CTSM),
University of Maryland, College Park, Maryland

8.1 INTRODUCTION: US FREIGHT RAIL AND TRAIN NETWORKS

Railroad networks constitute a critical component of transportation infrastructures for the nation's economy, environmental friendliness, and national security. The US freight rail system is made up of more than 140,000 miles of track and generates revenues of more than $80 billion a year (Association of American Railroads 2022). Each year, US freight trains are involved in transporting more than 59 tons of goods. This, in 2017, resulted in $219 billion dollars in economic production, 80,000 jobs, and $71 billion dollars in wages (United States Congress House Committee on Transportation and Infrastructure Subcommittee on Railroads 2021). The US Rail network also reduces congestion on highways and other roads; a US House of Representatives Committee found that a single freight train can remove 280 cars from the road (United States. Congress. House. Committee on Transportation and Infrastructure 2009). With one train that moves one ton nearly 500 miles in a single gallon, rail transportation is one of the most environmentally friendly methods of freight transportation. Finally, the US rail network is integral to the national security of more than 41,000 miles of the US freight rail system, which has been classified as critical for defense deployment (US Department of Defense, The Military Surface Deployment and Distribution Command Transportation Engineering Agency).

8.1.1 Network Vulnerabilities to a Changing Climate

Given the position of the railroad infrastructure in the US economy and national security, it would be prudent to identify any risks that may impact the operation of

DOI: 10.1201/9781003567332-8

this system. In the US, the rail system is divided between freight and passenger rail. Freight rail is operated by seven Class 1 rail companies (with operating revenues of $490 million or more each) and hundreds of other smaller rail companies, all overseen by the US Department of Transportation's Federal Railroad Administration (FRA). Each rail company owns, builds, and maintains its own infrastructure (Association of American Railroads 2022). Therefore, each company must individually deal with the effects of climate change on its own systems.

Every component of the railroad system is susceptible to risk due to climate change. Transportation systems, including the rail system, are negatively affected by a changing climate and associated extreme weather events (Liban et al. 2023). Without sufficient adaptation, climate change is expected to cost the rail industry at least $10 billion yearly (Chinowsky et al. 2019). To produce effective adaptation plans, the rail system must be evaluated for information on relevant climate events, potential responses, and the corresponding predicted impact (Garmabaki et al. 2021).

8.2 BACKGROUND: GRAPHS AND NETWORKS

When trying to understand the impact of a geographically interconnected system, such as a railroad, defining the system using networks is often a helpful tool. Networks enable the study of connection patterns in systems where a set of individual components or parts are linked together. By representing the system as an abstract series of nodes and edges, we can observe how this pattern influences and impacts system behavior and other characteristics (Newman 2010).

At its core, a network is a collection of nodes joined by edges. The "shape" of these connections (i.e., the network topology) can be represented mathematically via an adjacency matrix, defined as follows:

$$A_{ij} = \begin{cases} 1, & \text{if } i \text{ is linked to } j \\ 0, & \text{otherwise} \end{cases} \tag{8.1}$$

For undirected network and networks without self-edges, this matrix will be symmetrical and undirected, respectively. For weighted networks, the weight associated with two nodes i and j will be denoted as w_{ij}. The shortest path between any two nodes in a positively weighted network can be calculated using Dijkstra's algorithm.

Additionally, we can characterize each node using its degree. The node degree of each node is the number of edges connected to it, often excluding self-edges. Node degrees are always an integer, as edges are simply counted with link weights ignored (Newman 2010).

In this chapter, we use link and node-betweenness centrality as a metric for node and link importance. Link-betweenness centrality can be calculated using the following:

$$c_B(e) = \sum_{i,j \in N} \frac{d(i,j \mid e)}{d(i,j)} \tag{8.2}$$

Similarly, node-betweenness centrality can be calculated using the following:

$$c_B(n) = \sum_{i,j \in N} \frac{\sigma(i,j \mid n)}{\sigma(i,j)} \tag{8.3}$$

In the above equations, N represents the set of nodes in the network; $d(i,j)$ represents the number of shortest paths from node i to j, while $d(i,j \mid e)$ and $d(i,j \mid n)$ represent the number of shortest paths from node i to j that pass through link e and node n respectively (Brandes 2008).

To quantitatively describe the performance of the network as a whole, we use network-connectedness efficiency (Latora and Marchiori 2001). We compute efficiency using the following equation:

$$\bar{E}(c) = \frac{1}{n(n-1)} \sum_{i=1}^{n} \sum_{j=1}^{n} \frac{\bar{w}_{ji}}{d_{ji}}, \quad j \neq i \tag{8.4}$$

where $\bar{E}(c)$ is the efficiency, n is the number of nodes in the network, d_{ji} represents the count of links along the shortest path from node j to node i, and $\bar{w}_{ji}$ is the average of node and link weights along the shortest paths.

8.3 A NETWORK-BASED METHODOLOGY FOR IMPACTS AND ADAPTATION ANALYTICS

The objective of this chapter is to **develop and illustrate a methodology for predicting the impact of climate on a rail network**. The output will be a relative indication of the risk exposure for various areas of the network that takes into account network topology, usage, and climate hazards. This methodology is compatible with any level of input/data resolution, but is most applicable when the impacts of the hazard of interest vary by geography (Figure 8.1).

8.3.1 System Definition

One of the first steps in determining the risk exposure of a system is to define the scope and limits of that system (Ayyub 2014). The US freight rail system consists of many components, entities, and individuals – some of these components are laid out non-exhaustively in Figure 8.2.

When looking at the impact of climate hazards on the railway system, it is also useful to use a network to describe how the system is laid out geographically. For rail networks, network definitions based on track geography (as edges) and points of interest (as nodes) are often available from sources such as the North American Atlas Database (Federal Railroad Administration 2024), the United Kingdom Ordanance Survey (Ordnance Survey 2024), or Geofabrik / Open Street Maps (Geofabrik GmbH Karlsruhe). However, most of these sources describe the physical layout of the track

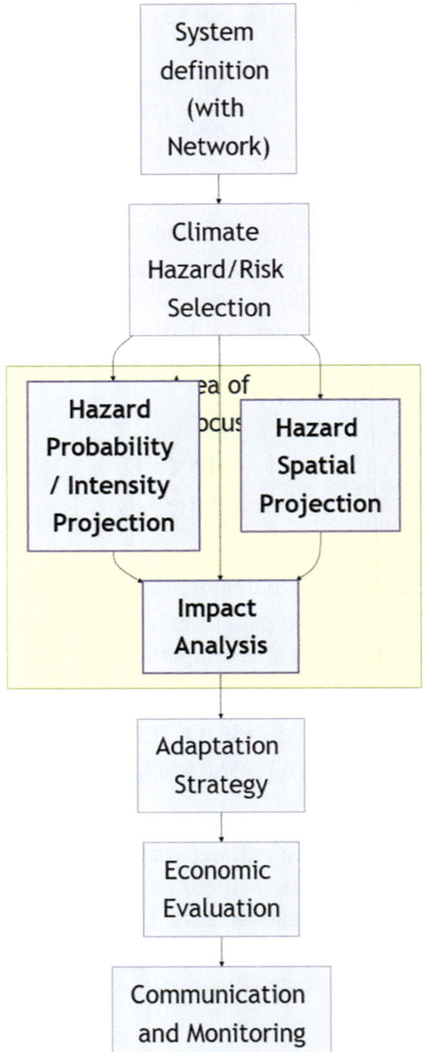

FIGURE 8.1 Methodology for treating climate risks to a rail system.

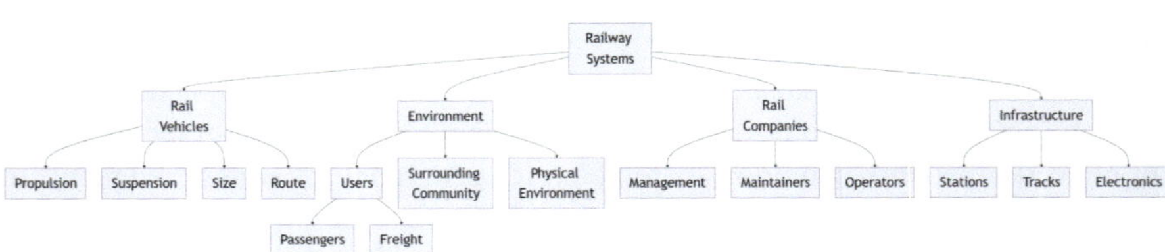

FIGURE 8.2 Breakdown structure of a railroad system. (Based on Blainey and Preston 2022.)

and other infrastructure in far too granular detail, especially compared to other data sources such as climate projections. As such, it is often desirable to aggregate the nodes and edges into a simplified network. In such a simplified network, the nodes represent geographically related collections of the original nodes, while the edges represent the interfaces between each collection. The details and challenges that arise from this aggregation are discussed in a later section.

Climate change presents a risk to railway systems through a variety of mechanisms, including extreme temperatures, precipitation, wind, and changes in coastal water levels. Each hazard can have an impact on multiple elements of the system throughout the system, including slopes, bridges, tunnels, tracks, and electrical systems. Then these impacts have some direct and indirect consequences (Palin et al. 2021).

For example, increasing precipitation from storms causes the following:

1. Increased incidents of blocked tracks, switches, and short circuits.
2. Increased occurrences of and damage from landslides.
3. Greater likelihood for the flooding of bridges, tunnels, and low-lying rails (Liban et al. 2023).

These impacts then cause degraded performance or complete failure of the track, switch, signal, and other systems. This then increases the risk of delays and accidents in trains (Palin et al. 2021).

Based on this example, we can see that even a single climate hazard can have many potential risk paths. Therefore, to simplify the next impact analysis performed in the next step, we want to narrow our scope to consider a feasible number of risk paths.

In addition to a qualitative determination of a hazard and its impact, we should also look to quantitatively determine the size, scope, and nature of the risk of interest. In the context of climate change and rail systems, this can be grouped into a few categories:

1. **The Probability of Occurrence**: How likely is a particular hazard to happen? How often do we expect it, with what level of confidence?
2. **The Intensity of Occurrence**: For each hazard event that does occur, how much of a deviation from the norm do we expect it to be?
3. **The Spatial Dimension**: How does geographical location affect the probability and intensity of this hazard?

A wide range of climate projections and models exist with various resolutions, most notably those involved in Coupled Model Intercomparison Project (CIMP5) and CIMP6 (Ma et al. 2024). These models provide projected climate variables at various resolutions. For instance, the Seamless System for Prediction and Earth System Research (SPEAR) model provides variables like the near-surface air temperature averages, minimums and maximums, wind, and humidity (Delworth et al. 2020). These values can be spatially aggregated (usually averaged) and joined with the same process that we simplify the network, then used to weight the various nodes and edges.

8.3.2 Impact Analysis

Once the hazard and the corresponding impact have been determined, we then shift to asking questions about the consequences. Firstly, we determine the measures of performance (MOPs) that we are interested in. For a railroad, a few potential MOPs include:

1. Amount of train delays (often aggregated as total delay minutes for the system).
2. Frequency of accidents.
3. Overall operational costs.

After selecting one or more appropriate MOPs, we then want to determine what the relationship among the hazard, its impacts, and the selected MOPs is. Since most climate hazards are extreme versions of phenomena observed today, we can often extrapolate this relationship based on historical data and climate projections, if we can successfully isolate the current impact of the associated hazard. In addition to using performance measures, we often end up using economic value as a metric to assess impact. We do this by establishing a relationship between an MOP and the operational or maintenance costs.

When working with networks, it is also useful to use network-specific metrics to determine the relative impact on various areas of the network. Using node importance metrics like betweenness centrality on a weighted network can be a good way to achieve this.

8.3.3 Adaptation Analysis

Following an impact analysis, we can then look at the various adaptation (or risk reduction/treatment) methods available and evaluate the efficiency of each. The available adaptation measures vary depending on the hazard and the system being considered. However, they can generally be grouped into the following categories:

1. **No Adaptation**: Where no response is taken to a potential risk. Often useful to consider as a baseline.
2. **Reactive Adaptation**: Where steps are taken during/after a risk has materialized.
3. **Proactive Adaptation**: Where steps are taken before a risk materializes.

Generally speaking, proactive adaptation will result in lower costs related to climate change damage compared to reactive adaptation or lack of adaptation (Wasley et al. 2023).

When we then attempt to evaluate each adaptation measure, we execute a cost–benefit analysis that takes into account the chosen MOPs. If we also perform an economic evaluation, we can quantify the reduction in risk given the associated improvement in the relevant MOPs and compare it to the cost of the said adaptation measure, giving a net present value for a particular adaptation measure (Andersson-Sköld et al. 2021).

8.4 USE CASE: TEMPERATURE EXTREMES AS A CLIMATE HAZARD

As an example, to illustrate an implementation of this methodology, we will take a look at the impact of extreme high temperatures on railroads, particularly as they affect rail buckling. Extreme temperatures do have effects on other rail systems, including catenary sag and electrical component failure – we will not be considering these risk pathways as part of this analysis.

Rail buckling occurs when the temperature-based expansion of a railway track causes the compressive force experienced by the track to increase. This then leads to lateral misalignment, an unstable equilibrium which can then laterally buckle out suddenly. When a rail vehicle goes over a railway track, it causes vertical movement in the track, termed "uplift waves". These waves can trigger buckling because they provide the energy needed to alter the equilibrium by reducing the lateral friction provided by the ballast surrounding the tracks. Whether or not buckling occurs is often modeled probabilistically and is affected by track neutral temperature design, rail temperatures, and total energy imparted by the vehicle (United States: Transportation Department: Federal Railroad Administration et al. 2013).

If a section of the rail buckles, we need to be concerned that a train that crosses the buckled area will get derailed, especially since the buckling can be triggered by crossing trains. Around 34 buckling-related derailments occur in the US per year, and in 2002 the industry cost around $17 million dollars in loss and damage to track and train cars (United States: Transportation Department: Federal Railroad Administration et al. 2013). Each buckling-related derailment has an average of 12 cargo cars derailed (Liu et al. 2012).

8.4.1 Data Sources Used

To simulate the below use case, we used the following sources of data:

- **North American Rail Network Lines** (Federal Railroad Administration 2024): A geographic reference for North American rail lines, part of the National Transportation Atlas Database..
- **Freight Analysis Framework (FAF) Zones** (US. Department of Transportation. Bureau of Transportation Statistics 2017): A geographic reference for the boundaries of each FAF Zone.
- **Freight Analysis Framework Tonnages**: A regional database of tonnage (and value), grouped by origin-destination pair, commodity type, and transportation mode. For this simulation, we will be using a summation of all commodity types and filtering the dataset down to commodities transported by rail.
- **Seamless System for Prediction and Earth System Research (SPEAR)** (Delworth et al. 2020): A climate model used for seasonal to multidecadal projections. This simulation will only be using the temperature at surface variable to extrapolate the temperature of rail tracks.

8.4.2 Current Adaptation Practices to Temperature Increase

As the track temperature is roughly proportional to the surface air temperature, higher daily maximum surface air temperatures result in less energy needed to cause buckling. To mitigate this, rail companies will often impose artificial speed limits during periods of high temperatures. This means that less energy will be imparted to the tracks through the uplift waves, reducing the chance of rail buckling (Dobney et al. 2009). For instance, at temperatures of 85°F–115°F, BNSF will reduce the speed of freight trains from 50 to 40 miles per hour, while CSX will reduce all speed limits by 20 miles per hour (Chinowsky et al. 2019).

Unfortunately, this reduction in speed causes a delay to the train, measured in "delay minutes", which can then be associated with direct and indirect economic impacts to determine a delay minute cost. For a freight train, this cost is dependent on many factors, especially the operating situation (bulk, manifest, or intermodal) (Lovett et al. 2015).

Multiple papers have attempted to estimate the cost of rising temperatures of climate change due to rail buckling, in both European (Dobney et al. 2009, Villalba Sanchis et al. 2020) and US (Neumann et al. 2021, Chinowsky et al. 2019) settings. These studies often consider multiple adaptation scenarios:

1. **No Adaptation**: Rail buckling becomes more common as temperatures rise, with no response from rail companies. In this scenario, the cost of a buckling incident is considered, multiplied by the expected rate of incidents.
2. **Reactive Adaptation**: Rail companies introduce speed limits during the hottest portion of the day, incurring costs via delays.
3. **Proactive Adaptation**: Rail companies install track temperature sensors, enabling a risk-based approach to speed orders.

8.4.3 System Definition: Creating a Simplified Network

To reduce this problem to a feasible state, we will first construct a network using 132 nodes corresponding to the zones defined in the Freight Analysis Framework (FAF) (US. Department of Transportation. Bureau of Transportation Statistics 2017). This will allow us to utilize the origin/destination freight flow projected data from the FAF, while also ensuring that the network is simple enough to analyze. Edges on this network will be created by searching for edges in the NTAD Rail Line dataset that start in one zone and ends in the other. Each pair of zones that has at least one rail line crossing the zone boundary will result in a link being added between those two nodes.

To ensure that the topology of the network aligns with the physical structure of the network, modifications may need to be applied to the FAF Zones and thus the network connectivity. Some features of the resultant network that require modification are as follows:

- Areas that are made of multiple subareas that are not connected physically. These types of zones allow traffic going through the area to "skip" through the network. These types of areas can be corrected by simply breaking the areas up into individual zones. For instance, in the Figure 8.3, the "Rest of California" zone (Purple) is noncontiguous and can be broken up into three separate zones.

- Areas that are completely surrounded in terms of rail connections. For instance, if you have area A mostly encircling area B, only being interrupted by area C, however the only rails leaving A head into area B, area A will only be accessible from B. These can be identified as nodes of degree 1, and are especially concerning when performing analysis of goods entering and leaving the network. An example in the Figure 8.3 is the surrounding of the "Fresno-Medera" zone (Dark Blue) by the "Rest of California" Zone (Purple). Unfortunately, there is currently no good way to deal with such zones.

Once these modifications are made, the tonnage data then needs to be modified to reflect the new areas. Flows are recalculated proportionally to the relative land area of the new zones. In the future, the proportion of rail nodes or population density could be used.

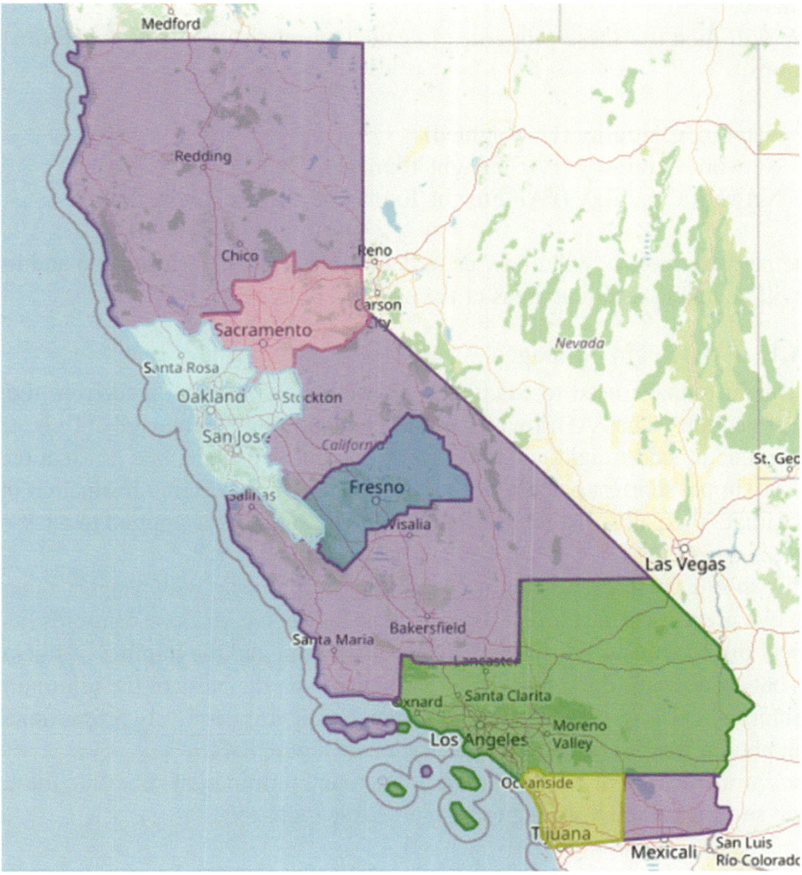

FIGURE 8.3 Example of freight analysis framework zones.

TABLE 8.1
List of Properties Used to Calculate Link Weights

Property	Description
Shortest-path betweenness centrality	The centrality of each link as calculated using NetworkX
FAF freight tonnage for 2050	Calculated
Temperature difference at surface	Derived from the SPEAR model

8.4.4 Climate Hazard Identification: Network Links Properties and Weighting

Once the topology of the network is solidified, weighting will be applied to the network based on a multitude of factors, including, but not limited to, the physical structure of the rail network from the US Department of Transportation's National Transportation Atlas Database (US Department of Transportation 2024).

To derive the network weights, the following properties will either be derived or calculated:

We will then use these properties to calculate relative network metrics in three iterations:

- Network Centrality (Unweighted).
- Network Centrality (FAF Freight Tonnage).
- Network Centrality (FAF Freight Tonnage * Surface Temperature).

These network metrics will allow us to determine relative performance and impact that will be had to different areas of the network (Table 8.1).

8.4.4.1 FAF Freight Tonnage

The freight tonnage used to calculate link weights will be sourced from the FAF 2050 projections. The FAF provides a dataset containing the tonnage of freight, economic value of said freight, commodity, the origin FAF Zone, and destination FAF Zone. A simple shortest path algorithm is applied to each origin–destination pair to approximate the paths that freight is transported, and flow is applied to each of the edges along this path.

This shortest path calculation uses a weighted approach, with weights calculated using two properties: the straight-line distance between FAF Zone centroids (i.e., the physical distance between two nodes in the network) and the number of tracks that cross between zones. Each property is normalized on the basis of the minimum and maximum of each property, where the distance weight is positively proportional and the number of tracks is negatively proportional (Figure 8.4).

For intrazone freight (where the origination and termination zones are the same), we had several potential strategies to apply these tonnages:

- **Apply to Every Connected Link**: Apply the tonnage to each connected link. Although simple, this will result in the intrazone freight applying the tonnage twice to routes that transit through the node.

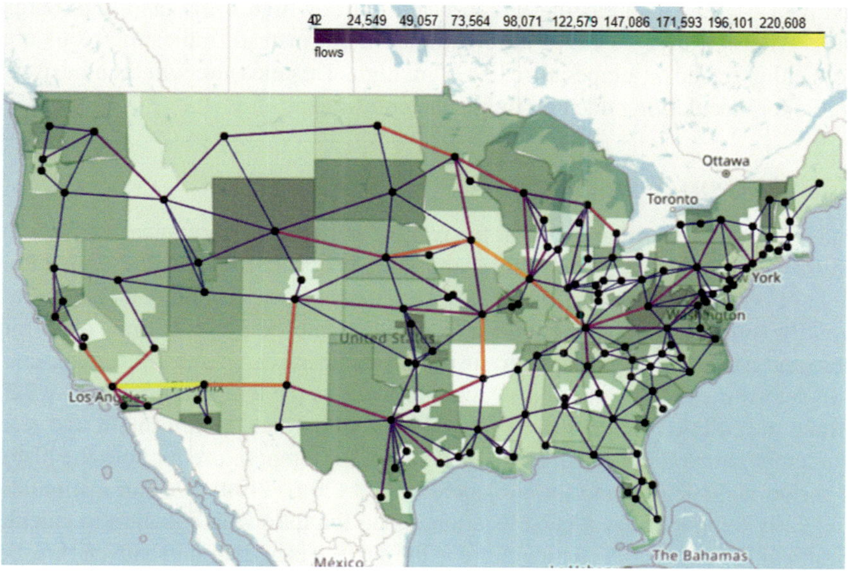

FIGURE 8.4 Map of freight tonnage transported.

- **Apply Half to Every Connected Link**: This will correct things so that the tonnage will only be applied once to the routes that transit through the node. This also has the effect of "averaging" the tonnage on each link based on the origin- and destination-node tonnages. However, high-degree nodes will still have an enormous impact on the overall freight in the network.
- **Introduced as Node Weights**: This is identical to applying half the tonnage to every connected link.
- Divide tonnage by node degree before applying to every connected link.

After some analysis of these options, we decided to divide each intrazone tonnage by the node degree before applying it to each connected link, "spreading" the tonnage across the various edges.

8.4.4.2 Temperature Increases

At this juncture, we will overlay projected near-surface temperatures from the SPEAR model (Delworth et al. 2020) to predict the impact of the temperature change on freight flows. These temperature readings will be averaged over 2001–2010 and compared with similarly averaged temperatures from 2091 to 2100, effectively giving us an idea of temperature increases over a 90-year period.

These temperature variations will then be applied to the simplified FAF Network by averaging average temperature deviations in each zone, and to the edges by averaging the temperature differences exhibited by the connected nodes.

8.4.4.3 Combining Weights

Once we have both sets of weights, we will normalize and combine them by multiplying the two weights associated with each link and node. This will give us a good measure for the areas that are affected both by high traffic and high temperatures. Alternatively, we could have summed them instead, indicating to us areas where either traffic or temperature levels are high. This measure will then be used as weights for calculating node and link betweenness centrality.

8.4.5 Network-Wide Performance

In addition to performing analysis for relative network performance, we will also analyze whole network performance and the impact that increasing temperatures will have on the network as a whole. For this analysis, we assume that freight trains normally run at 60 mph, and that rail operators impose a speed limit of 40 mph where and when surface air temperatures rise above 32°C.

To determine the amount of time and areas where the temperature is above 32°C, we take the SPEAR model's daily simulation and extract the minimum and maximum temperature at the surface. We then model the temperature for both the historical period (1991–2001) and the projected period (2091–2100). We run a simulation using daily temperatures at each location, using the following equation to calculate the ratio of the day that the temperature at the surface is higher than a given threshold (32°C in our specific use case):

$$R = \arccos\left(2 * \frac{t_T - t_{\min}}{t_{\max} - t_{\min}} - 1\right) / \pi \tag{8.5}$$

Where R is the ratio of the day that the temperature at surface is higher than t_T, $t_{\min}$ and $t_{\max}$ are the minimum and maximum temperatures for that day, and t_T is the threshold temperature.

Once we have calculated the above ratio for all nodes in our network, we then assign weights to each link using the following equation:

$$w_{ij} = \frac{\frac{S_L}{S_N} * t_T + (1 - t_T)}{D_{ij} * F_{ij}} \tag{8.6}$$

where w_{ij} is the assigned weight for link ij, F_{ij} is flow traffic for link ij, D_{ij} is the physical distance that link ij traverses, S_L and S_N are the imposed speed limit and normal speed respectively.

Using these weights, we can calculate the efficiency of our network for both the historical and projected cases.

8.4.6 Results

8.4.6.1 FAF Zone Network

Incorporating the FAF Zone definitions and the NA Rail Lines results in a network that originally has 131 nodes and 255 edges. However, given that the original zones have some non-contiguity, noncontiguous zones are subdivided such that each contiguous subzone above 25% of the zone's total area becomes its own zone. This results in ten new zones, bringing the FAF Network to 139 nodes and 275 edges. Further refinement could be applied to break up surrounded areas, however, this was skipped due to algorithmic complexity (Figure 8.5 and Table 8.2).

Compared to the aggregate network presented in Hamed et al. (2022), this FAF Network has more nodes that are more sparsely connected, as indicated by the network density and mean node degree. Interestingly, we observe a higher clustering metric, which can be verified by inspection. This is owing to the aggregate network being constructed more as a mesh network, while the FAF Network ends up looking more like a tree network in some areas (Table 8.3).

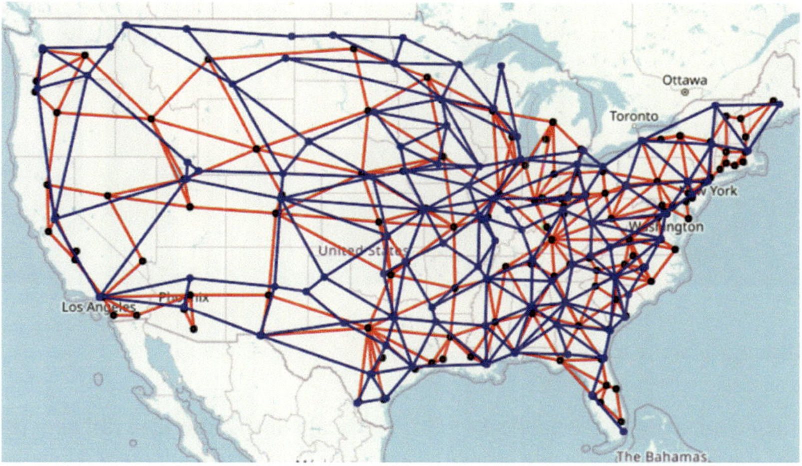

FIGURE 8.5 Overlay of freight analysis framework network (red) on top of aggregate network (blue).

TABLE 8.2
Network Comparison

	NTAD	FAF	Aggregate
Node count	250436	139	100
Link count	289309	275	226
Network density	0.000009	0.029	0.046
Mean node degree	2.310	3.957	4.520
Node degree standard deviation	0.836978	2.288	1.590
Clustering	0.04333	0.445	0.396
Characteristic path length	N/A	5.747	5.117
Mean connectivity	N/A	2.181	3.310
Mean betweenness centrality	N/A	0.607	0.090

TABLE 8.3
Freight Analysis Framework Network Baseline Centrality Top 5 Nodes

FAF Zone	Centrality
Remainder of Kentucky	0.465
Remainder of Illinois	0.358
Remainder of Pennsylvania	0.245
Remainder of Missouri	0.226
Remainder of Texas	0.161

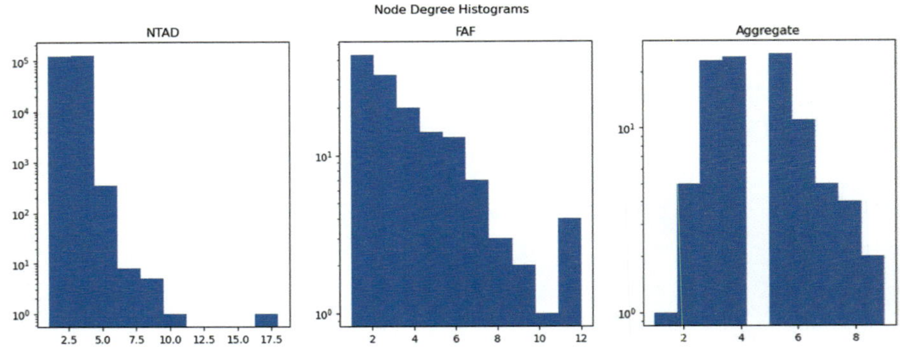

FIGURE 8.6 Network node degree histograms.

When comparing these networks to the physical North American rail network, we see that the FAF and aggregate networks are more similar to each other. However, we also observe a similar node degree distribution between the FAF and NTAD network (Figure 8.6).

In addition to comparing and contrasting this network to other similar networks, we can also calculate a baseline centrality metric. When we do this, we find, unsurprisingly, that the nodes with the highest centrality are geographically central and have high node degrees, as shown in Figure 8.7.

8.4.6.2 Freight Analysis Framework Freight Traffic Weighted Network Results

Based on the aforementioned FAF Freight tonnage algorithm, freight tonnages were calculated for each of the path segments in the FAF Network. When observing the map and output tables, we can see that the LA–Arizona area are the nodes with the largest amount of freight flow by far. This is a first-look indication that our simulation is accurate, since this corresponds to the large amount of traffic originating from the Los Angeles port (Figure 8.8 and Tables 8.4).

We then assign these flows as weights used to calculate network centrality, and compare them to the baseline, as shown in Table 8.5 and Figure 8.9. Most of the nodes do not shift much in centrality; however, a good number decrease in centrality, demonstrating a network that has a few important nodes.

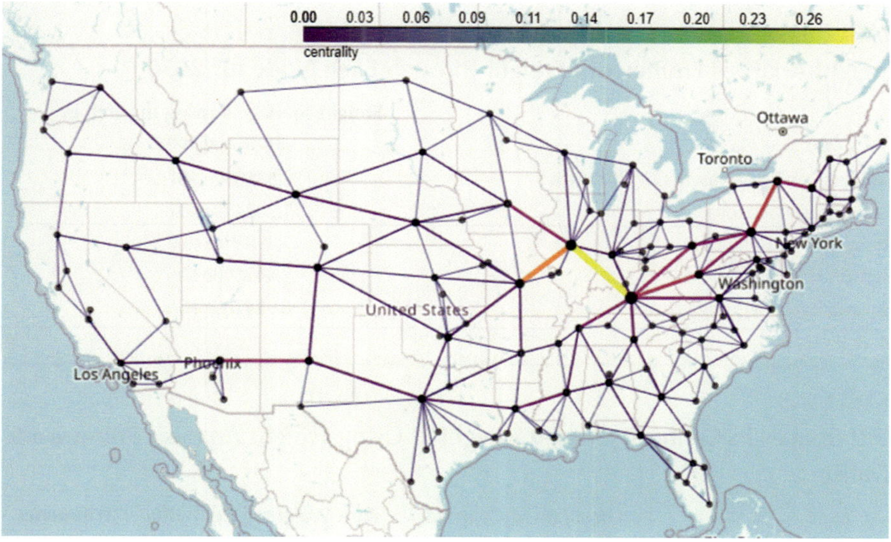

FIGURE 8.7 Freight analysis framework network baseline centrality.

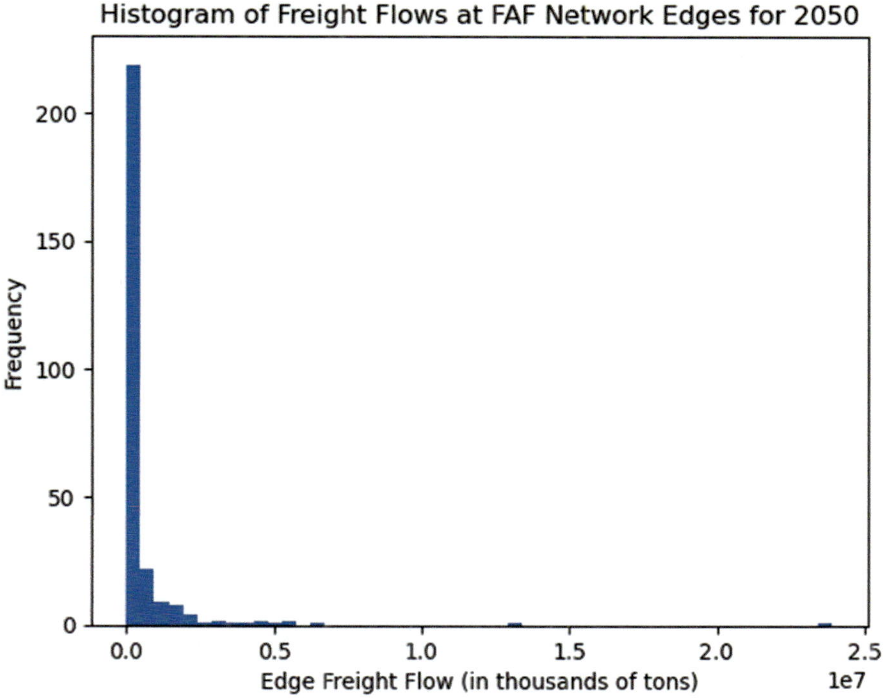

FIGURE 8.8 Freight analysis framework freight tonnage histogram.

TABLE 8.4

Freight Analysis Framework Freight through Each Node for 2050

FAF Zone	Freight Tonnage Passing through in thousand tons
Los Angeles-Long Beach, CA CFS Area	2.348032e+07
Remainder of Arizona	1.282123e+07
Remainder of Kentucky	7.499376e+06
Remainder of Illinois	7.442690e+06
Remainder of California	6.984260e+06

TABLE 8.5

Differences in Centrality When Weighted Using Freight Analysis Framework Traffic

FAF Zone	Unweighted Centrality	Flow Weighted Centrality	Difference
Remainder of Kentucky	0.465	0.315	0.149
Remainder of Illinois	0.358	0.203	0.155
Remainder of Pennsylvania	0.245	0.000	0.245
Remainder of Missouri	0.226	0.194	0.031
Remainder of Texas	0.161	0.293	−0.138

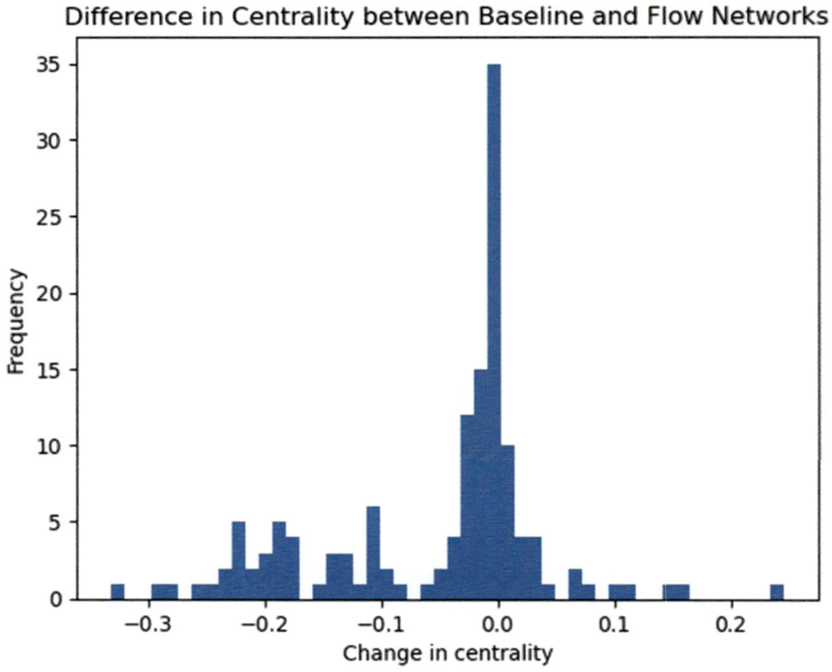

FIGURE 8.9 Difference in centrality between unweighted and traffic-weighted networks.

We can repeat this process using projected temperature data, overlaying data from National Oceanic and Atmospheric Administration (NOAA's) SPEAR model onto the network. When we do so, we see a temperature increase of anywhere from 3.5°C to 7°C (Figure 8.10).

When we recalculate our network centrality using both temperature and freight traffic as weights, we find that our network centrality has shifted again, as shown in Table 8.6 and Figure 8.11.

Looking at Figure 8.12, Tables 8.8 and 8.7, we will see that the link from the Los Angeles zone to Arizona is again considered central, but also that there are central nodes/edges in the Midwest/East Coast. This is probably due to a number of factors:

1. The large amount of traffic passing through associated nodes.
2. The large relative temperature increase.
3. The high node degrees.

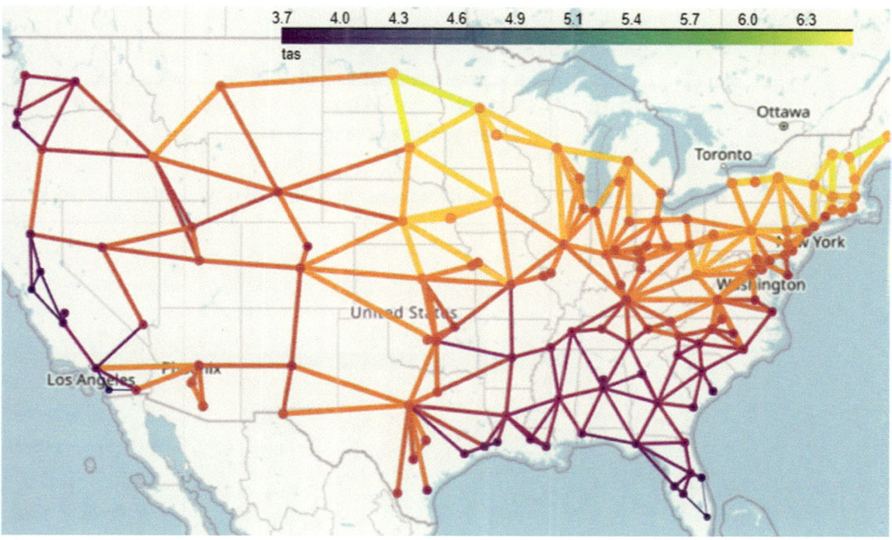

FIGURE 8.10 Map of temperature differences overlaid on the Freight Analysis Framework Zone Network.

TABLE 8.6

Combined Weighted Network Centrality

FAF Zone	Traffic Weighted Centrality	Combined Weighted Centrality	Difference
Remainder of Kentucky	0.315	0.500	−0.035131
Remainder of Illinois	0.203	0.416	−0.058024
Remainder of Pennsylvania	0.000	0.246	−0.000491
Remainder of Missouri	0.194	0.124	0.101542
Remainder of Texas	0.293	0.127	0.033953

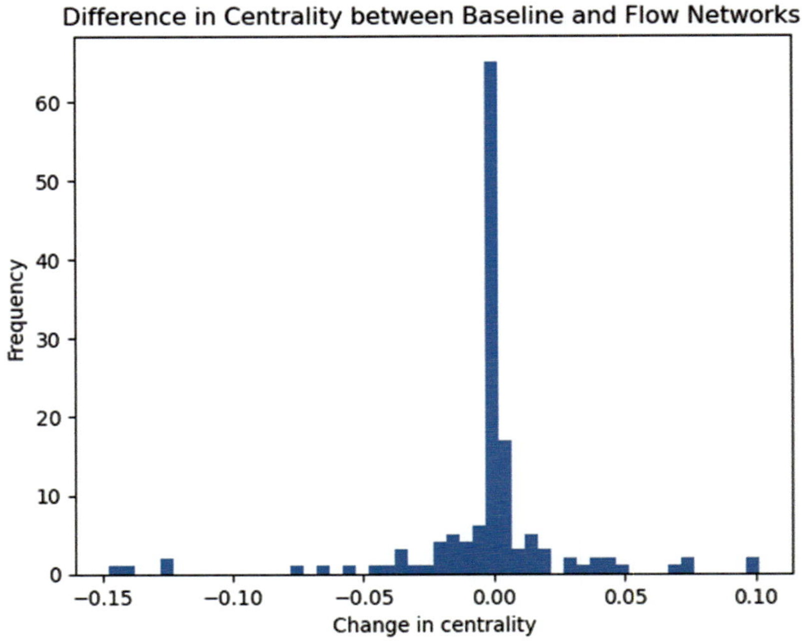

FIGURE 8.11 Histogram of centrality differences in combined weighted network vs. unweighted network.

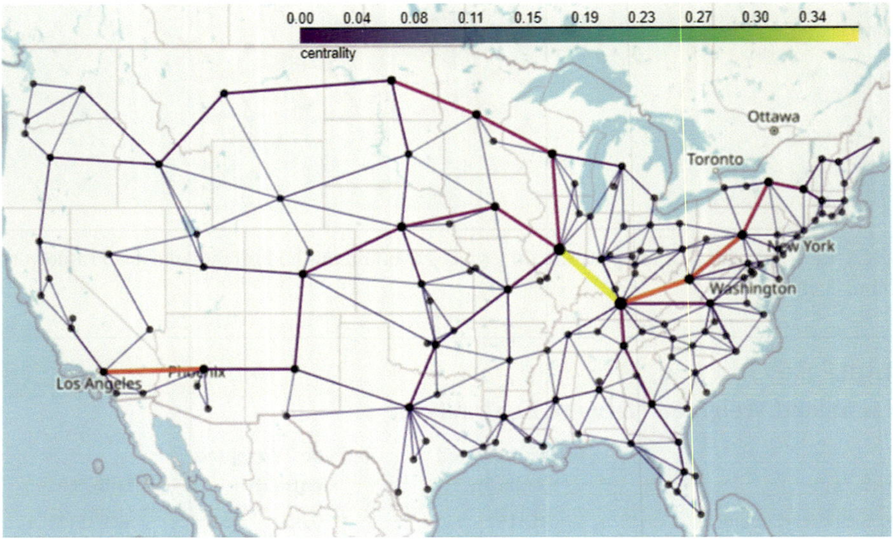

FIGURE 8.12 Freight analysis framework network combined temperature and traffic weighted centrality.

TABLE 8.7
Combined Weighted Network Node Centrality, 5 Most Central

FAF Zone	Node Centrality
Remainder of Kentucky	0.500159
Remainder of Illinois	0.416958
West Virginia	0.272929
Remainder of Pennsylvania	0.246483
Remainder of Nebraska	0.183011

TABLE 8.8
Combined Weighted Network Link Centrality, 5 Most Central

FAF Zone From	FAF Zone To	Link Centrality
Remainder of Illinois	Remainder of Kentucky	0.372745
Remainder of Kentucky	West Virginia	0.237775
Remainder of Arizona	Los Angeles, CA CFS Area	0.229330
Remainder of Pennsylvania	West Virginia	0.218069
Remainder of New York	Remainder of Pennsylvania	0.168543

TABLE 8.9
Whole-Network Results

Simulation Case	Historical Ratio	Projected Ratio	Network Impact (Efficiency)
Average	0.89%	3.6%	−2.7%
Maximum	8.00%	25.1%	−18.3%

8.4.6.3 Whole Network Performance

When we then turn to a holistic view of our network performance, we find the results listed in Table 8.9. We find that, on average, temperatures went above 32 an additional 2.7% of the time. This, in turn, decreased our network efficiency by around 2.3%, a similar value to the corresponding increase in the above time ratio. This demonstrates, at least at an initial glance, a potential resiliency that the multiple redundant paths give the network.

Consider, however, that the above figures stem from an average of 10 years that does not take into account any seasons. If we use the maximum instead of the average, perhaps representing a particularly hot summer day, we see that temperatures go above 32°C an additional 17% of the time. This, in turn, decreased our network efficiency by 18%, suggesting a point where the redundant paths of our network are unable to maintain their resiliency to adverse conditions.

8.5 DISCUSSION

This methodology for analyzing the US freight network clearly shows reasonable shifts in centrality and efficiency, based on the input weights. However, there is still a need to verify the weight generation process and the output results through other means. For FAF freight traffic weights, we can check against other data sources like the published waybill data, to ensure the model's correctness.

This methodology provides two key benefits over more frequently used operational methods:

1. While the available studies tend to take into account rising temperatures, few tend to consider potential changes in rail traffic and infrastructure. Although predictions of this nature are difficult to make and predict, changes in relative congestion in various areas of the network are important to consider. The proposed method provides the ability to scale our analysis based on the resolution and precision of the available data.
2. While speed limits on a specific portion of the network should not cause congestion within the limited area, since all vehicles within have roughly the same speed, the limits of a limited area of speed are a different story. Where there is some congestion, trains approaching the boundary will tend to cluster as the trains ahead slow down. As such, the impact of a speed limit can expand beyond the area of the limit. The proposed methodology takes into account the potential concentration and cascading effects of speed limits.

In the future, we could augment this methodology to take into account multiple potential hazards, thus giving a fuller picture of the overall system risk due to climate change.

Additionally, we could also increase the specificity of the study to provide higher-fidelity results. Specifically, we should continue to add details pertaining to seasons, traffic patterns, and company-specific adaptation measures to determine the types of extremes against which the network is able to demonstrate resilience.

8.6 CONCLUSION

As the US railroad network is critical piece of infrastructure, it is prudent to ensure that we adequately explore the impact of climate change on it. We should take into account both the scale and intensity of the impact particular hazards may have, but also the cascading effects that degradations in one region will affect the performance of another. The developed network-based methodology focuses on providing an understanding of the system performance as an interconnected gestalt. Through it, we find that temperature increases are expected to cause a corresponding decrease in network performance (around 2.7% year-round and 18% on the hottest days).

ACKNOWLEDGMENT

The authors acknowledge the partial financial support of the NOAA) for this research. The results and scientific opinions in this publication are those of the authors and are not necessarily that of NOAA or the US Government. The authors are grateful for the technical advice received from Yujie Mao, Sherief Elsibaie and Yating Zhang.

REFERENCES

Andersson-Sköld, Y., Nordin, L., Nyberg, E., and Johannesson, M. (2021). "A framework for identification, assessment and prioritization of climate change adaptation measures for roads and railways." *International Journal of Environmental Research and Public Health*, 18(23), 12314.

Association of American Railroads (2022). "Overview of America's Freight Railroads." (March)". https://www.aar.org/rail-industry-overview/

Ayyub, B. M. (2014). *Risk analysis in engineering and economics*, 2nd edition. CRC Press, Boca Raton, FL.

Blainey, S. and Preston, J. (2022). *Sustainable railway engineering and operations*. Transport and Sustainability; Volume 14. Emerald Publishing, London, England.

Brandes, U. (2008). "On variants of shortest-path betweenness centrality and their generic computation." *Social Networks*, 30(2), 136–145.

Chinowsky, P., Helman, J., Gulati, S., Neumann, J., and Martinich, J. (2019). "Impacts of climate change on operation of the US rail network." *Transport Policy*, 75, 183–191.

Delworth, T. L., Cooke, W. F., Adcroft, A., Bushuk, M., Chen, J., Dunne, K. A., Ginoux, P., Gudgel, R., Hallberg, R. W., Harris, L., Harrison, M. J., Johnson, N., Kapnick, S. B., Lin, S., Lu, F., Malyshev, S., Milly, P. C., Murakami, H., Naik, V., Pascale, S., Paynter, D., Rosati, A., Schwarzkopf, M., Shevliakova, E., Underwood, S., Wittenberg, A. T., Xiang, B., Yang, X., Zeng, F., Zhang, H., Zhang, L., and Zhao, M. (2020). "SPEAR: the next generation GFDL modeling system for seasonal to multidecadal prediction and projection." *Journal of Advances in Modeling Earth Systems*, 12(3), e2019MS001895.

Dobney, K., Baker, C. J., Quinn, A. D., and Chapman, L. (2009). "Quantifying the effects of high summer temperatures due to climate change on buckling and rail related delays in south-east United Kingdom." *Meteorological Applications*, 16(2), 245–251.

Federal Railroad Administration (2024). "North American Rail Network Lines." (October).

Garmabaki, A. H. S., Thaduri, A., Famurewa, S., and Kumar, U. (2021). "Adapting railway maintenance to climate change." *Sustainability*, 13(24), 13856.

Geofabrik GmbH Karlsruhe. (2024). "GEOFABRIK." https://www.geofabrik.de/

Hamed, M., Mao, Y., Ayyub, B. M., Elsibaie, M., and Omar, T. (2022). "Connectedness efficiency analysis of weighted U.S. Freight Railroad Networks." *ASCE-ASME Journal of Risk and Uncertainty in Engineering Systems, Part B: Mechanical Engineering*, 8(4), 041202.

Latora, V. and Marchiori, M. (2001). "Efficient behavior of small-world networks." *Physical Review Letters*, 87(19), 198701 Number: 19 Publisher: American Physical Society.

Liban, C. B., Kafalenos, R., Alessa, L., Anenberg, S., Chester, M., DeFlorio, J., Dóñez, F. J., Flannery, A., Sanio, M. R., Scott, B. A., Stoner, A. M. K., Crimmins, A. R., Avery, C. W., Easterling, D. R., Kunkel, K. E., Stewart, B. C., and Maycock, T. K. (2023). "Chapter 13: Transportation. Fifth National Climate Assessment." U.S. Global Change Research Program. https://toolkit.climate.gov/NCA5_Ch13_Transportation.

Liu, X., Saat, M. R., and Barkan, C. P. L. (2012). "Analysis of causes of major train derailment and their effect on accident rates." *Transportation Research Record: Journal of the Transportation Research Board*, 2289(1), 154–163.

Lovett, A. H., Dick, C. T., and Barkan, C. P. L. (2015). "Determining Freight Train Delay Costs on Railroad Lines in North America". https://railtec.illinois.edu/proceedings/determining-freight-train-delay-costs-on-railroad-lines-in-north-america/

Ma, D., Bai, Z., Xu, Y.-P., Gu, H., and Gao, C. (2024). "Assessing streamflow and sediment responses to future climate change over the Upper Mekong River Basin: a comparison between CMIP5 and CMIP6 models." *Journal of Hydrology: Regional Studies*, 52, 101685.

Neumann, J. E., Chinowsky, P., Helman, J., Black, M., Fant, C., Strzepek, K., and Martinich, J. (2021). "Climate effects on US infrastructure: the economics of adaptation for rail, roads, and coastal development." *Climatic Change*, 167(3-4), 44.

Newman, M. E. J. (2010). *Networks: an introduction*. Oxford University Press, Oxford; New York.

Ordnance Survey (2024). "Rail Ordnance Survey." https://www.data.gov.uk/dataset/5fa93846-ebda-44b7-b478-4fe0ee3c8595/rail4." (December).

Palin, E. J., Stipanovic Oslakovic, I., Gavin, K., and Quinn, A. (2021). "Implications of climate change for railway infrastructure." *WIREs Climate Change*, 12(5), e728.

United States. Congress. House. Committee on Transportation and Infrastructure. (2009). *Freight and passenger rail : present and future roles, performance, benefits, and needs : hearing before the Subcommittee on Railroads, Pipelines, and Hazardous Materials of the Committee on Transportation and Infrastructure, House of Representatives, One Hundred Eleventh Congress, first session*, January 28, 2009. U.S. G.P.O., Washington.

United States Congress House Committee on Transportation and Infrastructure Subcommittee on Railroads, P. (2021). *Full steam ahead for rail: why rail is more relevant than ever for eco-nomic and environmental progress: remote hearing before the Subcommittee on Railroads, Pipelines, and Hazardous Materials of the Committee on Transportation and Infrastructure, House of Representatives, One Hundred Seventeenth Congress, first session*, March 10, 2021. U.S. Government Publishing Office, Washington.

United States. Department of Transportation. Bureau of Transportation Statistics. (2017). "FreightAnalysis Framework (FAF) FAF5: HiLoForecasts [supporting datasets]." https://rosap.ntl.bts.gov/view/dot/72145. (January).

United States: Transportation Department: Federal Railroad Administration, Kish, A., Kish, A., Samavedam, G., United States: Federal Railroad Administration: Office of Research and Development, John A. Volpe National Transportation Systems Center (U.S.), and Foster-Miller (2013). *Track buckling prevention: theory, safety, concepts, and applications*. Transportation Department. https://purl.fdlp.gov/GPO/gpo48099. (March). Accession Number: GOVPUB-TD3-PURL-gpo48099 Source: DGPO.

US Department of Defense, The Military Surface Deployment and Distribution Command Transportation Engineering Agency. (2025). "Office of the Special Assistant for Transportation Engineering Railroads for National Defense." https://api.army.mil/e2/c/downloads/2025/08/25/be0ea552/bulletin-16-01-who-is-sddctea.pdf

US Department of Transportation (2024). "North American Rail Network." https://geodata.bts.gov/datasets/usdot::north-american-rail-network-nodes/about. (July).

Villalba Sanchis, I., Insa Franco, R., Martínez Fernández, P., Salvador Zuriaga, P., and Font Torres, J. B. (2020). "Risk of increasing temperature due to climate change on high-speed rail network in Spain." *Transportation Research Part D: Transport and Environment*, 82, 102312.

Wasley, E., Dahl, T. A., Simpson, C. F., Fischer, L. W., Helgeson, J. F., Kenney, M. A., Parris, A., Siders, A., Tate, E., Ulibarri, N., Crimmins, A. R., Avery, C. W., Easterling, D. R., Kunkel, K. E., Stewart, B. C., and Maycock, T. K. (2023). "Chapter 31: Adaptation. Fifth National Climate Assessment." U.S. Global Change Research Program. https://toolkit.climate.gov/sites/default/files/2025-07/NCA5_Ch31_Adaptation.pdf

9 Disaster Recovery Modeling for Residential Buildings and Its Policy Implications

Wanting (Lisa) Wang[a] *and John W. van de Lindt*[b]
[a]Department of Civil and Environmental Engineering, Old Dominion University, Norfolk, Virginia
[b]Department of Civil and Environmental Engineering, Colorado State University, Fort Collins, Colorado

9.1 INTRODUCTION

Community resilience is the ability of a community to prepare for natural or human-made hazards, adapt to changing conditions, and withstand and recover rapidly from disruptions (PPD-8 2011, PPD-21 2013, NIST 2015). Resilience modeling and analysis can support risk-informed decisions including assisting stakeholders and decision-makers in understanding the different dimensions of the challenge, establishing short-term and long-term community goals and objectives, and executing their community planning. Community resilience assessment addresses the quantitative measures of robustness, rapidity, resourcefulness, and redundancy from a multidimensional perspective (Bruneau et al. 2003, Koliou et al. 2017). However, numerous studies have focused on assessing the residual functional capacity of the built environment and, in particular, lifeline systems following a disaster. Limited studies have been performed to explore the continuous recovery of physical infrastructure systems and most model trajectories, making it difficult to model interventions at different points in time during the recovery process.

The Federal Emergency Management Agency (FEMA), which is an agency within the United States Department of Homeland Security, first proposed the National Disaster Recovery Framework in 2011 as guidance, in which they explained how government agencies, nongovernmental organizations (NGOs), and the private sector could (should) organize and utilize existing resources to promote effective recovery by phase and support disaster-impacted states (FEMA 2011). In more recent years, researchers have developed comprehensive restoration models to explore the restoration process specific to critical infrastructure systems like buildings (e.g., Koliou and van de Lindt 2020, Bonstrom and Corotis 2016), electric power networks (e.g., Ouyang and Duenas-Osorio 2014), water and wastewater (e.g., Tomar et al. 2020), oil and gas (e.g., Ameri and van de Lindt 2019), and the interaction between networks of

DOI: 10.1201/9781003567332-9

infrastructure systems (e.g., He and Cha 2018, Smith et al. 2019). When assessing the restoration time or sequences of the community involved with an interdependent network, the community topology associated with graph theory and critical paths is one of the most straightforward and efficient approaches to solve the problem (He and Cha 2018, Ramachandran et al. 2015, Masoomi and van de Lindt 2018b). Moreover, some research efforts have focused on modeling housing recovery and re-occupancy of displaced households in the aftermath of a disastrous event (Zhang and Peacock 2009, Hamideh et al. 2018, Lin et al. 2019). Others were devoted to examining the effects of household dislocation on business recovery reflected in resilience metrics such as revenue recovery and customer retention (Watson et al. 2020, Aghababaei et al. 2020a).

Data-driven models have played an important role in quantifying certain aspects of community recovery. For example, some researchers have measured building recovery progress after hazard events via longitudinal field investigations (van de Lindt et al. 2020) and aerial imagery techniques (Kikitsu and Sarkar 2015, Crawford et al. 2019, Aghababaei et al. 2020b). Hamideh et al. (2018) investigated the effect of damage, tenure, minority population, and income on the housing recovery trajectory using a regression model. Burton et al. (2019) considered integrating household decision-making into recovery models with real-world data collected from Los Angeles households combined with multinomial logistic regression. The results showed that the household decisions were significantly impacted by physical damage to the residential community as well as other household socioeconomic characteristics, such as neighborhood evacuation level, household income, and insurance coverage.

Simulation-based modeling is another process that helps researchers, engineers, and decision-makers better understand community recovery under different conditions. Such model simulation results can further help leaders and their stakeholders develop community plans to mitigate the adverse effects of hazard events and accelerate the restoration process. Solving an optimization problem for repair priority sequences or community resource allocation is one of the most common research topics examining recovery (Sun et al. 2020, Zhang et al. 2017, Nozhati et al. 2020). Hassan and Mahmoud (2020) used a Markov chain process coupled with a success tree to perform a dynamic optimization to estimate the recovery of the healthcare facilities, considering the availability of repair crews. González et al. (2016) developed the minimum-cost reconstruction strategy of a partially destroyed infrastructure network system with the limitation of budget, resources, and operationality.

A few studies have estimated residential and business downtime due to a variety of delays before the initiation of building repairs. Lin and Wang (2017a) utilized a discrete/continuous-time Markov Chain and transition probability matrix to predict the recovery trajectory involved with delay across the domain of a community for earthquake hazards using a building portfolio approach that clustered buildings by type. De Iuliis et al. (2019) applied a fuzzy logic hierarchical scheme to evaluate the delay time considering repair sequences and repair crews for a residential building. Aghababaei et al. (2020b) adjusted the The Resilience-based Earthquake Design Initiative (REDi™) framework and used real-world Small Business Administration (SBA) loans and permitting time data to determine the delay time, filling the gap between the analytical and empirical fragilities.

Moreover, policies are essential to effectively influence the speed and quality of the recovery process when they are developed related to public resources with potential causes of delays considered (Ingram et al. 2006, Joshi and Aoki 2014, Sutley and Hamideh 2018, Richmond 2019). For example, recovery fund policies from the FEMA restricted any money used to repair damages caused by deferred maintenance (Sapat and Esnard 2016). Hirayama (2000) examined the temporary and permanent housing recovery policy implemented in Kobe, a city heavily damaged by the Great Hanshin Earthquake in 1995, and identified a significant gap in housing recovery for the urban area after the earthquake due to socioeconomic and spatial polarization. Drennan et al. (2016) confirmed the effectiveness of intergovernmental policy and funding arrangements on reinforcing the management of natural disasters through case studies of two of the most expensive and deadly natural disasters in Australia. However, the work presented in this chapter differs from past studies in that no studies to date have performed a quantitative probabilistic analysis to determine the community recovery when impacted by policy changes implemented by federal, state, or local governments. Another significant contribution of this chapter is, for the first time, a methodology that provides recovery modeling as a time-stepping algorithm (e.g., monthly, quarterly, and yearly) is presented with full propagation of uncertainties, thereby allowing investigation of changes in policies at different points in time during the recovery process. Finally, the methodology presented herein applies an existing housing unit allocation algorithm (Rosenheim 2020) that allows the analyst to include detailed household characteristics and to track how the recovery of housing impacts the people living in each building. This last feature allows the prediction of delays based on the household income allocated to each housing unit based on race, ethnicity, and household size at the census block level as well as census-tract-level income distributions. Figure 9.1 shows the schematic description of community resilience following a disruptive hazard event in terms of robustness (system-level residual functional capacity to withstand an extreme event) and rapidity (the time required to recover the desired functionality). Where $Q(t)$ is the functionality performance, which is measured as a dimensionless function of time, t_r is the recovery time for the system to be fully restored, and t_o is the time of occurrence of disasters (Bruneau et al. 2003, Koliou and van de Lindt 2017). Resilience (R) was defined mathematically by Bruneau et al. (2003), illustrated in Eq. (9.1), and can also be shown as the area under the curve between t_o and t_r in Figure 9.1.

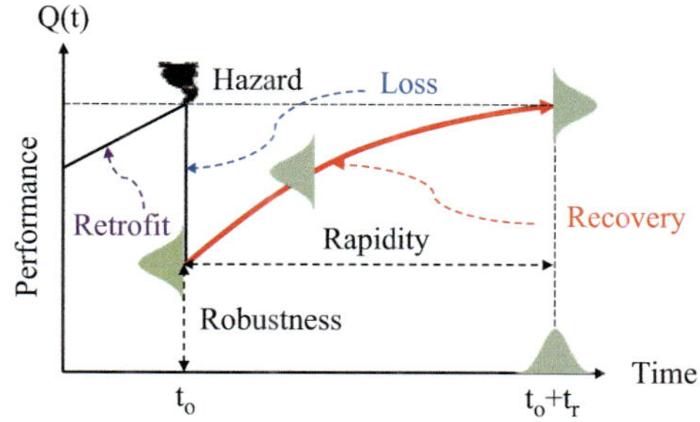

FIGURE 9.1 The schematic description of community resilience.

$$R = \int_{t_0}^{t_0+t_r} Q(t)\,dt \tag{9.1}$$

9.2 PROBABILISTIC RESIDENTIAL RECOVERY MODELING

Total recovery time for a damaged building induced by a hazard event is not limited to the time required to simply repair a building. Recovery modeling in this chapter is considered a two-step recovery: downtime due to delays and downtime due to repairs. The recovery modeling overall was a probabilistic quantitative analysis which uses multilayer Monte Carlo simulation. The first layer of Monte Carlo simulation included N1 scenarios to probabilistically estimate the delay time, which is directly related to building damage from the hazard. For each realization of delay time within the MCS, N2 realizations of repair time were generated and then combined with this delay time realization to form the multilayer MCS. The delay assessment methodology was developed based on the approach in the REDi framework (Almufti and Willford 2013). It should be mentioned that the REDi framework was initially designed for loss assessment of post-earthquake events, but the methodology developed in this chapter modified the framework to generalize it for all events, with an illustrative example for wind-induced events presented later in this chapter. The functions comprehensively incorporate the delay variables resulting from typical impeding factors, which include issues such as post-disaster inspection, securing an engineer/contractor, financial assistance, and a building permit. The financing delay is assumed to be directly related to the financial resources available to a household. Thus, the methodology proposes to use annual household income as a sociodemographic characteristic indicator to predict the funding options for households to finance their residential building repairs. Repair time was estimated by fragility functions, and recovery time was the combination of delay time plus repair time. Weibull distributions can be used to fit the Monte Carlo simulation results for delay time and recovery time for each building. It was assumed that no households would out-migrate but rather stay in the community to complete the repairs following a hazard event. Relaxing and addressing this assumption is beyond the scope of the current chapter, but its importance is duly noted and will be addressed in future work.

9.2.1 DELAY TIME ASSESSMENT

Several delays inevitably occur before the initiation of building repairs, which will increase the time needed for damaged buildings to reach the next (and eventually complete) recovery state. Eq. (9.2) shows the delay variables considered in this chapter, where i is an index for each realization within the Monte Carlo simulation, j is a specific building in the community; T_{Delay} is the modeled delay time, and T_{INSP}, T_{ENGM}, T_{FINA}, T_{CONM}, T_{PERM} are the times for the post-event inspection, to contact engineers, to secure funding, to secure contractors, and procure construction permitting, respectively.

$$T_{\text{Delay},\,ij} = T_{\text{INSP},\,ij} + \max\left\{T_{\text{ENGM},\,ij}, T_{\text{FINA},\,ij}, T_{\text{CONM},\,ij}\right\} + T_{\text{PERM},\,ij} \tag{9.2}$$

This chapter assumes the following sequence of events associated with modeled delay time. First, a building inspector performs a thorough visual inspection of the damaged buildings to assess the extent of the damage and protect the safety of the building occupants. The building owner is then told to obtain a structural inspection if the structural integrity of the building is in question. In addition, the jurisdiction, tenants, or insurance companies have a right to request an inspection if they deem the event may have caused damage to a building. Upon receiving the inspection report, the homeowner will typically approach a professional contractor to prepare for the repair and, if applicable by jurisdiction, identify an engineer or equivalent to review/redesign and provide drawings for the building. Simultaneously, the financial loss estimate for the building may be provided by the insurance inspection report and/or the contractor, and the owner must proceed to secure any funding resources needed. Delays during this period need to be combined, and the time required for each component cannot simply be added, since some tasks are usually performed in parallel. The duration used for the engineer to review/redesign may depend on the structural damage level and may also be related to the height and footprint area of the building. Finally, if the building exhibits structural damage and then structural repairs are needed, a permit needs to be approved by the local building jurisdiction.

The duration of most of the delay variables is well correlated with the extent of the damage. Unlike most of the delay variables, financing delays rely on the household's capability to access government funding, the bank, insurance payments, and charitable organizations. Note that federal, other state governments, and NGOs like the American Red Cross and local churches are not considered in this chapter due to the uncertainty of their assistance for each event. Instead, financial resources, including insurance, loans, and savings, are considered herein. More than 85% of homeowners have homeowner insurance in the United States, which typically covers wind damage (Croll 2021). The SBA, which is a United States government agency that provides support and loans to entrepreneurs and small businesses, has invested a large sum of money in disaster loans to repair or replace damaged components. In theory, businesses and homeowners can borrow within limits, which depend on their qualifications. Additionally, owners may qualify for private loans from banks if they meet specific minimum requirements. The personal loan amount is related to financial impact factors such as property value, credit report, and market conditions. However, specific populations, such as homeowners with low incomes and minorities, historically have less access to such public assistance (Peacock et al. 1997). For these households, any savings or family assistance may become the only option for financing repair.

The impeding factor functions are quantitatively described in the format of lognormal cumulative distribution functions with a high degree of uncertainty, as illustrated in Table 9,1 and Eq. (9.3) (Almufti and Willford 2013). Note that β is the standard deviation of the lognormal function, and the columns of the median (θ_1) and coefficient of variation (CoV) shown in Table 9.1 were referenced from the REDi framework, while the modified median (θ_2) values are defined in the present

TABLE 9.1
Statistics of Delay Time Associated with Building Damage (Unit: Weeks)

Sequence	Delay Time Impeding Factor	Building Damage Conditions	Median (θ_1)	Modified Median (θ_2)	Coefficient of Variation (CoV)
Delay phase 1	Inspection ($T_{\mathrm{INSP},\,ij}$)	Slight	0	0	-
		Above slight	0.5	0.1	1.08
Delay phase 2	Engineering mobilization and review/redesign ($T_{\mathrm{ENGM},\,ij}$)	Slight	6	-	0.07
		Moderate/extensive	12	-	0.03
		Complete	50	-	0.01
	Financing ($T_{\mathrm{FINA},\,ij}$)	Insurance	6	3	0.19
		Private loans	15	7.5	0.05
		SBA-backed loans	48	24	0.01
		Savings/others	48	24	0.01
	Contractor mobilization ($T_{\mathrm{CONM},\,ij}$)	Slight	7	1	0.09
		Above slight	19	11	0.02
Delay phase 3	Permitting ($T_{\mathrm{PERM},\,ij}$)	Slight	1	0	0.86
		Above slight	8	0	0.04

chapter to represent shortened delay time impacted by different policies, which will be expanded on later in the policy lever section. It is assumed that the coefficients of variation for the delay time remain the same with or without the impact of post-disaster policies. The variable of $T_{n,\,ij}$ shown in Eq. (9.3), is consistent with the delay variables such as post-event inspection defined in Eq. (9.2). It is worth noting that these delay time estimates may be merely applicable to the buildings located in the United States since other countries may have different emergency response plans. Lin and Wang (2017b) combined delay time components with specific conditions of buildings damaged at different levels (see Table 9.1). The estimated delay time is directly related to the initial building damage level in the present chapter, which is modeled using fragilities for each building, as illustrated conceptually in Figure 9.2.

$$F_R\left(T_{n,ij}\right)=\Phi\left[\frac{\ln\left(T_{n,ij}\right)-\ln(\theta)}{\beta}\right] \tag{9.3}$$

For residential buildings, a household's income has been considered a sociodemographic indicator that can describe how households can access different financial resources (Carvalho et al. 2016). Eq. (9.4) presents an estimation of the time used for financing delays, where n refers to four specific funding options considered herein, including homeowners' insurance, SBA-backed loans, private loans, and savings/others. The indices i and j are still representative of each realization in the Monte Carlo simulation and building identification itself, respectively; $P\left[T_{n,ij}\right]$ is the probability of

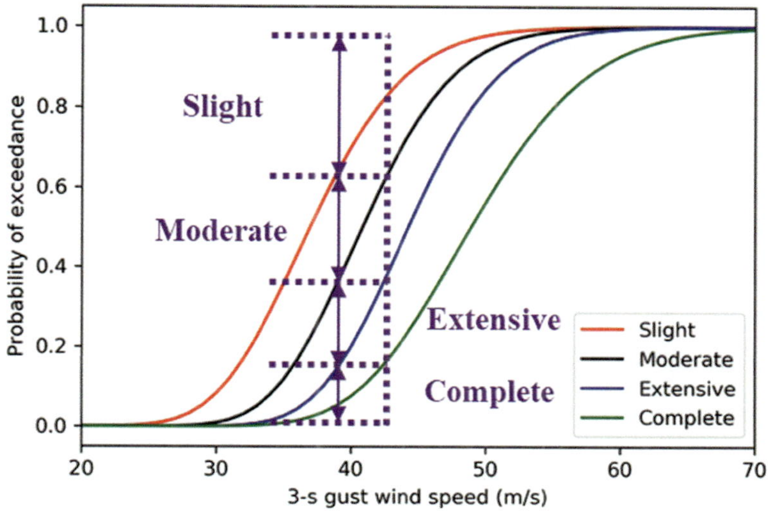

FIGURE 9.2 Fragility curve of a wood-frame residential building subjected to a hazard (tornado).

the households to access one of the financial options as discussed, and $T_{n,ij}$ is the estimated time derived from cumulative lognormal functions. It is assumed in the model that each household does not assess two or more financial resources, and the probability of different household groups to approach a single financial resource is different.

$$T_{\text{FINA},\,ij} = \sum_{n=1}^{4} P\left[T_{n,ij}\right] \cdot T_{n,ij} \tag{9.4}$$

The income distributions at the census tract level can be derived from the dataset developed by the American Community Survey (ACS), which is a continuous survey led by the U.S. Census Bureau. However, issues such as missing data are unavoidable due to a lack of response by households or incomplete reports provided by the household. Therefore, when the household income information cannot be fully measured/estimated by household surveys or the specific algorithm, it is necessary to predict (extrapolate) the missing data to estimate financing delay times for all housing units affected by the hazard exposure. This chapter introduces what is believed to be a feasible and relatively straightforward predictive approach to use a Gaussian model to predict missing data combined with the existing procured income data. The approach incorporated housing unit de-aggregation from the census block level, as shown in Eq. (9.5), where $I_{\text{phh,imk}}$ is the predicted household income and $I_{\text{hh,mk}}$ is the known household income of the kth housing unit in the mth census block for the ith realization of the Monte Carlo simulation. The first level of the Gaussian function is used to obtain the statistics of the household income distribution at the census block-level using all the known U.S. Census household income data. The second level of the Gaussian function is then used to predict the missing household income for specific households using the determined Gaussian distribution census block in which the household with missing census data is located. This step uses Monte Carlo simulation to make the prediction probabilistic. The motivation for this process is that, currently, census block-level is the minimum geographic unit publicly reported by the U.S. Census Bureau to maintain individual privacy.

$$I_{\text{phh,imk}} = \begin{cases} I_{\text{hh,mk}} & I \in I_{\text{hh,mk}} \\ \Phi\left(\Phi^{-1}\left(I_{\text{hh,mk}}\right)\right) & I \notin I_{\text{hh,mk}} \end{cases} \tag{9.5}$$

9.2.2 Repair Time Assessment

To this point in this chapter, the methodology has been presented in a general sense for any hazard event. Although any repair fragilities could be implemented in the step described in this section, the authors move to a hazard-specific explanation, namely a tornado hazard, to enable clarity for the example presented later. In this chapter, repair time was developed following the steps of combining isolated component damage fragility and consequence functions (i.e., repair estimates) introduced in the FEMA P-58 (FEMA 2012) methodology and then assembling the repair fragility functions from building component levels to system levels (Koliou and van de Lindt 2020). Table 9.2 presents tornado repair fragility functions of residential buildings for different damage states to reach full functionality ($Q = 100\%$), and Table 9.3 demonstrates the defined damage combination and functionality level for wood-frame residential buildings. Please refer to Koliou and van de Lindt (2020) for details on the definitions of performance level and operational status for all building archetypes in detail. Note that the fragilities only refer to the building repair time due to damage without considering external parameters such as crew availability, which means the time used for delay and repair in this chapter does not overlap.

9.2.3 Recovery Time Assessment

Following the immediate damage, the total time required for each building to be fully restored can be determined by combining delay time and repair time. It is assumed that the two stages of building recovery (delay and repair) are statistically

TABLE 9.2
Repair Fragilities for Wood-Frame Residential Buildings (Days)

Damage States	Mean of the Natural Logarithm (μ)	Standard Deviation (σ)
DS1	3.09	0.51
DS2	3.52	0.55
DS3	4.62	0.55
DS4	5.19	0.52

TABLE 9.3
Damage Combination and Functionality Level for Wood-Frame Residential Buildings

Functionality Q (%)	Damage Combination	Description of Damage Combination: Roof Cover Failure	Window/ Door Failures	Roof Sheathing Failure	Roof Truss Failure
100	-	-	-	-	-
75	DS1	>2% and ≤15%	1	No	No
50	DS2	>15% and ≤50%	2 or 3	1–3 panels	No
25	DS3	>50%	>3	>3 panels and ≤35% by area	No
<25	DS4	Typically>50%	Typically>3	>35% by area	Yes

independent, but it is noted that correlation could be included using a copula approach if evidence of correlation becomes known. The recovery results can be presented as histograms for a single building, a single sector (i.e., household, business, healthcare, and education), and even the entire community. A three-parameter Weibull distribution was used to fit the recovery time statistics for each building for modeling efficiency.

To track the community recovery performance and building recovery over time, the model steps through time by quarter, as shown in Eqs. (9.6) and (9.7), where $B_{rj}^{i}(t)$ is the building recovery performance indicator of the jth building at the rth quarter in the ith realization. $R_r(t)$ is the time by quarter (90 days, 180 days, etc.) and $R_{j,\text{th}}^{i}$ is the calculated recovery time of buildings as thresholds. $\text{RP}_{rj}(t)$ is the quarterly recovery probability, continuously increasing over the analysis period based on Monte Carlo simulation. For each time step, the probability that each building is fully recovered and therefore the percentage of residential buildings that are fully restored can be calculated.

$$B_{rj}^{i}(t) = \begin{cases} 1 & R_r(t) \geq R_{j,\text{th}}^{i} \\ 0 & R_r(t) < R_{j,\text{th}}^{i} \end{cases}, \quad i \in 1,2,\ldots,n \tag{9.6}$$

$$\text{RP}_{rj}(t) = \frac{\sum_{i=1}^{n} B_{rj}^{i}(t)}{n}, \quad i \in 1,2,\ldots,n \tag{9.7}$$

9.3 POLICY LEVERS

Post-disaster policies (policies that take effect in the event of a disaster) aim to create strategic plans to help expertly guide the completion of repair across the community. These policies can be related to delay time variables for completing building restoration or refer to the assessment of available resources to a household (Sutley and Hamideh 2018). In general, flexibility in existing policies and programs is more critical than implementing rigid policies to effect recovery following a natural or human-induced disaster (Richmond 2019). Excessive government involvement might result in increased costs and interruptions in the recovery process. After the 2011 Joplin tornado occurred, city and state officials published a series of policies. The policies, which included decisions such as relaxing regulations and avoiding micro-management, were implemented to facilitate the recovery. Regulatory relief allowed local businesses to maintain operations and quickly recover the Joplin Public Schools and let them reopen on time for the next academic year (Smith and Sutter 2013) with some relocation. This chapter proposed a series of policies referring to the real environment after the 2011 Joplin tornado, expanded the introduced methodology, and quantitively measured how the community recovery was impacted by these policies. Table 9.4 provides selected policies that positively impact community recovery and shows examples of their combinations as potential policy cases for a community.

Recall the relationship between delay time and building damage conditions shown in Table 9.1. Building retrofits improve the performance of buildings and reduce damage for buildings within the tornado path (Amini and van de Lindt 2014), and therefore delay times such as for inspection, and contractor mobilization will be shorter since damage is less significant. In general, where damage is reduced for some portion of the buildings, both delay time and repair time are shortened accordingly. In this chapter, wood-frame residential building retrofits are reflected in the different construction product combinations such as roof covering, roof sheathing nailing patterns, and roof-to-wall connections. See Wang et al. (2021) for a description of wind retrofit strategies on residential buildings. In this chapter, the construction product combination of asphalt shingles, 8d nails with 150 mm (6 in) edge nail spacing and 300 mm (12 in) field nail spacing for roof sheathing, and two 16d toenails for the wall-to-roof truss connection was considered to be the base case without any retrofits. The retrofit case consisted of clay tiles, 8d nails with 150 mm (6 in) edge and field spacing for the roof sheathing, and two H2.5 hurricane clips connecting the roof truss and wall. Additionally, if the city or county can relax building permit requirements for wood-frame residential buildings, the recovery process can be moderately expedited. It is assumed that the city can choose to relax certain regulations and allow homeowners to schedule repairs without obtaining a permit. Another approach is to hire extra building inspectors, which allows time reduction for home inspectors when there is a demand surge after a hazard event. A typical inspection of a house under 95 m^2 (1,000 ft^2) takes 2–3 hours, so one inspector can perform three to five inspections per day. For contractor mobilization, the bidding process for procurement typically takes 30 days for heavily damaged buildings since the losses are significant and competitive bids need to be sought. Provided that the bidding is waived by the insurance company (or others) for procurement, only the procurement and management of items (e.g., labor, equipment, and material) and their delivery to the site is needed. Upon receiving claims from homeowners following a tornado (or any event), an insurer decides whether to accept or deny the claim. Having financing in place shortens the time needed to begin the next delay phase, thus the duration to

TABLE 9.4
Policies and Their Combinations as Policy Cases That Can Facilitate Community Recovery Progress

Index	Policy	Case 1	Case 2	Case 3	Case 4	Case 5	Case 6	Reference
1	Building retrofits		X				X	Sutley and Hamideh (2018)
2	Relax building permits issuance			X		X	X	Sutley and Hamideh (2018)
3	Hire extra building inspectors			X		X	X	Smith and Sutter (2013)
4	Waive bidding rules for procurement			X		X	X	Smith and Sutter (2013)
5	Have financing in place				X	X	X	Lindell and Prater (2003)

obtain financial assistance was assumed to be cut in half to model when financing is in place in this chapter. Currently, the cost of different policy combinations and the budget limits for government investments are not considered and are beyond the scope of this chapter.

9.4 ILLUSTRATIVE EXAMPLE

Tornadoes, which are low-probability, high-consequence natural hazard events, result in numerous damage and casualties associated with high socioeconomic and environmental costs through an entire community (Simmons et al. 2013, Standohar-Alfano and van de Lindt 2015, Masoomi and van de Lindt 2018a, Changnon 2009). Over the last two decades, tornadoes have caused almost 1,500 fatalities, 20,000 injuries, and \$30.3 billion property damage loss in the U.S., according to the National Weather Service (NOAA 2020). The Enhanced Fujita scale (EF scale) is used to assign a tornado intensity rating (from EF0 to EF5) based on damage assessments and estimated wind speeds since 2007 (Standohar-Alfano and van de Lindt 2015). The 2011 Joplin tornado was a devastating EF5 multiple-vortex tornado and ranked as the deadliest and costliest single tornado in U.S. history. A comprehensive field investigation was performed to make recommendations for other communities related to tornadoes (Kuligowski et al. 2014), and a number of researchers performed analyses related to this event (e.g., Attary et al. 2018, Prevatt et al. 2012, Roueche and Prevatt 2013).

As a small- to medium-size community in the United States, Joplin, in the Midwest state of Missouri, was selected to illustrate the introduced approach for a two-step recovery model for residential buildings that is able to step through time, thereby allowing the analyst to explore the impact of policy levers on community recovery with the goal of eventually providing the results to support decision-making for stakeholders and community planners. Wang et al. (2020) performed a functionality analysis using the Joplin testbed to consider the interdependency between buildings and electric power networks and examined the effect of wind events on a basic set of physical and socioeconomic resilience metrics. The detailed building dataset and the simulated EF5 tornado hazard used herein were not illustrated again, and the interested reader, is referred to Wang et al. (2020). It is known that tornadoes have small footprints and, although a very large tornado, the Joplin tornado only hit a small geographical area of Joplin, and many buildings fell outside the path of the tornado and remaining undamaged. Wood-frame residential buildings are the most vulnerable building sector to wind events, and therefore, the recovery modeling of residential buildings located within the tornado path is the focus herein.

9.4.1 Probabilistic Initial Building Damage

Tornado fragilities together with the simulated wind speeds at each location, were combined to estimate the building damage for each building across the community (see Figure 9.2). Please refer to Masoomi et al. (2018) regarding tornado fragilities for five residential building archetypes and their descriptions (e.g., the number of stories and building sizes/area). Figure 9.2 shows the fragility functions of a typical wood-frame residential building (archetype 4) as an example. Table 9.5 indicates building spatial damage results in a typical single realization within Monte Carlo simulation. Note that the damage levels, as indicated in Table 9.5, are as follows: 1 is slight, 2 is moderate, 3 is extensive, and 4 is complete. It can be observed from Table 9.5 that more than 85% of residential buildings within the tornado path were initially damaged at the complete level in this single realization. For each realization, probabilistic initial building damage results were then used to estimate the delay time and repair time.

9.4.2 Financial Resources

Low-income and ethnic minority families are more vulnerable to the risks of disasters and struggle most to recover. One of the reasons is that low-income households are more likely to reside in neighborhoods that are more susceptible to environmental shocks, and they do not have many choices to relocate to safer areas (Krause and Reeves 2017). However, this is less true for tornadoes than for many hazards. For the socioeconomic groups described above that have a lower probability

TABLE 9.5
A Typical Scenario of Initial Building Damage

Building Specific Conditions	Building Count
Slight	174 (3.3%)
Moderate	181 (3.4%)
Extensive	234 (4.4%)
Complete	4,738 (88.9%)
Grand total	5,327 (100%)

of being able to access some funding sources such as insurance and loans, savings and other community-level resources may be options to perform repairs/reconstruction. In this chapter, five household groups are differentiated by annual household income as shown in Table 9.6 to span the wealth distribution across the community. It is assumed that households belonging to income groups over household income group (HH3) are homeowners. HH1 is mostly a group for renters, and HH2 has a balance of low-income homeowners and renters. In Missouri, about 90% of homeowners are insured (Insurance Agency Plano 2014), and the number of people with renter's insurance is far less, at 27% (Joplin Globe 2011). In general, for homeowners with an annual income of over US$30,000, an increase in household income of US$11,000 increases the likelihood of insurance purchases by approximately one percent (Landry et al. 2020). Overall, insurance is the dominant resource for households impacted by wind damage, while SBA-backed loans and private loans are also important resources for low-income households to consider. The reason is that homeowners (and renters) insurance can be quite a burden for low-income families if not required, and the premiums may account for too significant a percentage of their monthly household income. In theory, SBA-backed loans are preferable compared to private loans since they have lower interest, even though the processing time might be longer. Unfortunately, SBA loans go to those who would have qualified for commercial loans, not the households in the greatest need, because the ability to repay is always a crucial issue. Therefore, the probability of low-income households receiving SBA loans is limited as well (Lindell and Prater 2003).

Table 9.6 shows the deterministic household restoration financial resource distribution based on information from the literature described above. The percentages shown in Table 9.6 are the probabilities of households funding their repairs through the different financial resources. For example, households with an annual income below $15,000 will only have a low probability of obtaining funding supported by insurance and SBA-backed loans. To survive and recover from the disaster, this group of households is more likely to depend on their savings or other means such as family/friends. Conversely, households with an annual income of over $100,000 will have homeowner's insurance.

TABLE 9.6
Household Restoration Financial Resource Distribution

Household Income Group	Insurance	SBA-Backed Loans	Private Loan	Savings/Others
HH1 (less than $15,000)	30%	5%	0%	65%
HH2 ($15,000 to $24,999)	50%	5%	5%	40%
HH3 ($25,000 to $74,999)	80%	10%	10%	0%
HH4 ($75,000 to $99,999)	85%	15%	0%	0%
HH5 (more than $100,000)	100%	0%	0%	0%

9.4.3 Housing Unit De-Aggregation

To employ annual household income as an indicator to determine household financial resources during the recovery process, household income group information needs to be collected or determined first. This information was determined through data merging at the housing unit level based on the ACS (2012) 5-year survey. The household income came from a combination of tables B19001 and B19101. This variable was designed to provide income comparable to income distributions associated with race/ethnicity . The allocation of income to a physical location does not represent the actual income for the household. Please refer to Rosenheim (2021) for the detailed household income dataset created by the housing unit allocation algorithm. The building dataset was developed for Joplin circa 2010 prior to the tornado, thus allowing some level of validation 10 years later. There are 7,201 housing units contained in 5,327 buildings in the path of the Joplin tornado in Joplin based on the 2010 Decennial Census.

After completing the household income prediction by applying the housing unit allocation algorithm, there are still 10.5% of housing units with missing household income information (see Table 9.7). The reason for missing values for household income might be that the household has income, but the characteristics (race/ethnicity) did not match between the 2010 Census and the 2012 ACS at the census tract scale. Future research is required to reduce the number of missing values, especially for minority households. It is important to mention that 656 housing units are vacant, and 6 housing units are group quarters such as nursing homes, but the household income prediction from the housing unit allocation algorithm does not consider such cases, introducing error into the approach.

Census geographic entities cover the entire United States, with the smallest being census blocks, which aggregate into census block groups and then census tracts. Residential buildings located in the tornado path were distributed across 10 census tracts, 28 census block groups, and 437 census blocks, respectively. Thus far, the

TABLE 9.7
Housing Unit Inventory within the Tornado Path

	Owner-Occupied	Renter-Occupied	Housing Unit Count	Household Income Missing
1-person household	1,035	1,177	2,212	147 (6.6%)
2-person household	1,135	842	2,177	293 (13.5%)
3-person household	483	487	970	146 (15.1%)
4-person household	371	325	696	97 (13.9%)
5-person household	160	138	298	43 (14.4%)
6-person household	66	56	122	22 (18.0%)
7-person household	37	27	64	9 (14.1%)
Group quarters	-	-	6	-
Vacant	-	-	656	-
Total housing units	3,487	3,052	7,201	757 (10.5%)

census block is the smallest geographic unit published by the U.S. Census Bureau to maintain data confidentiality. Therefore, housing units were de-aggregated to the census block level to predict the missing income data accurately. Housing unit details and buildings with single or multiple dwellings were tracked by unique identifiers (Housing unit ID and Globally Unique Identifier, respectively).

For each realization of the Monte Carlo simulation, known income groups assigned to each house remained the same to ensure recovery was tracked for each allocated household. The recovery modeling performed in this chapter conducted the analysis at the building level, while the allocation of income groups described their income distribution at the household level. For multifamily buildings, the income groups were aggregated from household levels to building levels by selecting the housing unit having the highest annual income group. It is assumed that housing units with the highest estimated income group had the best restoration financial resources, which could be used to recover the multiple dwelling units.

9.4.4 Recovery Modeling Results

The probabilistic residential recovery modeling described throughout this chapter was implemented for each of the policy cases described earlier. Figures 9.3 and 9.4 show delay time results and recovery time results for two typical residential buildings at different damage levels under Case 1, respectively. The results of Monte Carlo simulation realizations are shown in histograms as well as the fitted Weibull distribution curves and parameters. Recall that recovery time, which means the total time needed for buildings to be fully restored, consists of two independent steps: delay time and repair time. It is evident that delay time dominated throughout the recovery process for residential structures, but this is likely not the case for other types of buildings where repairs take much longer, for example, hospitals, schools, and office buildings. For example, the mean recovery time for a typical building shown in Figure 9.3 was 97.2 weeks, and the mean delay time accounted for 62.6 weeks. Households need at least several months to as much as 2 years before initiating the repair process to schedule inspection, contact engineers and contractors, and so forth. Additionally, since both delay time and repair time are directly related to initial building damage levels, buildings damaged at a higher level will, on average, take more time to achieve full recovery (see Figures 9.3 and 9.4). It is worth noting that the ability to utilize Monte Carlo simulation and produce statistical distributions for each building within a community enables full uncertainty propagation into the broader community models.

Figure 9.5 shows the building recovery when different policy cases (combinations) are implemented community wide. The times shown in Figure 9.5 are the mean recovery time, mean delay time, and mean repair time of all the in-path residential buildings in a single realization. All the other realizations within the Monte Carlo Simulation (MCS) have similar results for these parameters. Since Case 2 and Case 6 considered the policy of building retrofits, the repair time in these cases was shortened from 27 weeks to approximately 14 weeks, on average. Note that building retrofits were assumed to be implemented to all buildings right now, which is not realistic for a community. Additionally, all the policy cases decreased recovery time.

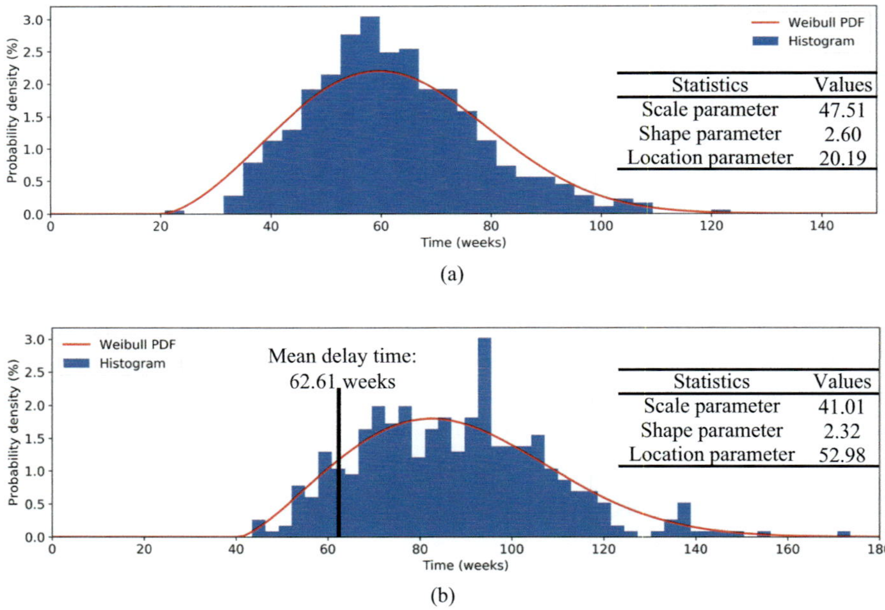

FIGURE 9.3 A typical residential building in the mean damage level of 3.998: (a) delay time results; (b) recovery time results.

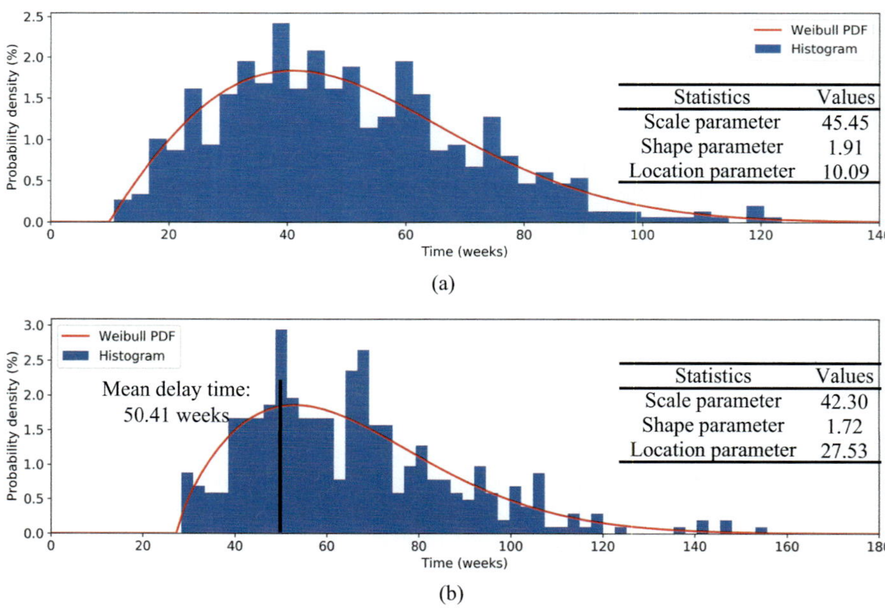

FIGURE 9.4 A typical residential building in the mean damage level of 2.744: (a) delay time results; (b) recovery time results.

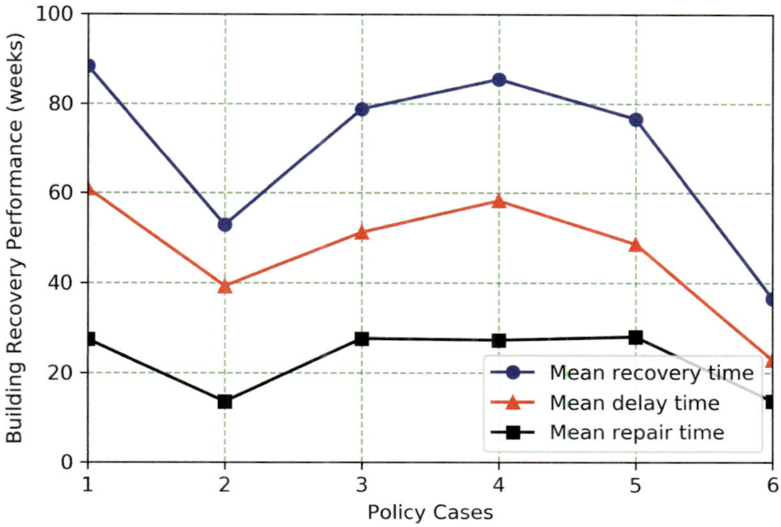

FIGURE 9.5 Building recovery performance impacted by different policy cases.

Overall, Case 6, incorporating all the proposed policies, enabled recovery 2.4 times faster (i.e., 88.4 weeks reduced to 36.6 weeks) than the base case of no new policies or mitigations/retrofitting. If only examining the impact of a single policy, mandatory retrofits requested by Case 2 facilitated the recovery by 1.7 times than the base case. Other cases related to the relaxation of regulations (e.g., bidding and inspection) and having financing in place are also shown slightly to speed up the recovery. However, the resulting decrease in the time resulting from delay variables did not offset the time required for engineers to review and redesign due to the damage. Thus, Case 3, Case 4, and Case 5 also shortened the recovery time, but the effects were not noticeable.

Quarterly building recovery at a community level under the six policy cases was calculated and compared. Figure 9.6 illustrates the building post-disaster recovery probability at a community level for Case 6 as an example. After 90 days following the tornado, almost all the residential buildings that were within the EF2 and higher wind regions still had a probability of less than 0.2 for completion of recovery, and only a few buildings in the EF1 region had a possibility of getting back to normal. For Case 6, which included all the proposed policies and mitigation strategies, the percentage of buildings 100% recovered was 51.5%, 85.4%, and 98.6% after 1.5 years, 3 years, and 4 years, respectively. However, if only Case 1 is considered, the percentage of buildings achieving recovery in these same time periods was only 0.3%, 29.47%, and 88.49%.

In addition to the community-level recovery presented in Figure 9.6, the probability of the mean building recovery times, defined as the average of the percentage of buildings fully recovered, is as shown in Eq. (9.8). Figure 9.7 illustrates the probability for the mean building recovery times for several of the cases explored in this chapter. The significant difference among cases was the probability at the short-term

FIGURE 9.6 Time-stepping building recovery performance probability under Case 6: (a) time = 1 quarter; (b) time = 2 quarters; (c) time = 6 quarters.

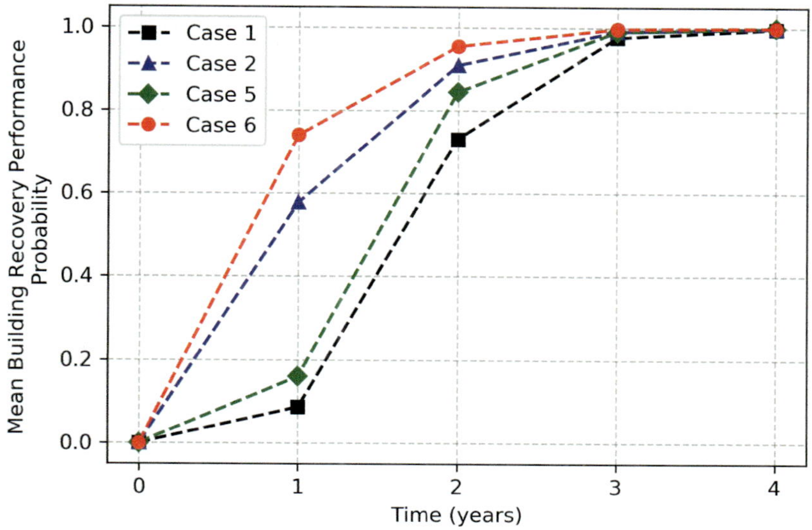

FIGURE 9.7 Time-stepping building recovery trajectory impacted by different policy cases.

and intermediate recovery stages. The probabilities of mean building recovery times were 0.09, 0.58, 0.16, and 0.74 in the first year after the tornado, and were increased to 0.73, 0.91, 0.84, and 0.95 in the second year under Case 1, Case 2, Case 5, and Case 6, respectively. It is worth mentioning that the restoration curves for Case 3 and Case 4 were close to that of Case 1 and were not included in the plot to provide clarity.

$$RP_{r\_mean}(t) = \frac{\sum_{j=1}^{N} RP_{rj}(t)}{N} \tag{9.8}$$

9.5 CONCLUDING REMARKS

Post-disaster recovery planning can help decision-makers mitigate risk and facilitate the recovery process for their communities and thus enhance community resilience. However, limited studies have performed only a qualitative analysis to evaluate the rapidity of community recovery impacted by dynamic policies. This chapter considered typical delay variables as well as repair fragilities for residential buildings for wind events. A time-stepping methodology to model residential building recovery was introduced and approximately validated based on existing longitudinal field investigation data. The recovery statistics were tracked for each building through the entire process from initial damage to full restoration. Based on the work described herein, the following conclusions can be reached:

1. Delay time dominated throughout the recovery modeling process. For a typical building located in the EF5 region of the simulated Joplin tornado, the mean delay time was 62.6 weeks, which accounted for 64.4% of the mean

recovery time totaling 97.2 weeks. This ratio of times may differ for other types of buildings.
2. All the proposed policy cases affected the delay process to different extents. Based on the typical results of Monte Carlo simulation in a single realization as an example, when considering the effect of building retrofits, the repair time was shortened from 27 weeks to 14 weeks, and the overall recovery process was expedited by 1.7 times. Case 6 involved all the proposed policies and mitigation strategies, enabled the recovery process to occur 2.4 times faster than the base case. Other cases related to the relaxation of regulations and financial support assessment were also shown slightly to speed up the recovery, but the results were not as significant.
3. The percentage of buildings in the study area achieving full recovery after 1.5 years, 3 years, and 4 years was 0.3%, 29.47%, and 88.49% according to Case 1, which was improved to 51.49%, 85.41%, and 98.61% with the implementation of Case 6.

In summary, short-term post-disaster activities, like the proposed policies, can effectively guide and facilitate the recovery process. The resilience-based analysis methodology presented in this chapter can support decision-making and community planning for stakeholders by being included in a broader community-level model for decision-making. The methodology, while demonstrated for tornado hazard, should be extensible to most hazards provided household income has an effect on delay time for that hazard.

ACKNOWLEDGMENTS

The authors gratefully acknowledge the support of this research by the Center for Risk-Based Community Resilience Planning funded through a cooperative agreement between the US National Institute of Standards and Technology and Colorado State University (NIST Financial Assistance Award Numbers: 70NANB15H044 and 70NANB20H008).

REFERENCES

Aghababaei, Mohammad, Maria Koliou, Maria Watson, and Yu Xiao. "Quantifying post-disaster business recovery through Bayesian methods." *Structure and Infrastructure Engineering* 17, no. 6 (2020a): 1–19.

Aghababaei, Mohammad, Maria Koliou, Stephanie Pilkington, Hussam Mahmoud, John W. van de Lindt, Andrew Curtis, Steve Smith, Jayakrishnan Ajayakumar, and Maria Watson. "Validation of Time-Dependent Repair Recovery of the Building Stock Following the 2011 Joplin Tornado." *Natural Hazards Review* 21, no. 4 (2020b): 04020038.

Almufti, Ibrahim, and Michael Willford. REDiTM Rating System: Resilience-Based Earthquake Design Initiative for the Next Generation of Buildings." Arup Co (2013).

Ameri, Mohammad R., and John W. van de Lindt. "Seismic performance and recovery modeling of natural gas networks at the community level using building demand." *Journal of Performance of Constructed Facilities* 33, no. 4 (2019): 04019043.

Amini, M. Omar, and John W. van de Lindt. "Quantitative insight into rational tornado design wind speeds for residential wood-frame structures using fragility approach." *Journal of Structural Engineering* 140, no. 7 (2014): 04014033.

Attary, Navid, John W. van de Lindt, Hussam Mahmoud, Steve Smith, Christopher M. Navarro, Yong Wook Kim, and Jong Sung Lee. "Hindcasting community-level building damage for the 2011 Joplin EF5 tornado." *Natural Hazards* 93, no. 3 (2018): 1295–1316.

Bonstrom, Holly, and Ross B. Corotis. "First-order reliability approach to quantify and improve building portfolio resilience." *Journal of Structural Engineering* 142, no. 8 (2016): C4014001.

Bruneau, Michel, Stephanie E. Chang, Ronald T. Eguchi, George C. Lee, Thomas D. O'Rourke, Andrei M. Reinhorn, Masanobu Shinozuka, Kathleen Tierney, William A. Wallace, and Detlof Von Winterfeldt. "A framework to quantitatively assess and enhance the seismic resilience of communities." *Earthquake Spectra* 19, no. 4 (2003): 733–752.

Burton, Henry, Hua Kang, Scott Miles, Ali Nejat, and Zhengxiang Yi. "A framework and case study for integrating household decision-making into post-earthquake recovery models." *International Journal of Disaster Risk Reduction* 37 (2019): 101167.

Carvalho, Leandro S., Stephan Meier, and Stephanie W. Wang. "Poverty and economic decision-making: Evidence from changes in financial resources at payday." *American Economic Review* 106, no. 2 (2016): 260–284.

Changnon, Stanley A. "Tornado losses in the United States." *Natural Hazards Review* 10, no. 4 (2009): 145–150.

Crawford, P. Shane, Alexander Hainen, Sebastain Awondo, Andrew Graettinger, and Lawrence Powell. "Development and Testing of a Resilience Rapidity Model for Disasters." *Journal of Architectural Engineering* 25, no. 2 (2019): 04019013.

Croll, Maxime. "Home Insurance Facts and Statistics: Coverage & Claims." ValuePenguin, March 15, 2021. https://www.valuepenguin.com/home-insurance-statistics

De Iuliis, Melissa, Omar Kammouh, Gian Paolo Cimellaro, and Solomon Tesfamariam. "Downtime estimation of building structures using fuzzy logic." *International Journal of Disaster Risk Reduction* 34 (2019): 196–208.

Drennan, Lex, Jim McGowan, and Anne Tiernan. "Integrating recovery within a resilience framework: Empirical insights and policy implications from regional Australia." *Politics and Governance* 4, no. 4 (2016): 74–86.

Federal Emergency Management Agency (FEMA). *Seismic Performance Assessment of Buildings (Volume 1: Methodology)*. FEMA, *Washington, DC,* 2012.

FEMA. (2011). "National Disaster Recovery Framework." 1–116.

González, Andrés D., Leonardo Dueñas-Osorio, Mauricio Sánchez-Silva, and Andrés L. Medaglia. "The interdependent network design problem for optimal infrastructure system restoration." *Computer-Aided Civil and Infrastructure Engineering* 31, no. 5 (2016): 334–350.

Hamideh, Sara, Walter G. Peacock, and Shannon Van Zandt. "Housing recovery after disasters: Primary versus seasonal/vacation housing markets in coastal communities." *Natural Hazards Review* 19, no. 2 (2018): 04018003.

Hassan, Emad M., and Hussam Mahmoud. "An integrated socio-technical approach for post-earthquake recovery of interdependent healthcare system." *Reliability Engineering & System Safety* 201 (2020): 106953.

He, Xian, and Eun Jeong Cha. "Modeling the damage and recovery of interdependent critical infrastructure systems from natural hazards." *Reliability Engineering & System Safety* 177 (2018): 162–175.

Hirayama, Yosuke. "Collapse and reconstruction: Housing recovery policy in Kobe after the Hanshin Great Earthquake." *Housing Studies* 15, no. 1 (2000): 111–128.

Ingram, Jane C., Guillermo Franco, Cristina Rumbaitis-del Rio, and Bjian Khazai. "Post-disaster recovery dilemmas: Challenges in balancing short-term and long-term needs for vulnerability reduction." *Environmental Science & Policy* 9, no. 7–8 (2006): 607–613.

Joshi, Abhay, and Misa Aoki. "The role of social capital and public policy in disaster recovery: A case study of Tamil Nadu State, India." *International Journal of Disaster Risk Reduction* 7 (2014): 100–108.

Kennedynews@joplinglobe.com, Wally Kennedy. "Insurance Payout from Tornado to Be Largest in State History." *Joplin Globe*, July 26, 2011. www.joplinglobe.com/news/local_news/insurance-payout-from-tornado-to-be-largest-in-state-history/article_3556cdc9-5f09-53f9-875f-9f65bbe76d88.html

Kikitsu, Hitomitsu, and Partha P. Sarkar. "Building damage, wind speed estimation, and post disaster recovery in an EF5 Tornado." *Natural Hazards Review* 16, no. 2 (2015): 04014019.

Koliou, Maria, and John W. van de Lindt. "Development of building restoration functions for use in community recovery planning to tornadoes." *Natural Hazards Review* 21, no. 2 (2020): 04020004.

Koliou, Maria, John W. van de Lindt, Therese P. McAllister, Bruce R. Ellingwood, Maria Dillard, and Harvey Cutler. "State of the research in community resilience: Progress and challenges." *Sustainable and Resilient Infrastructure*, 5, no. 3 (2017): 131–151.

Krause, Eleanor, and Richard V. Reeves. "Hurricanes Hit the Poor the Hardest." *Social Mobility Memos (Blog)*, September 18, 2017.

Kuligowski, Erica D., Franklin T. Lombardo, Long T. Phan, Marc L. Levitan, and David P. Jorgensen. "Final Report, National Institute of Standards and Technology (NIST) Technical Investigation of the May 22, 2011, Tornado in Joplin, Missouri." (2014).

Landry, Craig E., Sarah Anderson, Elena Krasovskaia, and Dylan Turner. (2020). "Willingness to Pay for Multi-Peril Hazard Insurance." Available at SSRN 3662668.

Lin, Peihui, and Naiyu Wang. "Stochastic post-disaster functionality recovery of community building portfolios I: Modeling." *Structural Safety* 69 (2017a): 96–105.

Lin, Peihui, and Naiyu Wang. "Stochastic post-disaster functionality recovery of community building portfolios II: Application." *Structural Safety* 69 (2017b): 106–117.

Lin, Peihui, Nathanael P. Rosenheim, Naiyu Wang, and Walter Gillis Peacock. "A Stochastic Approach to Model Household Re-occupancy in a Community Following a Natural Hazard." (2019).

Lindell, Michael K., and Carla S. Prater. "Assessing community impacts of natural disasters." *Natural Hazards Review* 4, no. 4 (2003): 176–185.

Masoomi, Hassan, and John W. van de Lindt. "Fatality and injury prediction model for tornadoes." *Natural Hazards Review* 19, no. 3 (2018a): 04018009.

Masoomi, Hassan, and John W. van de Lindt. "Restoration and functionality assessment of a community subjected to tornado hazard." *Structure and Infrastructure Engineering* 14, no. 3 (2018b): 275–291.

Masoomi, Hassan, Mohammad R. Ameri, and John W. van de Lindt. "Wind performance enhancement strategies for residential wood-frame buildings." *Journal of Performance of Constructed Facilities* 32, no. 3 (2018): 04018024.

NIST. (2015). "Community Resilience Planning Guide for Buildings and Infrastructure Systems." *NIST Special Publication 1190*, National Institute of Standards and Technology, *Gaithersburg, MD*.

NOAA (National Oceanic and Atmospheric Administration). (2020). "Weather Related Fatality and Injury Statistics." Accessed August, 2020. https://www.weather.gov/hazstat/

Nozhati, Saeed, Yugandhar Sarkale, Edwin KP Chong, and Bruce R. Ellingwood. "Optimal stochastic dynamic scheduling for managing community recovery from natural hazards." *Reliability Engineering & System Safety* 193 (2020): 106627.

Ouyang, Min, and Leonardo Duenas-Osorio. "Multi-dimensional hurricane resilience assessment of electric power systems." *Structural Safety* 48 (2014): 15–24.

Peacock, Walter Gillis, Betty Hearn Morrow, and Hugh Gladwin, eds. *Hurricane Andrew: Ethnicity, Gender, and the Sociology of Disasters*. Psychology Press, 1997.

PPD-21 (2013). "Presidential Policy Directive/PPD-21." *The White House*, February 12, 2013. https://www.cisa.gov/sites/default/files/2023-01/ppd-21-critical-infrastructure-and-resilience-508_0.pdf

PPD-8 (2011). "Presidential Policy Directive, PPD-8 – National Preparedness." *The White House*, March 30, 2011. https://www.dhs.gov/presidential-policy-directive-8-national-preparedness

Prevatt, David O., David B. Roueche, John W. Van De Lindt, Shiling Pei, Thang Dao, William Coulbourne, Andrew J. Graettinger, Rakesh Gupta, and David Grau. "Building damage observations and EF classifications from the Tuscaloosa, AL, and Joplin, MO, Tornadoes." In *Structures Congress 2012*, Chicago, Illinois, pp. 999–1010, 2012.

Ramachandran, Varun, Suzanna K. Long, Tom Shoberg, Steven Corns, and Hector J. Carlo. "Framework for modeling urban restoration resilience time in the aftermath of an extreme event." *Natural Hazards Review* 16, no. 4 (2015): 04015005.

Richmond, Joseph Lee. "The Relationship between Policy and Disaster Economic Recovery: A Case Study of the 2011 Joplin, Missouri Tornado." PhD diss., Northeastern University, 2019.

Rosenheim, Nathanael. "Detailed Household and Housing Unit Characteristics: Alpha Release of Housing Unit Inventories." (2021). Version 3. DesignSafe-CI. https://doi.org/10.17603/ds2-jwf6-s535

Roueche, David B., and David O. Prevatt. "Residential damage patterns following the 2011 Tuscaloosa, AL and Joplin, MO tornadoes." *Journal of Disaster Research* 8, no. 6 (2013): 1061–1067.

Sapat, Alka, and Ann-Margaret Esnard, eds. *Coming Home after Disaster: Multiple Dimensions of Housing Recovery*. CRC Press, Boca Raton, FL, 2016.

Simmons, Kevin M., Daniel Sutter, and Roger Pielke. "Normalized tornado damage in the United States: 1950–2011." *Environmental Hazards* 12, no. 2 (2013): 132–147.

Smith, Andrew M., Márton Pósfai, Martin Rohden, Andrés D. González, Leonardo Dueñas-Osorio, and Raissa M. D'Souza. "Competitive percolation strategies for network recovery." *Scientific Reports* 9, no. 1 (2019): 1–12.

Smith, Daniel J., and Daniel Sutter. "Response and recovery after the Joplin tornado: Lessons applied and lessons learned." *The Independent Review* 18, no. 2 (2013): 165–188.

Standohar-Alfano, Christine D., and John W. van de Lindt. "Empirically based probabilistic tornado hazard analysis of the United States using 1973–2011 data." *Natural Hazards Review* 16, no. 1 (2015): 04014013.

Sun, Wenjuan, Paolo Bocchini, and Brian D. Davison. "Model for estimating the impact of interdependencies on system recovery." *Journal of Infrastructure Systems* 26, no. 3 (2020): 04020031.

Sutley, Elaina J., and Sara Hamideh. "An interdisciplinary system dynamics model for post-disaster housing recovery." *Sustainable and Resilient Infrastructure* 3, no. 3 (2018): 109–127.

Tomar, Agam, Henry V. Burton, Ali Mosleh, and Ji Yun Lee. "Hindcasting the Functional Loss and Restoration of the Napa Water System Following the 2014 Earthquake Using Discrete-Event Simulation." *Journal of Infrastructure Systems* 26, no. 4 (2020): 04020035.

Insurance Agency Plano (2014). "Tornados, Floods and without Home Insurance as Reported by AP." *Insurance Agency Plano TX*, November 25, 2014. www.tristarins.com/tornados-floods-and-without-home-insurance-as-reported-by-ap/

van de Lindt, J. W., W. G. Peacock, J. Mitrani-Reiser, N. Rosenheim, D. Deniz, M. Dillard, T. Tomiczek, M. Koliou, A. Graettinger, P. S. Crawford, K. Harrison. "Community Resilience-Focused Technical Investigation of the 2016 Lumberton, North Carolina Flood: An Interdisciplinary Approach." *ASCE Natural Hazards Review*, 21, no. 3 (2020). https://ascelibrary.org/doi/full/10.1061/%28ASCE%29NH.1527-6996.0000387.

Wang, Wanting, John W. van de Lindt, Nathanael Rosenheim, Harvey Cutler, Jong Sung Lee, and Maria Koliou. "Community resilience assessment of an EF-5 Tornado using the IN-CORE modeling environment." In *7th International Symposium on Life-Cycle Civil Engineering (IALCCE 2020)*, Shanghai, China, 2020.

Wang, Wanting, John W. van de Lindt, Nathanael Rosenheim, Harvey Cutler, Brad Hartman, Jong Sung Lee, and Diego Calderon. "Effect of residential building wind retrofits on social and economic community-level resilience metrics." *Journal of Infrastructure Systems* 27, no. 4 (2021): 04021034.

Watson, Maria, Yu Xiao, Jennifer Helgeson, and Maria Dillard. "Importance of Households in Business Disaster Recovery." *Natural Hazards Review* 21, no. 4 (2020): 05020008.

Zhang, Weili, Naiyu Wang, and Charles Nicholson. "Resilience-based post-disaster recovery strategies for road-bridge networks." *Structure and Infrastructure Engineering* 13, no. 11 (2017): 1404–1413.

Zhang, Yang, and Walter Gillis Peacock. "Planning for housing recovery? Lessons learned from Hurricane Andrew." *Journal of the American Planning Association* 76, no. 1 (2009): 5–24.

10 Fragility Assessment of Building Portfolios under Natural Hazards Using Machine Learning and Poisson Binomial Distribution

De-Cheng Feng[a], *Jia-Yi Ding*[a], *and Carmine Galasso*[b]

[a]School of Civil Engineering, Southeast University, Nanjing, China
[b]Department of Civil, Environmental & Geomatic Engineering, University College London, London, United Kingdom

10.1 INTRODUCTION

Recently, there has been a global trend of rapid urbanization and continuous population growth. This has resulted in increasingly dense urban populations and building concentrations, especially in certain developing countries that are exposed to a multitude of natural hazards (e.g., Ref. [1]). Specifically, the devastating effects of major natural hazard events have been observed on the built environment, which encompasses both building structures and infrastructure systems, as well as on socio-economic conditions (e.g., Ref. [2]). In this context, regional-scale risk and resilience assessment of building portfolios subjected to hazard-induced loading has been widely investigated (e.g., Refs. [3–7]), with fragility analysis being a crucial step for hazard performance assessment of structural components or systems. Generally, urban areas comprise numerous buildings with diverse geometries, materials, and structural characteristics (e.g., lateral-load resisting systems and details) [8]. Hence, different building types, or building classes, are typically characterized by a blend of certain attributes. These may include the systems and materials used for resisting lateral loads, the height or number of stories, the age of construction, or the

DOI: 10.1201/9781003567332-10

time of major renovations. Each building type is assigned specific fragility relationships, quantifying the conditional probability that a structure will reach or exceed a specified level of damage (or damage state (DS)) for a given value of a considered hazard-specific intensity measure (IM). These relationships are often established based on an 'average' representative (archetype) building of the class, also known as an 'index' building, and are applied to all buildings within that class. However, even though a building might be classified under a specific type (for instance, a high-rise with a reinforced concrete, moment-resisting frame), it could still exhibit variations in the number of bays and stories, architectural geometries, and structural details. These differences can lead to significant variability in the extent of damage following a natural hazard. While fragility models for individual or representative buildings are relatively straightforward, developing fragility models on a regional scale for portfolios of buildings can pose a challenge due to the need to account for variability from one building to another.

Fragility models for individual buildings can be established using several methods. One such approach involves simulating structural responses through mechanics-based numerical models. These simulations can incorporate a range of analyses, each differing in complexity and computational demands, including nonlinear time history analysis (NLTHA).

Nevertheless, modeling and performing NLTHA for a large number of buildings in a region involves significant computational costs, particularly when using refined finite element models. To reduce the computational cost of NLTHA, several numerical models of representative buildings (for each building type) can be simulated to establish hazard damage and fragility databases, as described above (e.g., Refs. [7,9,10]). However, this approach may inadequately consider building-to-building variability in the hazard fragility assessment of building portfolios, as discussed in Ref. [10]. Alternatively, simplified numerical models have been used to simulate the nonlinear dynamic response of individual buildings in regional investigations, such as the fish-bone models for steel moment frames [11], the multi-degree-of-freedom (MDoF) shear model for multistory reinforced concrete (RC) frames [12], the MDoF flexure-shear model for high-rise RC frame-shear walls [13], inelastic single-degree of freedom for realistic RC frames [14,15], and the reduced order model for hysteretic structures [16] and 3D RC frame buildings [17]. Both simplified MDoF shear and flexure-shear models have demonstrated relatively high accuracy and moderate computational efficiency in practical applications (e.g., Refs. [18–20]) and thus are employed in this study.

On the other hand, conventional fragility analysis methods based on the assumption of logarithmic linear probabilistic hazard demand models (e.g., Refs. [21,22]) face challenges in handling high-dimensional input variables (including multiple IMs to adequately characterize natural hazards (e.g., Refs. [23,24]), along with structural attribute parameters (e.g., Ref. [9]) for building portfolios subjected to natural hazard loads). A surrogate (or statistical) model (or metamodel) can be established to relate hazard IMs and structural attributes to engineering demand parameters (EDPs) for building's response estimation and subsequent fragility analysis. For this purpose, various machine learning (ML) approaches have been employed to fit predictive models based on relatively small sets of samples for approximating expensive numerical model simulations. Such studies addressed the hazard fragility analysis of buckling restrained braced frames using artificial neural networks [25], the hazard fragility assessment of RC frame buildings employing adaptive boosting and extreme gradient boosting [26], and the fragility of existing steel frames considering the effects of masonry fillers using feedforward neural networks [27], among others (e.g., Refs. [28,29]). Some scholars have also comprehensively compared various ML algorithms used for hazard fragility assessment of RC and steel buildings (e.g., Refs. [30,31]). Although these ML models offer the advantages of low cost and high accuracy in predicting responses and deriving fragility models, they only provide deterministic predictions of responses and the resulting fragility models are often applicable to individual buildings. Regional-scale hazard fragility analysis requires establishing ML models for each different building separately (i.e., employing very large training datasets).

A wealth of promising research has been conducted in the field of regional fragility assessment for buildings. For example, Gentile and Galasso [10] employed Gaussian process regression to establish a surrogate model for simulation-based hazard risk assessment of building portfolios. Bayesian neural networks were utilized to develop surrogate models and fragility analysis for each building in a region in Ref. [32]. Zhang et al. [7] performed 3D high-fidelity NLTHAs on 16,030 RC buildings in Istanbul, Turkey, considering soil–structure interaction effects to evaluate regional-scale hazard fragility, loss, and resilience. While these studies offer valuable insights into the hazard damage sustained by individual buildings within a region, they fall short of providing straightforward answers to questions such as the number of unsafe buildings or the count of household displaced post-disaster. These are critical pieces of information that various stakeholders and decision-makers might be particularly interested in (e.g., Ref. [33]). Therefore, from a regional standpoint, the focus should be on the hazard fragility assessment of the entire building portfolio after obtaining the fragility model for each building in the region. Specifically, this study investigates a regional-scale hazard fragility model to describe the functionality failure probability of building portfolios under different hazard intensity levels. The ratio of the number of unsafe buildings to the total number of buildings within the building portfolio represents a given portfolio's functionality failure probability (e.g., Refs. [3,33,34]). Building portfolio fragility models, as proposed in this study, can support scalable community resilience assessment [3].

10.2 RESEARCH SCOPE AND OBJECTIVES

10.2.1 Considering Independent Functionality for Regional Buildings

Community-wide infrastructure systems, including transportation, water, and power networks, may display substantial interdependencies. These are primarily physical, where the functioning of one infrastructure depends on the services rendered by others, and geographical, where infrastructure failures may be interconnected due to their spatial proximity, especially in the aftermath of a hazardous event (e.g., Refs. [35,36]). This study primarily concentrates on evaluating the functionality of a building (for instance, determining its safety for occupancy) based on damages induced by natural hazards, whether structural or nonstructural. However, it does not consider the loss of functionality in undamaged buildings due to disruptions in essential utilities like water and power networks (e.g., Ref. [37]). As such, the functionality of buildings is considered independent of other buildings and systems. The investigation of the interdependence of operations will be addressed in future research.

10.2.2 Strategies for Regional Fragility Analysis of Buildings

In the realm of the probabilistic performance assessment of buildings under natural hazards, the goal is to derive a fragility model tailored to a particular structure. This process encompasses the (nonlinear) computational modeling of the considered structure, performing NLTHAs for various hazards, and computing the damage probabilities as a function of various IM levels. However, when it comes to regional buildings, this approach becomes unfeasible because of the necessity to create numerical models for each building in the region and execute hundreds of NLTHAs for each building to develop building-specific fragility models, as illustrated in Figure 10.1a. Even with ML-based surrogate models that approximate NLTHAs, this strategy still demands costly computational resources due to establishing a surrogate model for each building. An alternative strategy involves identifying specific archetype or

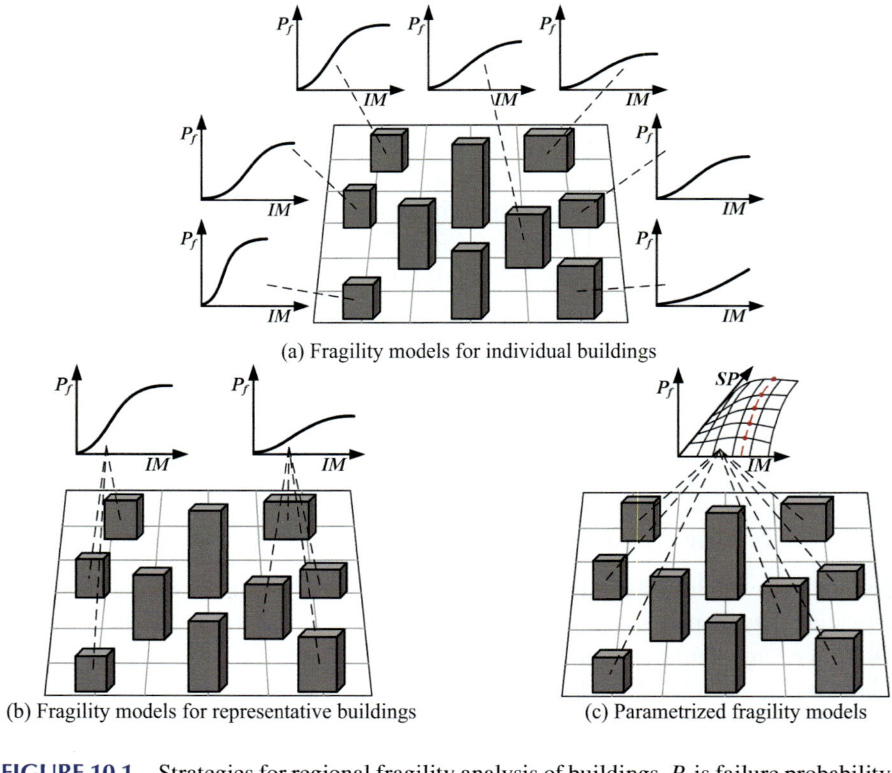

FIGURE 10.1 Strategies for regional fragility analysis of buildings. P_f is failure probability, and IM is the natural hazard intensity measure.

'index' buildings to represent various building types. This is performed based on a limited set of attributes, including the building material, the lateral resisting system, the height, and the code level (determined by the age of construction). As a result, it becomes sufficient to only develop fragility models for these representative buildings within a region, as depicted in Figure 10.1b. Identifying representative buildings involves a complex trade-off, balancing the consideration of the building-to-building variability versus reducing the number of archetype buildings to consider.

Figure 10.1c illustrates the concept of parametrized fragility models, as adopted in this study, indicating the damage probability of a building conditioned on natural hazard IMs and structural design parameters (e.g., building types, age of construction, number of stories, story heights, number of bays, and bay lengths). The methodology includes the following steps: (1) generation of random samples to formulate the building models, drawing upon available statistics related to the structural design parameters of the building portfolio under consideration; (2) selection of a variety of natural hazard scenarios that accurately represent the site conditions of interest and their random pairing with the building models; (3) development of a regional-scale (or parametrized) fragility model, leveraging the hazard damage database of the considered buildings. For step (3), a probabilistic ML method is further extended to the parametrized fragility models of building portfolios. The probabilistic ML method has been applied by the authors to the fragility model of a specific building (input features include concrete compressive strength, rebar yield strength, and rebar diameter), as described in detail in Ref. [38]. Such a probabilistic ML model can directly predict the probability distribution of structural responses, for example, their mean and standard deviation, considering the aleatory uncertainties associated with the input variables (i.e., natural hazard IMs and structural design parameters). This method eliminates the need to train individual surrogate models with ground motion IMs as input variables for each building. As a result, this approach could offer a straightforward and efficient way of assessing the hazard fragility of all buildings within a region.

Upon review, it is clear that these three strategies for regional fragility analysis of buildings primarily provide decision-makers with the probability of damage to each building under a specific hazard scenario. However, decision-makers may have a more pressing need for quantitative data on the performance of building portfolios under natural hazards. This could include, for instance, the number or percentage of unsafe residential buildings in the study area or the probability of half the school buildings losing functionality. Moreover, just as the safety of each building in the area is an uncertain event with a specific probability (e.g., damage probability), the number of unsafe buildings with uncertainty cannot be directly counted. Although simulation-based approaches (e.g., Monte Carlo simulation) can calculate the expected value and variance of the number of unsafe buildings based on thousands of realizations (each with the DSs of each building may different), this may require high computational costs. Therefore, this study aims to derive the proportion of unsafe buildings (which can serve as the functionality failure probability for building portfolios, e.g., Refs. [3,33,34]) in the probability space from the damage probability for each building in a region. As a result, the regional-scale hazard fragility models for building portfolios can provide valuable insights to decision-makers for managing risk at the community level and formulating disaster mitigation and resilience-enhancing plans.

10.2.3 Research Gaps and Contributions

Addressing these critical research gaps, this chapter employs the probabilistic ML method and the Poisson binomial distribution to delve into the regional-scale hazard fragility assessment of building portfolios. Simplified MDoF models are utilized to simulate the dynamic (nonlinear) response of various buildings (e.g., RC frame and RC frame-shear walls), thereby generating a regional hazard damage database. Natural hazard IMs and structural design parameters act as input variables, whereas EDPs representing damage to structural or nonstructural components serve as output variables. Subsequently, the probabilistic ML models (i.e., NGBoost in Ref. [39]), capable of providing the probability distribution of building responses, are trained and tested to calculate the damage exceedance probability of each building in the region under a given natural hazard IM. Lastly, the Poisson binomial distribution is introduced to derive the functionality failure probability and fragility model of building portfolios based on the damage probabilities of individual buildings. The proposed methodology offers the following features:

1. Numerical models of buildings, which account for building-to-building variability, can be efficiently and swiftly constructed using simplified MDoF models based on structural design parameters.
2. Probabilistic ML models can quantify the inherent randomness in the building responses and can directly provide the probability distribution parameters required for parametrized fragility models.
3. The regional-scale fragility model of building portfolios, considering independent functionality, is derived from the damage probability of each building based on the Poisson binomial distribution.

10.3 NUMERICAL MODELS OF BUILDING PORTFOLIOS

This study considers the widely prevalent RC frame and RC frame-shear wall buildings in urban areas using the simplified MDoF shear and the MDoF flexure-shear models, which are proposed in Refs. [12,13], respectively. Given the limited availability of building information at the regional scale, basic structural details such as building types, age of construction, number of stories, story heights, number of bays, and bay lengths are considered for modeling the target buildings. The simplified MDoF models also apply to masonry walls, steel moment frames, and wood frames in regional buildings (e.g., Ref. [18]).

10.3.1 Modeling of Multistory RC Frame

When modeling a multistory RC frame using the simplified MDoF shear model considered here, the following convenient assumptions are made: the floors are rigid with no rotational displacements, the building primarily experiences shear deformations, the mass of each floor is concentrated on the floor plan, and there is a regular floor layout. Therefore, each floor can be simplified as a mass point, with adjacent mass points connected by shear springs, as depicted in Figure 10.2a.

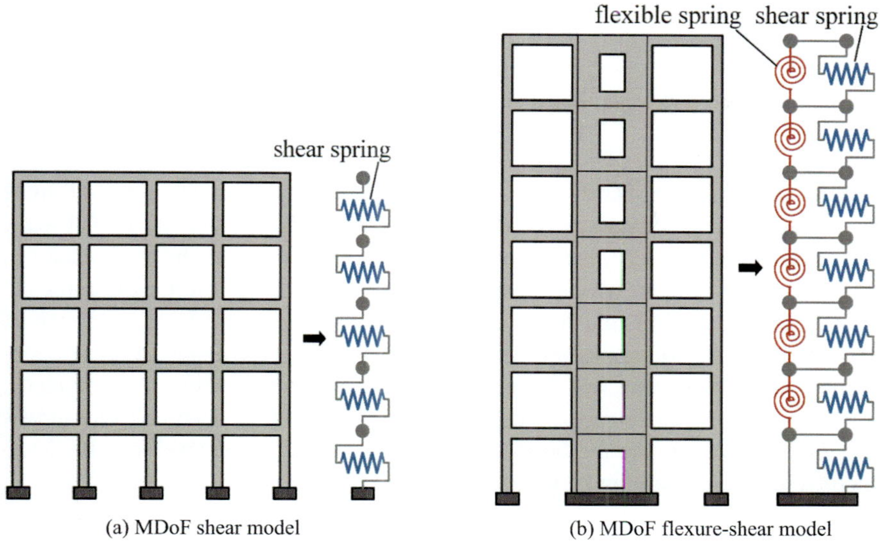

FIGURE 10.2 Simplified numerical model of buildings.

The trilinear backbone curve recommended by HAZUS (Hazards U.S.) [40] and the single-parameter pinching hysteresis model proposed by Steelman and Hajjar [41] are adopted to characterize the shear springs.

The trilinear backbone curve of each shear spring, as shown in Figure 10.3, can be identified through initial lateral stiffness k_0, shear yield strength F_y, hardening coefficient η, shear peak strength F_p, and ultimate inter-story drift Δ_u. For a target building with known basic structural information, the initial inter-story stiffness of each shear spring can be calculated as follows:

$$k_0 = 4\pi^2 \lambda \frac{m}{T_1^2} \tag{10.1}$$

where λ, related to the type and number of stories of the building, can be found in Table 5-5 of HAZUS [40]; m is the mass of the floor, including the slab, beams, and half of the columns; and T_1 is the fundamental period of the target building, which can be calculated as $T_1 = 0.25 + 0.00053H^2 / \sqrt[3]{W}$ for RC frames [42], where H and W are the height and width (i.e., the product of number of bays and bay lengths) of the building. The parameters on the backbone curve of the shear spring corresponding to the i-th floor can be determined as follows:

$$F_y = A_y \cdot \alpha \cdot m \cdot g \cdot N_S \cdot \rho, \ \rho = 1 - \frac{i(i-1)}{(N+1)N} \tag{10.2}$$

$$\eta = \frac{\left(A_u - A_y\right) D_y}{\left(D_u - D_y\right) A_y} \tag{10.3}$$

$$F_p = \frac{A_u}{A_y} F_y \tag{10.4}$$

$$\Delta_c = \delta_c h_i \tag{10.5}$$

where N_S is the number of stories; (A_y, D_y), (A_u, D_u) are the yield capacity and ultimate capacity points, respectively, of the capacity curve listed in Tables 5-7 to 5-10 (corresponding to the four seismic design levels) of HAZUS [40]; α is the mode factor in Table 5-5 [40]; g is the gravity acceleration; and δ_c is the drift ratio for the complete DS [40]. Note that the seismic design level (i.e., high-code, moderate-code, low-code, and precode) can be found in Table 5-35 of HAZUS [40], according to the age of construction and seismic zone. For this pinching hysteresis model, only one parameter τ, describing the degree of structural degradation, is needed to determine its hysteresis behavior. The τ, related to the building type and seismic design level, can be determined from the degradation factor in Table 5-33 of HAZUS [40].

10.3.2 Modeling of High-Rise RC Frame-Shear Walls

In the MDoF flexure-shear model for RC frame-shear wall buildings considered in this study, a flexural spring simulating the flexural deformation of the shear wall and a shear spring simulating the shear deformation of the frame are both utilized to connect two adjacent stories, as depicted in Figure 10.2b. The trilinear backbone curve and single-parameter pinching hysteresis model in Figure 10.3 are also employed to characterize the flexible spring. The T_1 and the second period T_2 can be determined following the guidelines provided by ASCE 2016 [43] and Taghavi and Miranda [44]:

$$T_1 = C_t H^x \tag{10.6}$$

$$T_2 = 0.27T_1 \tag{10.7}$$

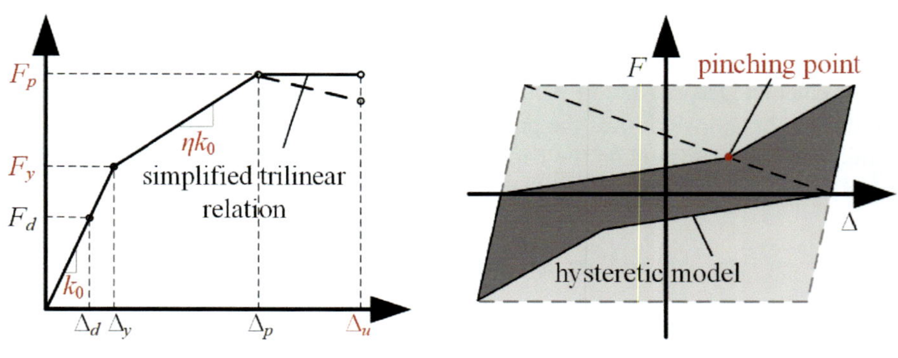

FIGURE 10.3 The trilinear backbone curves and single-parameter pinching model of the simplified MDoF model.

Then, the shear stiffness for the shear spring and the flexural stiffness for the flexural spring can be determined using floor mass m, T_1, and T_2, according to Xiong et al. [13]. Subsequently, the design strength values F_d are computed using elastic design spectrum and lateral seismic force pattern [18,45]. Therefore, yield strength F_y, peak strength F_p, yield deformations Δ_y, and peak deformation Δ_p are calculated as follows:

$$F_y = \beta_y F_d \tag{10.8}$$

$$\Delta_y = \frac{F_y}{k_0} \tag{10.9}$$

$$F_p = \beta_p F_y \tag{10.10}$$

$$\Delta_p = \frac{F_p}{\eta k_0} \tag{10.11}$$

where β_y and β_p represent the yield and peak overstrength ratio of a building, respectively, and can be determined using the recommendations in Table 5-5 of HAZUS [40]; k_0 is shear stiffness for a shear spring or flexural stiffness for a flexible spring; and η denotes a stiffness reduction factor provided in ACI318-11 [46].

Following these specifications, all parameters of the MDoF shear model and MDoF shear model can be determined based on the building's basic information. Note that the simplified MDoF models have been extensively utilized in numerical simulations of regional buildings, and the detailed validation is outside the scope of this study but can be found in Refs. [18,31,32].

10.4 FRAGILITY ASSESSMENT OF BUILDING PORTFOLIOS

In the context of this study, the ratio of the number of unsafe buildings z to the total number of buildings N is used to represent the functionality failure probability in hazard fragility assessment of building portfolios (e.g., Refs. [3,33,34]). The safety of a building can be determined by assessing whether its structural response exceeds a specific DS threshold [3]. Therefore, the probability that the number of unsafe buildings z in a region, conditioned on natural hazard IM, can be expressed as:

$$P(z \mid x_{\text{IM}}) = \int_z \int_{y_{\text{DS}}} f_{Z|\text{DS}}(z \mid y_{\text{DS}}) f_{\text{DS|IM}}(y \mid x_{\text{IM}})\, dy\, dz \tag{10.12}$$

where $f_{\text{DS|IM}}$ $(y|x_{\text{IM}})$ represents the probability density function (PDF) of structural response y of a building conditioned on a given natural hazard IM x_{IM}, which is detailed in Section 10.4.1; and $f_{Z|\text{DS}}$ $(z|y_{\text{DS}})$ is the PDF of the number of unsafe buildings z conditioned on a specified DS y_{DS}, which is described in Section 10.4.2. It is important to recognize that although the regional-scale fragility model discussed here focuses on building portfolios impacted by hazard-induced loading, the underlying fragility assessment method has the potential to be applied to other types of infrastructure (such as bridges and tunnels) exposed to various natural hazards (such as floods and strong winds).

10.4.1 Exceedance Probability of Individual Buildings Conditioned on IM

The fragility of an individual building is defined as the conditional probability that its structural response reaches or exceeds thresholds of specific (discrete) DSs for a given natural hazard IM value. According to HAZUS [40], among other risk calculation frameworks, the DSs of buildings under hazard-induced loading can be classified as slight, moderate, extensive, and complete; corresponding thresholds in terms of EDPs (e.g., the maximum inter-story drift ratio, MIDR, and peak floor acceleration, PFA) for each DS can also be found in HAZUS [40].

Then, once the conditional distribution parameters of the structural response (e.g., mean μ_{lny} and standard deviation σ_{lny}) for each IM are obtained, the exceedance probability of individual building can be calculated as

$$P_B\left(y > y_{\text{DS}} \mid x_{\text{IM}}\right) = \int_{y_{\text{DS}}} f_{\text{DS}|\text{IM}}\left(y \mid x_{\text{IM}}\right) dy = 1 - \Phi\left(\frac{\ln\left(y_{\text{DS}}\right) - \mu_{\ln y}}{\sigma_{\ln y}}\right) \tag{10.13}$$

where y_{DS} is the threshold corresponding to different damage levels; ln(·) represents the natural logarithm; and $\Phi(\cdot)$ denotes the standard normal cumulative distribution function (CDF). In this study, μ_{lny} and σ_{lny} of structural responses can be provided by the NGBoost model; the detailed steps can be found in Ref. [38].

10.4.2 Probabilistic Model of Building Portfolios Conditioned on IM

The functionality failure probability of building portfolios, which is the proportion of unsafe buildings [3,34,47] as introduced above, can be derived based on the damage exceedance probabilities of all buildings in the area. Firstly, following the approach of Lin and Wang [3] (but other decisional rules can be similarly used), a building without damage (DS_0) or in slight (DS_1) or moderate (DS_2) damage is considered safe to occupy immediately following a hazard scenario; a building becomes unsafe to occupy (UO) when it exhibits extensive (DS_3) or complete (DS_4) damage. The probability that a building is UO can be obtained from Eq. (10.13):

$$\begin{aligned} P_{B,\text{UO}} &= P_B\left(y > y_{\text{DS}_3} \mid x_{\text{IM}} \cap y > y_{\text{DS}_4} \mid x_{\text{IM}}\right) \\ &= P_B\left(y > y_{\text{DS}_3} \mid x_{\text{IM}}\right) = 1 - \Phi\left(\frac{\ln\left(y_{\text{DS}_3}\right) - \mu_{\ln y}}{\sigma_{\ln y}}\right) \end{aligned} \tag{10.14}$$

Then, the ratio of the number of buildings that are UO z to the total number of buildings N can be expressed as

$$P(z \mid x_{\text{IM}}) = \frac{z}{N} = \frac{\sum_{i=1}^{N} \text{IO}_i}{N} \tag{10.15}$$

$$\text{IO}_i = \begin{cases} 0, & \text{safe to occupy with } P = 1 - P_{Bi,\text{UO}} \\ 1, & \text{unsafe to occupy with } P = P_{Bi,\text{UO}} \end{cases} \tag{10.16}$$

Due to the independent probability events of whether each building in a region can be safely occupied, the number or proportion of unsafe buildings in a region is uncertain and cannot be directly counted.

Now, let us consider N independent trials where each building is either safe or UO, with a given probability of success (i.e., the probability that a building is UO). The number of successes z in these N independent yes/no experiments follows a Poisson binomial distribution – a discrete probability distribution that arises from a sum of independent Bernoulli trials. Unlike the ordinary binomial distribution, where the success probabilities are identical, the Poisson binomial distribution allows for varying success probabilities.

The probability of having z successful trials out of a total of N can be expressed as:

$$P(z) = \sum_{A \in Fz} \prod_{Bi \in A} P_{Bi} \prod_{Bi \in A^c} (1 - P_{Bi}) \tag{10.17}$$

where z denotes the number of unsafe buildings; P_{Bi} is the probability that the i-th building is deemed UO; F_z is defined as the set of all subsets of size z; and A^c represents the complement of set A. As a variable following the Poisson binomial distribution is a sum of N independent Bernoulli distributed variables, its mean and variance will merely be sums of the mean and variance of the N Bernoulli distributions:

$$\mu = \sum_{i=1}^{N} P_{Bi} \tag{10.18}$$

$$\sigma^2 = \sum_{i=1}^{N} P_{Bi}(1 - P_{Bi}) \tag{10.19}$$

Combining Eqs. (10.15), (10.16), and (10.17), the PDF and CDF of the proportion of unsafe buildings ($P(z|x_{\text{IM}}) = z/N$) are given by:

$$P(z = Z \mid x_{\text{IM}}) = \sum_{A \in F_z} \prod_{Bi \in A} P_{Bi} \prod_{Bi \in A^c} (1 - P_{Bi}) \tag{10.20}$$

$$P(Z < z \mid x_{\text{IM}}) = \sum_{l=0}^{z} \sum_{A \in F_z} \prod_{Bi \in A} P_{Bi} \prod_{Bi \in A^c} (1 - P_{Bi}) \tag{10.21}$$

Recursive formulae are available to compute the exact CDF values of the Poisson binomial distribution. However, F_z in Eq. (10.21) will contain $N!/((N-Z)!Z!)$ elements, and the summation over these elements is infeasible to compute due to the large total number of buildings N in a region. An efficient algorithm derived from the discrete Fourier transform of characteristic functions can be employed to evaluate the distribution function [48].

10.4.3 Fragility Assessment of Building Portfolios

According to Eqs. (10.15), (10.16), and (10.18), the proportion of unsafe buildings within a region under a given hazard scenario is expected to be:

$$P(z = \mu \mid x_{\mathrm{IM}}) = \mu / N = \sum_{i=1}^{N} P_{Bi} / N \tag{10.22}$$

where, μ and N are the mean number of unsafe buildings and the total number of buildings in the region, respectively; and P_{Bi} is the probability of the i-th building being unsafe. Subsequently, different natural hazard IMs can be considered to obtain the regional-scale fragility model of a building portfolio. In addition, the probability of the proportion of unsafe buildings in a region under a given natural hazard IM not exceeding a certain threshold (e.g., half of buildings being unsafe) can be evaluated based on the CDF of the building portfolio functionality failure in Eq. (10.21).

10.5 ILLUSTRATIVE APPLICATION

10.5.1 Building Inventory for a Virtual City

This section considers a virtual city in China to demonstrate the proposed fragility assessment of building portfolios employing the NGBoost algorithm and Poisson binomial distribution. The city consists of 1,000 buildings with various structural design parameters, including building types, age of construction, number of stories, story heights, number of bays, and bay lengths. Table 10.1 lists the possible values of these design parameters as per GB 50010-2010 [45] or based on expert judgment. Specifically, GB 50010-2010 classifies buildings into different seismic design categories based on their intended use, location, and importance. Each category corresponds to a specific fortification intensity. The specific site conditions (such as soil type) and the classification of the design earthquake also play a role. Latin hypercube sampling generates random samples from Table 10.1, facilitating the construction of 600 RC frame buildings and 400 RC frame-shear wall buildings in this city. It should be noted that these buildings need to satisfy the design requirements of GB 50010-2010 [45] and GB 50009-2012 [42] in terms of axial compression ratio, lateral resistance moment, and aspect ratio. Additionally, RC frames are mostly low to mid-rise buildings (i.e., with seven floors or less), whereas RC frame-shear walls are mostly high-rise buildings. For cities in China, post-2002

TABLE 10.1
The Possible Values of Key Design Parameters.

Parameters	Possible Values	References
Building types	RC frames and RC frame-shear walls	\
Age of construction	Pre-1989, 1990–2001, and Post-2002	GB 50010-2010 [45]
Number of stories	1–16	JGJ 3-2010 [45]
Story height (m)	3.6, 4.2	Cao et al. [49]
Number of bays in *x* direction[a]	1–6	\
Number of bays in *y* direction[a]	1–6	\
Bay length in *x* direction (m)[b]	6, 6.5, 7, 7.5, 8	GB 50010-2010 [45] and ACI 318 [46]
Bay length in *y* direction (m)[b]	6, 6.5, 7, 7.5, 8	GB 50010-2010 [45] and ACI 318 [46]

[a] The number of bays of a specific building needs to meet the design requirements of Ref. [42].

[b] The bay lengths are determined according to the design guidelines for medium span RC beams in Ref. [45, 46].

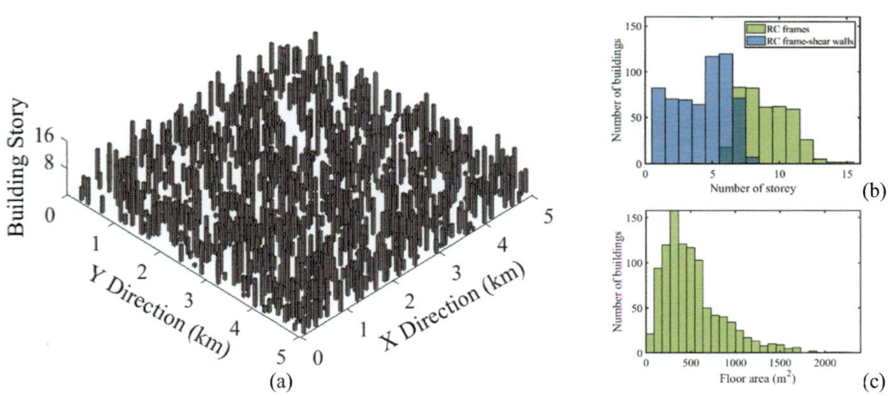

FIGURE 10.4 Building information in the virtual city.

buildings are the most prevalent, followed by those built between 1990 and 2001 and pre-1989.

The locations of 1,000 buildings are randomly generated over a rectangular region with uniform soil conditions spanning 5 km by 5 km, as depicted in Figure 10.4a. In addition, Figure 10.4b and c show the number of stories and floor area (calculated by the number of bays and bay length) of these buildings. Most RC frame buildings have less than seven stories, whereas RC frame-shear wall buildings have between 7 and 12 stories. This virtual city complies with a fortification level of seven degrees (i.e., the peak ground acceleration (PGA) corresponding to a 10% exceeding probability in 50 years is 0.15 g according to GB 50010-2010), a site condition of class II (or moderate soil conditions, i.e., the shear wave velocity between 250 and 500 m/s), and a first-group design earthquake (related to the magnitude and epicenter distance of recorded strong earthquakes).

10.5.2 Ground Motion Records and Intensity Measures

A total of 100 ground motion records (GMRs) is selected from the Pacific Earthquake Engineering Research Center NGA-West2 databases [50] for structural response analysis. Figure 10.5b shows that the mean response spectrum of the considered GMRs matches well the site-specific target response spectrum for a long range of periods, suggesting that the selected GMRs are suitable for the site-specific hazard conditions of the virtual city [51]. Additionally, Figure 10.5a shows the distributions of magnitude *M* and rupture distance *R* corresponding to these 100 earthquake events. These events include both non-pulse and pulse-like ground motions to consider a wide range of potential seismic events in probabilistic assessments of building portfolios. Among the 80 non-pulse ground motions, the average shear velocity V_{s30} ranged from 180 to 360 m/s for 40 ground motions recorded on soil sites, and it exceeded 360 m/s for 40 considered ground motions. For the 20 pulse-like ground motions, the V_{s30} ranged from 200 to 660 m/s, and the pulse periods ranged from 1.0 to 10 seconds with a mean of 3.7 seconds. The specific treatment for pulse effects in probability assessment of building portfolios can refer to the approach from Gentile and Galasso [52].

Engineering attention is directed toward the inelastic behavior exhibited by structures under strong earthquakes, whereas the original GMRs cover a lower range of intensities. Therefore, the 100 selected GMRs are scaled by factors of 0.5, 1.0, 1.5, and 2.0, yielding a total of 400 GMRs in this study, facilitating the representation of a broad spectrum of structural damages from elastic to highly inelastic (and collapse). To comprehensively characterize seismic features and variations, 28 candidate IMs were calculated for each GMR. The candidate IMs, listed in detail in Ref. [24], comprise 14 IMs related to acceleration, 9 IMs related to velocity, 3 to displacement, and 2 to time.

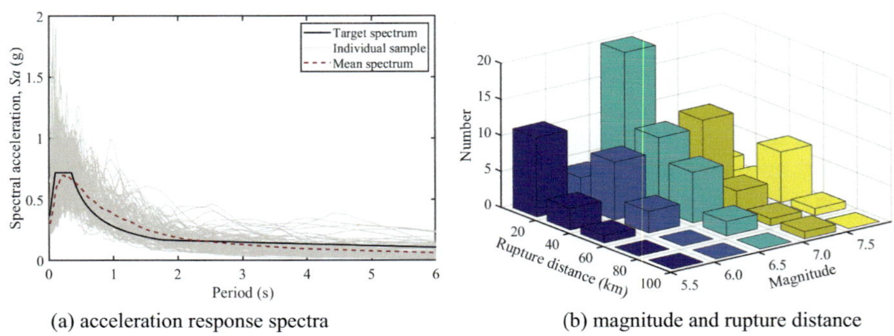

(a) acceleration response spectra (b) magnitude and rupture distance

FIGURE 10.5 The statistical properties of GMRs used in this study.

10.5.3 Damage Database for the Considered Buildings

To create a comprehensive database of seismic damage for the targeted building portfolio, the MDoF shear model and the MDoF flexure-shear model introduced above are specifically used, as detailed in Section 10.3. Specifically, 600 buildings, half RC frames, and half RC frame-shear walls, are randomly selected from the virtual city, and corresponding building models are developed based on the design parameter samples. Subsequently, NLTHAs are performed on these building models in the OpenSees software to acquire the building responses (i.e., MIDR and PFA), with GMRs randomly paired with the building models. Note that MIDR describes the DSs of structural and nonstructural drift-sensitive components, whereas PFA is used for calculating the DSs of nonstructural acceleration-sensitive components [40]. Ultimately, a seismic damage dataset is assembled. This dataset incorporates 600 candidate IM sets and samples of design parameters as the input variables. It also includes the corresponding MIDR and PFA values, which function as the output responses.

10.6 SEISMIC FRAGILITY ANALYSIS

10.6.1 NGBoost Model Training and Performance

Before constructing a surrogate model based on the developed seismic damage database, the optimal IMs selection method proposed by Ding et al. [38] is employed to identify the optimal input features. Due to space constraints, this study does not delve into the specific details of this analysis component. The results revealed that the selected optimal ground motion IMs are the geometric mean of the 5%-damped spectral acceleration in the period range of 0–3.0 seconds (*AvgSA*), peak ground velocity (*PGV*), and Housner intensity (*HI*) for MIDR, and *AvgSA*, *PGA*, and effective peak acceleration (*EPA*) for PFA. The definitions of these optimal IMs can be found in Ref. [24]. A database containing 11-dimensional inputs ($\boldsymbol{x}_{\mathrm{IM}}$, $\boldsymbol{x}_S$) (i.e., *AvgSA*, *PGV*, and *HI* as ground motion IMs, and eight structural design parameters in Table 10.1) and one-dimensional output y_M (i.e., the natural logarithm of MIDR, LnMIDR) is randomly split into training and testing sets at a ratio of 7:3.

NGBoost, as a probabilistic ML method developed in Ref. [39], is employed to construct a surrogate model based on the training set for predicting LnMIDRs of the considered buildings, referred to as NGBoost-MIDR. The lognormal distribution is initially selected as the response variable's distribution type according to judgment and tests. Throughout the model process, five key hyperparameters, including the number of weak learners, learning rate, maximum depth, minimum sample numbers for splitting, and minimum sample numbers for a leaf, were fine-tuned through a grid search method based on the coefficient of determination (R^2) as a score of the model's performance.

Note that the constructed NGBoost-MIDR model can predict the probability distribution of the response $P_\theta(y_M|\boldsymbol{x}_{\mathrm{IM}},\boldsymbol{x}_S)$ for each building under various ground motion IMs. However, to discuss the model's performance, the mean prediction accuracy of the NGBoost-MIDR model is demonstrated through the testing and training sets, as shown in Figure 10.6a. In addition to the coefficient of determination (R^2), several

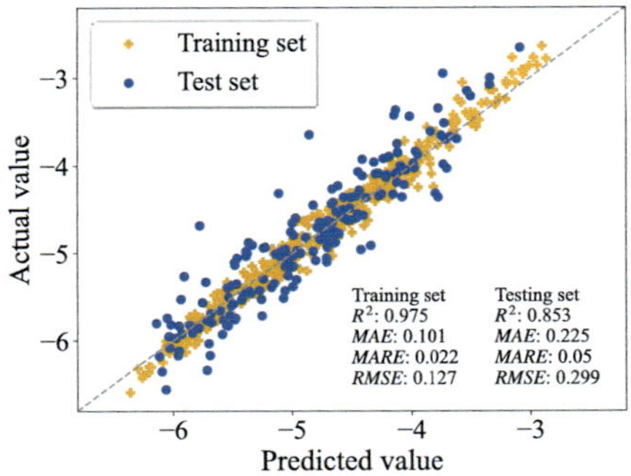

(a) LnMIDR

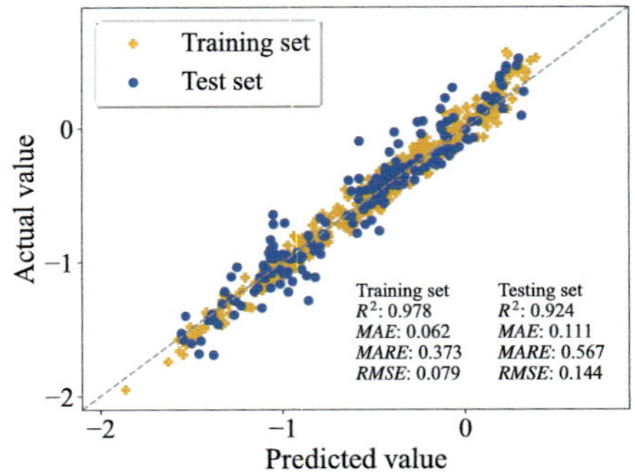

(b) LnPFA

FIGURE 10.6 The mean prediction performance of the NGBoost models.

other metrics are employed to evaluate the goodness-of-fit of the model. These include the Mean Absolute Error (*MAE*), Mean Absolute Relative Error (*MARE*), and Root Mean Squared Error (*RMSE*). Upon examination, it is observed that the actual and predicted data points align closely along the $y=x$ line. This is further confirmed by the near-zero values of *MAE*, *MARE*, and *RMSE*, and an R^2 value of 0.853 for the testing set. In simpler terms, these results indicate that the NGBoost-MIDR model demonstrates a high degree of prediction accuracy.

Similarly, another database consisting of 11-dimensional inputs ($\boldsymbol{x}_{\text{IM}}$, $\boldsymbol{x}_S$) (i.e., *AvgSA PGA*, and *EPA* as ground motion IMs, and eight structural design parameters in Table 10.1) and one-dimensional output y_P (i.e., the natural logarithm of PFA, LnPFA) is used to develop another NGBoost model for predicting LnPFAs of these buildings, referred to as NGBoost-PFA. From Figure 10.6b, it can be observed that the predicted points of the model closely match the actual ones, with *MAE*, *MARE*, and *RMSE* near 0 and R^2 near 1 over training and testing sets. The results demonstrate that both models are highly accurate and effectively predict building responses of structural and nonstructural components.

10.6.2 Seismic Fragility Analysis of Individual Buildings

In this section, the established NGBoost models (i.e., NGBoost-MIDR for damage of structural components and NGBoost-PFA for damage of nonstructural acceleration-sensitive components) are utilized to provide the $\mu_{\ln y}$ and $\sigma_{\ln y}$ of responses for each building under a given ground motion IM. Then, the probability of exceeding a specific DS is calculated according to Eq. (10.13). According to the considered seismic zone (corresponding to the fortification level in the China code) and the age of construction, the buildings in the city comply with low-code or precode seismic design levels. Tables 10.2 and 10.3 summarize the four DSs (i.e., the slight, moderate, extensive, and complete damage) thresholds of buildings from low-code and precode seismic design levels in HAZUS [40], corresponding to MIDR and PFA, respectively.

Figure 10.7a shows the conditional probabilities of each building in this virtual city suffering extensive damage under three different scenarios corresponding

TABLE 10.2
Threshold Values of MIDR Corresponding to Each Damage State [40].

	Low-Code				Precode			
Building Type	**Slight**	**Moderate**	**Extensive**	**Complete**	**Slight**	**Moderate**	**Extensive**	**Complete**
C1L	0.0050	0.0080	0.0200	0.0500	0.0040	0.0064	0.0160	0.0400
C1M	0.0033	0.0053	0.0133	0.0333	0.0027	0.0043	0.0107	0.0267
C1H	0.0025	0.0040	0.0100	0.0250	0.0020	0.0032	0.0080	0.0200
C2L	0.0040	0.0076	0.0197	0.0500	0.0032	0.0061	0.0158	0.0400
C2M	0.0027	0.0051	0.0132	0.0333	0.0021	0.0041	0.0105	0.0267
C2H	0.0020	0.0038	0.0099	0.0250	0.0016	0.0031	0.0079	0.0200

Note: the number of stories was 1 to 3, 4 to 7, and more than 8, representing low-, mid- and high-rise buildings, respectively.
C1L, C1M, and C1H are RC frames with low-, mid-, and high-rise, respectively.
C2L, C2M, and C2H are RC shear walls with low-, mid-, and high-rise, respectively.

TABLE 10.3
Threshold Values of PFA Corresponding to Each Damage State [40].

Low-Code				Precode			
Slight	**Moderate**	**Extensive**	**Complete**	**Slight**	**Moderate**	**Extensive**	**Complete**
0.20	0.40	0.80	1.60	0.20	0.40	0.80	1.60

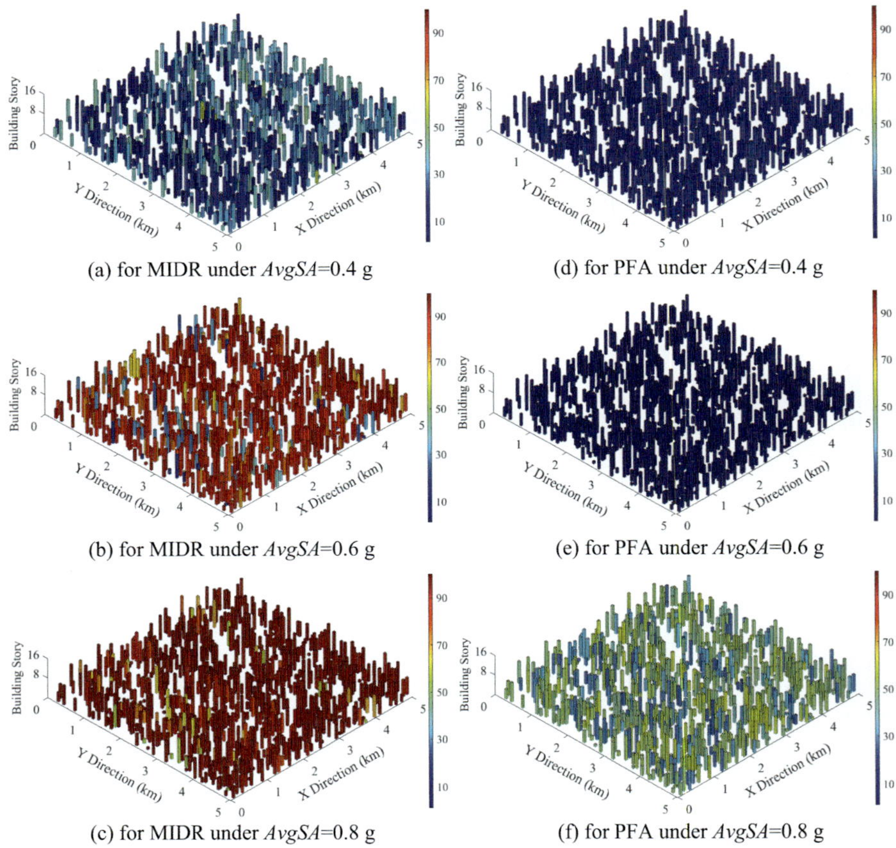

FIGURE 10.7 Conditional probability that all buildings in the considered virtual city reach extensive damage state.

to given ground motion IM values (i.e., $AvgSA = 0.4$, 0.5, and 0.6 g) based on the NGBoost-MIDR model for structural components. Damage probabilities under different *PGV* or *HI* or multiple IMs (i.e., *AvgSA*, *PGV*, and *HI*) are no longer shown here due to redundancy. The extensive damage probabilities of each building in the city under $AvgSA = 0.4$, 0.5, and 0.6 g based on NGBoost-PFA model for nonstructural acceleration-sensitive components are displayed in Figure 10.7b. It is important to note that extensive damage (i.e., DS_3) is considered because its corresponding threshold is adopted to determine whether buildings are safe for occupancy according to Eq. (10.14). From Figure 10.7, it can be seen that the probabilities of extensive damage to structural components under the same ground motion IM may be higher than that of nonstructural acceleration-sensitive components, even if their exceedance probability increases with the increase of IM.

Furthermore, six buildings with varied design parameters are chosen from the considered virtual city to investigate their seismic fragility models. Table 10.4 summarizes essential information about the six buildings. The first three correspond to multistory RC frame buildings, while the last three correspond to high-rise RC frame-shear wall buildings. The seismic fragilities of these six buildings based on MIDRs (indicating structural damage) and PFAs (indicating nonstructural acceleration-sensitive damage) are shown in Figures 10.8 and 10.9, respectively. Regardless of MIDR or PFA, the fragility curves of RC frame-shear wall buildings are relatively scattered compared to RC frame buildings. For instance, the former (e.g.,

TABLE 10.4
Information of Randomly Selected Buildings.

No.	Building Type	Age of Construction	Number of Stories	Floor Area (m^2)	Fundamental Period (seconds)
35	C1L	Pre-1989	2	540	0.36
92	C1M	Post-2002	5	682	0.68
500	C1H	1990–2001	8	336	0.95
677	C2H	1990–2001	9	540	1.25
693	C2M	Pre-1989	7	634	1.03
829	C2H	Post-2002	12	254	1.38

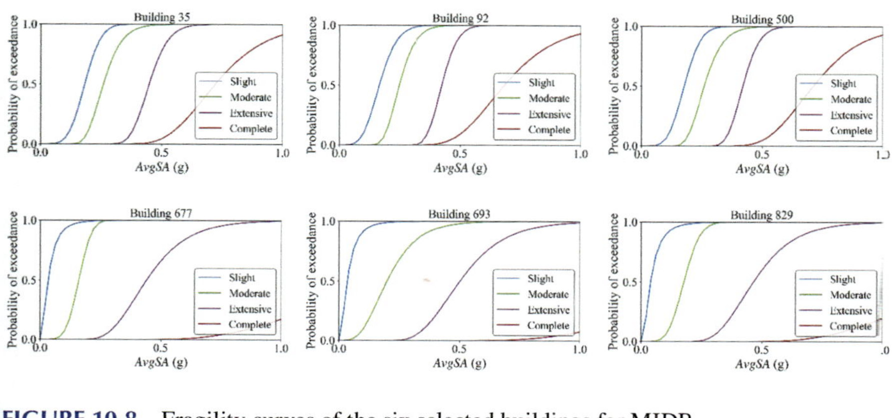

FIGURE 10.8 Fragility curves of the six selected buildings for MIDR.

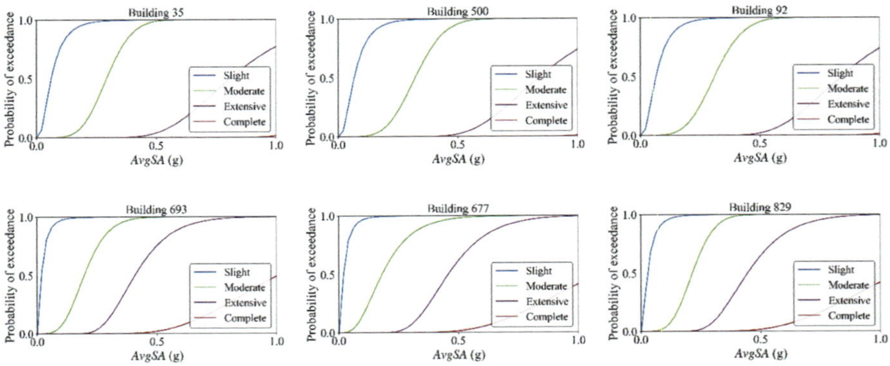

FIGURE 10.9 Fragility curves of the six selected buildings for PFA.

Buildings 677, 693, and 829) exhibit higher fragility curves for slight damage than the latter (e.g., Buildings 35, 92, and 500). The fragility curves for complete damage are significantly lower than the latter. This is because, compared to RC frames, RC frame-shear walls have excellent seismic performance (measured in terms of MIDR and PFA) for high-intensity earthquakes due to their lateral resistance, while the difference in seismic performance between them for low-intensity earthquakes may not be significant.

Comparing Figures 10.8 and 10.9, significant discrepancies in exceedance probabilities for MIDR among buildings with diverse design parameters under the same seismic scenario are observed (see Figure 10.8). In contrast, differences in exceedance probabilities for PFA are relatively minor (see Figure 10.9). The observed patterns may be attributed to the substantial influence of both seismic intensity level and design parameters on structural damage. Conversely, the extent of nonstructural damage appears to be predominantly determined by the seismic intensity level.

10.6.3 Seismic Fragility Analysis of Building Portfolio

The number of unsafe buildings in the city follows the Poisson binomial distribution as detailed in Section 10.4, and its probability distribution can be obtained using Eq. (10.20) based on the exceedance probability associated with extensive damage of each building. Figure 10.10a illustrates the probability distribution of the number of unsafe buildings for structural damage conditioned on different ground motion IMs (i.e., *AvgSA*), with μ and σ denoting the mean and standard deviation of the respective distribution. It is acceptable for the probability distribution of the number of unsafe buildings to approximate a normal distribution. This is especially true when the number of trials is significantly large (for instance, $N = 1{,}000$ in this study). Under such conditions, the Poisson binomial distribution converges to a normal distribution [48]. In Figure 10.10a, as the *AvgSA* increases, there is a marked rise in the expected

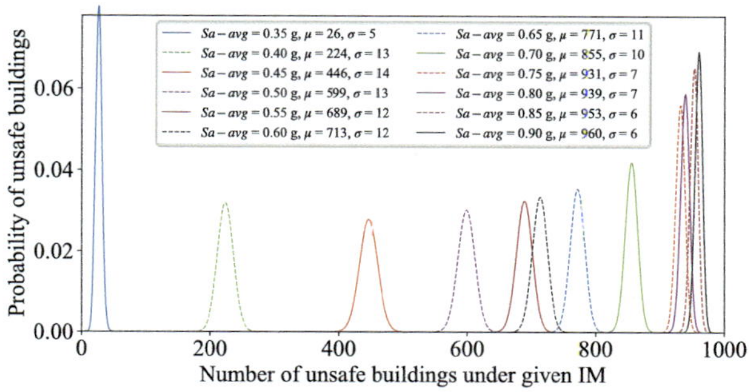

(a) based on MIDR

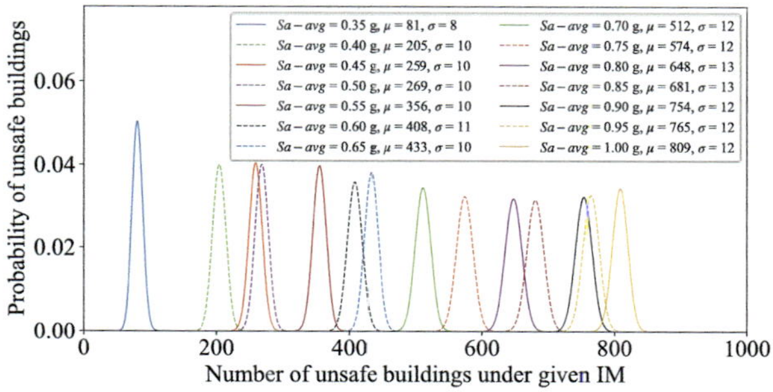

(b) based on PFA

FIGURE 10.10 Trends of the probability distribution of the number of unsafe buildings in the city.

number of unsafe buildings, and this increase plateaus once the *AvgSA* attains a value of 0.75 g. This pattern holds true for the nonstructural acceleration-sensitive damage observed in city buildings, even when conditioned on varying *AvgSA* values (see Figure 10.10b).

Furthermore, the expected proportion of unsafe buildings in the city under a given ground motion IM can be calculated using Eq. (10.22). The fragility curves, representing the proportion of unsafe buildings under different IMs, are plotted as Figure 10.11a based on MIDR for structural damage and Figure 10.11b based on PFA for nonstructural acceleration-sensitive damage. Due to the uncertainty in the number or proportion of unsafe buildings, as depicted in Figure 10.10, there is variability in the range of the city's fragility curves. This is exemplified by the 95% confidence interval of $\mu \pm 1.96\sigma$. Figure 10.11 indicates that the fluctuation range of regional-scale seismic fragility is relatively small. In other words, for the Poisson binomial distribution model, the proportion of successful trials in a series of independent Bernoulli tests fluctuates only slightly. This is a consequence of the law of large numbers, where the percentage (or proportion) of successes becomes more stable (closer to the mean) when the number of trials is significantly large. This suggests that the expected proportion of unsafe buildings in the city conditioned on different IMs used to evaluate regional-scale seismic fragility of building portfolio is deemed acceptable.

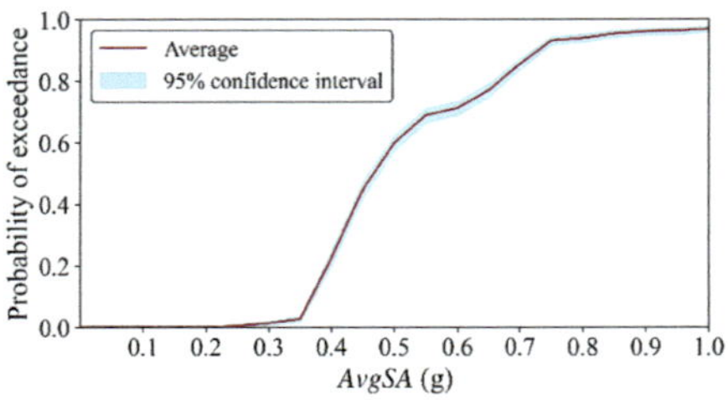

(a) based on MIDR

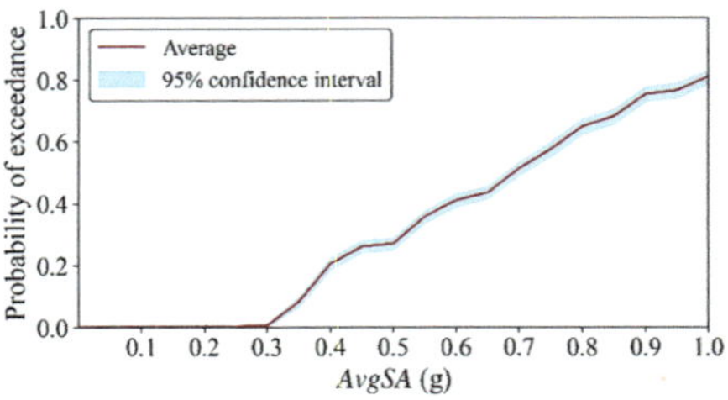

(b) based on PFA

FIGURE 10.11 Fragility assessments of the building portfolio in the city.

10.7 CONCLUSIONS

This study introduced a probabilistic ML method, that is, NGBoost, to predict the exceedance probability of damage for individual buildings in a region. The probabilistic model of building portfolio functionality failure was also developed based on the Poisson binomial distribution, providing a regional-scale fragility model for building portfolios. A virtual city consisting of 600 RC frame and 400 RC frame-shear wall buildings was analyzed to demonstrate this work. The main results are as follows:

1. The seismic damage database of regional-scale buildings, considering building-to-building variability, could be effectively and quickly obtained based on a simplified MDoF shear model for RC frames and simplified MDoF flexure-shear model for RC frame-shear walls.
2. The NGBoost model can directly predict the probability distribution parameters of individual building response for parametrized fragility model, obviating the need for deterministic ML methods to iteratively train models across different buildings.
3. The seismic fragility of building portfolios could be assessed using the Poisson binomial distribution rather than solely focusing on the damage probability of individual buildings in previous regional-scale research. Notably, the established fragility model of building portfolios encompasses structural and nonstructural damage, which holds great promise for assessing loss, risk, and resilience of building portfolios.

ACKNOWLEDGMENTS

The authors express their sincere gratitude for the financial support received from the Project of National Key Research and Development Program of China (Grant No. 2022YFC3803000), the National Natural Science Foundation of China (Grant No. 52311540017), and the Natural Science Foundation of Jiangsu Province (Grant No. BK20211564).

REFERENCES

1. C. Mesta, G. Cremen, C. Galasso, Urban growth modelling and social vulnerability assessment for a hazardous Kathmandu Valley, *Sci Rep* 12(1) (2022) 6152. https://doi.org/10.1038/s41598-022-09347-x
2. Y.N. Du, D.C. Feng, G. Wu, InSAR-based rapid damage assessment of urban building portfolios following the 2023 Turkey earthquake, *Int J Disaster Risk Reduc* 103 (2024) 104317. https://doi.org/10.1016/j.ijdrr.2024.104317
3. P. Lin, N. Wang, Building portfolio fragility functions to support scalable community resilience assessment, *Sustain Resil Infras* 1(3–4) (2016) 108–122. https://doi.org/10.1080/23789689.2016.1254997
4. M. Polese, M. Di Ludovico, M. Gaetani d'Aragona, A. Prota, G. Manfredi, Regional vulnerability and risk assessment accounting for local building typologies, *Int J Disaster Risk Reduc* 43 (2020) 101400. https://doi.org/10.1016/j.ijdrr.2019.101400
5. R. Fernández, J.F. Correal, D. D'Ayala, A.L. Medaglia, Towards disaster risk mitigation on large-scale school intervention programs, *Int J Disaster Risk Reduc* 90 (2023) 103655. https://doi.org/10.1016/j.ijdrr.2023.103655
6. I. Sianko, Z. Ozdemir, I. Hajirasouliha, K. Pilakoutas, Probabilistic seismic risk assessment framework: case study Adapazari, Turkey, *Bull Earthq Eng* 21(7) (2023) 3133–3162. https://doi.org/10.1007/s10518-023-01674-2
7. W.Y. Zhang, P.Y. Chen, J.G.F. Crempien, A. Kurtulus, P. Arduino, E. Taciroglu, Regional-scale seismic fragility, loss, and resilience assessment using physics-based simulated ground motions: an application to Istanbul, *Earthq Eng Struct D* 52(6) (2023) 1785–1804. https://doi.org/10.1002/eqe.3843
8. S.W. Li, C. Farrar, Y.C. Yang, Efficient regional seismic risk assessment via deep generative learning of surrogate models, *Earthq Eng Struct D* 52(11) (2023) 3435–3454. https://doi.org/10.1002/eqe.3849
9. H.Y. Zhang, X.W. Cheng, Y. Li, D.J. He, X.L. Du, Rapid seismic damage state assessment of RC frames using machine learning methods, *J Build Eng* 65 (2023) 105797 https://doi.org/10.1016/j.jobe.2022.105797
10. R. Gentile, C. Galasso, Gaussian process regression for seismic fragility assessment of building portfolios, *Struct Saf* 87 (2020) 101980. https://doi.org/10.1016/j.strusafe.2020.101980
11. M. Nakashima, K. Ogawa, K. Inoue, Generic frame model for simulation of earthquake responses of steel moment frames, *Earthq Eng Struct D* 31(3) (2002) 671–692. https://doi.org/10.1002/eqe.148
12. X.Z. Lu, B. Han, M. Hori, C. Xiong, Z. Xu, A coarse-grained parallel approach for seismic damage simulations of urban areas based on refined models and GPU/CPU cooperative computing, *Adv Eng Software* 70 (2014) 90–103. https://doi.org/10.1016/j.advengsoft.2014.01.010
13. C. Xiong, X.Z. Lu, H. Guan, Z. Xu, A nonlinear computational model for regional seismic simulation of tall buildings, *Bull Earthq Eng* 14(4) (2016) 1047–1069. https://doi.org/10.1007/s10518-016-9880-0
14. R. Gentile, C. Galasso, Surrogate probabilistic seismic demand modelling of inelastic single-degree-of-freedom systems for efficient earthquake risk applications, *Earthq Eng Struct D* 51(2) (2021) 492–511. https://doi.org/10.1002/eqe.3576
15. R. Gentile, C. Galasso, Simplicity versus accuracy trade-off in estimating seismic fragility of existing reinforced concrete buildings, *Soil Dyn Earthq Eng* 144 (2021) 106678. https://doi.org/10.1016/j.soildyn.2021.106678

16. D. Patsialis, A.A. Taflanidis, Reduced order modeling of hysteretic structural response and applications to seismic risk assessment, *Eng Struct* 209 (2020) 110135. https://doi.org/10.1016/j.engstruct.2019.110135
17. S. Ruggieri, A. Chatzidaki, D. Vamvatsikos, G. Uva, Reduced-order models for the seismic assessment of plan-irregular low-rise frame buildings, *Earthq Eng Struct D* 51(14) (2022) 3327–3346. https://doi.org/10.1002/eqe.3725
18. X.Z. Lu, H. Guan, *Earthquake disaster simulation of civil infrastructures: from tall buildings to urban areas* (2nd edition), Springer, New York (2021). https://doi.org/10.1007/978-981-15-9532-5
19. M.J. Xiang, J. Chen, J.X. Shen, Z.T. Wang, Regional-scale stochastic nonlinear seismic time history analysis of RC frame structures, *Bull Earthq Eng* 20(15) (2022) 8123–8149. https://doi.org/10.1007/s10518-022-01473-1
20. D.L. Gu, A. Kareem, X.Z. Lu, Q.L. Cheng, A computational framework for the simulation of wind effects on buildings in a cityscape, *J Wind Eng Ind Aerodyn* 234 (2023) 105347. https://doi.org/10.1016/j.jweia.2023.105347
21. C.A. Cornell, F. Jalayer, R.O. Hamburger, D.A. Foutch, Probabilistic basis for 2000 SAC Federal Emergency Management Agency steel moment frame guidelines, *J Struct Eng* 128(4) (2002) 526–533. https://doi.org/10.1061/(ASCE)0733-9445(2002)128:4(526)
22. A. Miano, F. Jalayer, H. Ebrahimian, A. Prota, Cloud to IDA: efficient fragility assessment with limited scaling, *Earthq Eng Struct D* 47(5) (2018) 1124–1147. https://doi.org/10.1002/eqe.3009
23. Y.J. Xu, X.Z. Lu, Y. Tian, Y.L. Huang, Real-time seismic damage prediction and comparison of various ground motion intensity measures based on machine learning, *J Earthq Eng* 26(8) (2020) 4259–4279. https://doi.org/10.1080/13632469.2020.1826371
24. J.Y. Ding, D.C. Feng, Feature selection of ground motion intensity measures for data-driven surrogate modeling of structures, *Earthq Eng Struct D* 53(3) (2024) 1216–1237. https://doi.org/10.1002/eqe.4068
25. Z.Y. Wang, N. Pedroni, I. Zentner, E. Zio, Seismic fragility analysis with artificial neural networks: application to nuclear power plant equipment, *Eng Struct* 162 (2018) 213–225. https://doi.org/10.1016/j.engstruct.2018.02.024
26. S.H. Hwang, S. Mangalathu, J. Shin, J.S. Jeon, Machine learning-based approaches for seismic demand and collapse of ductile reinforced concrete building frames, *J Build Eng* 34 (2021) 101905. https://doi.org/10.1016/j.jobe.2020.101905
27. J.R. Wu, L. Di Sarno, A machine-learning method for deriving state-dependent fragility curves of existing steel moment frames with masonry infills, *Eng Struct* 276 (2023) 115345. https://doi.org/10.1016/j.engstruct.2022.115345
28. D. Shahnazaryan, G.J. O'Reilly, Next-generation non-linear and collapse prediction models for short- to long-period systems via machine learning methods, *Eng Struct* 306 (2024) 117801. https://doi.org/10.1016/j.engstruct.2024.117801
29. S.Z. Chen, D.C. Feng, E. Taciroglu, Prior knowledge-infused neural network for efficient performance assessment of structures through few-shot incremental learning, *Comput-Aided Civ Inf* (2024) 1–18. https://doi.org/10.1111/mice.13175
30. F. Kazemi, N. Asgarkhani, R. Jankowski, Machine learning-based seismic fragility and seismic vulnerability assessment of reinforced concrete structures, *Soil Dyn Earthq Eng* 166 (2023) 107761. https://doi.org/10.1016/j.soildyn.2023.107761
31. S. Bhatta, J. Dang, Seismic damage prediction of RC buildings using machine learning, *Earthq Eng Struct D* 52 (2023) 3504–3527. https://doi.org/10.1002/eqe.3907
32. P.Y. Su, F. Xiong, Y. Lu, Q.D. Hu, B.W. Zhang, Efficient Bayesian metamodeling for fine-grained and robust fragility analysis of buildings at a regional scale, *Struct Saf* 102 (2023) 102324. https://doi.org/10.1016/j.strusafe.2023.102324
33. P. Heresi, E. Miranda, RPBEE: Performance-based earthquake engineering on a regional scale, *Earthq Spectra* 39(3) (2023) 1328–1351. https://doi.org/10.1177/87552930231179491

34. J. Zhang, M.Y. Zhang, Seismic robustness computational methodology of community building portfolio coupled with water supply network based on probability-cloud model, *Comput-Aided Civ Inf* 39 (2023) 1676–1698. https://doi.org/10.1111/mice.13088
35. S.M. Rinaldi, J.P. Peerenboom, T.K. Kelly, Identifying, understanding, and analyzing critical infrastructure interdependencies, *IEEE Contr Syst Mag* 21(6) (2001) 11–25. https://doi.org/10.1109/37.969131
36. A. Atef, D. Bristow, Risk assessment of infrastructure facilities considering spatial and operational interdependencies: temporal simulation model, *Struct Infrastruct Eng* 18(8) (2021) 1138–1151. https://doi.org/10.1080/15732479.2021.1877737
37. E.A. Opabola, C. Galasso, A probabilistic framework for post-disaster recovery modeling of buildings and electric power networks in developing countries, *Reliab Eng Syst Saf* 242 (2024) 109679. https://doi.org/10.1016/j.ress.2023.109679
38. J.Y. Ding, D.C. Feng, E. Brunesi, F. Parisi, G. Wu, Efficient seismic fragility analysis method utilizing ground motion clustering and probabilistic machine learning, *Eng Struct* 294 (2023) 116739. https://doi.org/10.1016/j.engstruct.2023.116739
39. T. Duan, A. Avati, D.Y. Ding, K.K. Thai, S. Basu, A. Ng, A. Schuler, NGBoost: natural gradient boosting for probabilistic prediction, *Int Conf Mach Learn* 119 (2019) 2640–3498. https://doi.org/10.48550/arXiv.1910.03225
40. FEMA, Hazus 5.1, Hazus earthquake model technical manual, FEMA (Federal Emergency Management Agency), Washington, DC (2022).
41. J.S. Steelman, J.F. Hajjar, Influence of inelastic seismic response modeling on regional loss estimation, *Eng Struct* 31(12) (2009) 2976–2987. https://doi.org/10.1016/j.engstruct.2009.07.026
42. MHURD, Load code for design of building structures GB 50009-2012, Ministry of Housing and Urban-Rural Development of the People's Republic of China, Beijing (2012).
43. ASCE, *Minimum design loads for buildings and other structures*, American Society of Civil Engineering, Reston, VA (2016).
44. S. Taghavi, E. Miranda, Approximate floor acceleration demands in multistory buildings. II: applications, *J Struct Eng* 131(2) (2005) 212–220. https://doi.org/10.1061/(ASCE)0733-9445(2005)131:2(212)
45. MHURD, Technical specification for concrete structures of tall building JGJ 3-2010, Ministry of Housing and Urban-Rural Development of the People's Republic of China, Beijing (2010).
46. ACI, Building code requirements for structural concrete and commentary ACI 318-19, American Concrete Institute, Farmington Hills, MI (2019).
47. P. Heresi, E. Miranda, Structure-to-structure damage correlation for scenario-based regional seismic risk assessment, *Struct Saf* 95 (2022) 102155. https://doi.org/10.1016/j.strusafe.2021.102155
48. Y. Hong, On computing the distribution function for the Poisson binomial distribution, *Comput Stat Data Anal* 59 (2013) 41–51. https://doi.org/10.1016/j.csda.2012.10.006
49. X.Y. Cao, D.C. Feng, G. Wu, Pushover-based probabilistic seismic capacity assessment of RCFs retrofitted with PBSPC BRBF sub-structures, *Eng Struct* 234 (2021) 111919. https://doi.org/10.1016/j.engstruct.2021.111919
50. PEER, PEER NGA database. NGA-West, NGA-West2, NGA-East (2019). https://peer.berkeley.edu/research/data-sciences (accessed March 10, 2024).
51. J.W. Baker, Lin, T., Shahi, S. K., N. Jayaram, New ground motion selection procedures and selected motions for the PEER transportation research program, PEER Report 2011-03, Pacific Earthquake Engineering Research Center, University of California, Berkeley, CA (2011).
52. R. Gentile, C. Galasso, Accounting for directivity-induced pulse-like ground motions in building portfolio loss assessment, *Bull Earthq Eng* 19(15) (2020) 6303–6328. https://doi.org/10.1007/s10518-020-00950-9

11 Predictive Modelling of Hail Impact Damage on Toughened Glazing Panels

Yiwen Cui[a], Nelson Lam[a], Shuangmin Shi[a], Emad Gad[b], and Lihai Zhang[a]

[a] Department of Infrastructure Engineering, The University of Melbourne, Parkville, VIC, Australia

[b] School of Engineering, Swinburne University of Technology, Hawthorn, VIC, Australia

11.1 INTRODUCTION

A hailstorm event can cause significant damage to the glazing façades of buildings. Panels impacted by hail may suffer surface defacement, often necessitating full replacement. In urban areas, post-storm replacement costs can reach tens of millions of dollars [1–3]. If hail impact is severe enough to perforate the glazing, allowing water ingress, extensive damage can be caused to both the building and its contents [4].

Accurate analytical methods for predicting façade damage are essential for setting fair insurance premiums, distributing risk, and supporting realistic hazard assessments that guide effective preventive measures. However, knowledge of hail impact damage prediction is not widely shared amongst building professionals. The highly transient and localised, nature of hail impact makes such predictions particularly challenging. To date, no analytical model has been developed to predict glazing failure under hail impact, and as a result, the failure mechanisms remain poorly understood.

This chapter presents a systematic modelling framework for predicting the risk of hail-inflicted failure of glazing installations. It begins by discussing the fundamental failure mechanisms of glazing panels when impacted by hail, and examining key influential factors such as microcrack accumulation over the service life, impact location, and impact velocity. A deterministic approach is then introduced to quantify the damage potential from individual hail events, followed by a probabilistic method for developing fragility curves that account for uncertainty and variability in the size of the hailstones, frequency of occurrences, and impact energy. Holistically, these methods offer a comprehensive basis for assessing hail-induced failure risks. The framework enables engineers and designers to evaluate the vulnerability of glazing

 DOI: 10.1201/9781003567332-11

systems under projected climate conditions and supports the development of more resilient, data-informed façade designs for hail-prone regions.

11.2 FUNDAMENTALS OF GLASS FAILURE UNDER HAIL IMPACT

Understanding the failure behaviour of glazing panels under hail impact is essential for developing accurate predictive models. This section outlines the fundamental failure mechanisms involved and highlights key contributing factors such as micro-cracks, impact location, impact velocity, and boundary conditions. These factors collectively influence the impact resistance of a glazing panel and the likelihood of failure in a hailstorm.

11.2.1 Wave-Controlled Phenomenon

When subject to impact by hail, the development of flexural strains in a glazing panel is primarily governed by the initial generation of flexural waves. Experimental observations show that strain intensity at the impact point is largely insensitive to panel size, impact location, or boundary conditions [5]. This wave-controlled phenomenon arises because of the very rapid development of flexural strains initiated at the point of contact in an extremely short period of time, well before any stress wave is reflected from the plate boundaries. As a result, the outcome of the impact is not dependent on the size, nor geometry, of the glazing panel. The risk of onset of fracture is instead controlled by the size of the ice sphere, impact velocity, panel thickness, and temperature.

Wave-controlled conditions cease to apply when the impact occurs near an edge, allowing reflected waves to influence the strain field during the loading phase. The critical offset distance of the point of contact from the edge of the panel, beyond which wave reflections have negligible contributions to additional stresses in the glass pane, can be estimated using the flexural wave propagation equation, as shown in Eq. (11.1) [5].

$$L(t) = \sqrt{\pi}\left(\frac{2D}{m}\right)^{1/4}\sqrt{t} \tag{11.1}$$

where $L(t)$ is the distance of wave travel for a given time (t), is the flexural rigidity of the panel, and m is the mass per unit area. The approximate amount of time (t) for the flexural strain reaching the peak value is taken as 0.1 ms according to observations from impact experiments undertaken by the authors. Inputting $t = 0.1$ ms into Eq. (11.1) would give estimates of the distance of flexural wave travel during the rise time (to peak flexural strains) for plates of different thicknesses as listed in Table 11.1.

11.2.2 Dominance by Flexural Failure

Hertzian fracture and flexural failure are the two predominant failure modes observed in glass subjected to the impact of a solid projectile, as illustrated in Figure 11.1. These modes are characterised by distinct crack patterns occurring on opposite surfaces of the glass pane.

TABLE 11.1
The Propagation Distance of Flexural Waves in a Glass Panel within the Impact Duration

Thickness	4 mm	6 mm	8 mm
$r_1(t)$	55 mm	65 mm	75 mm

Note: Values listed in the table have been rounded up to the nearest 5 mm.

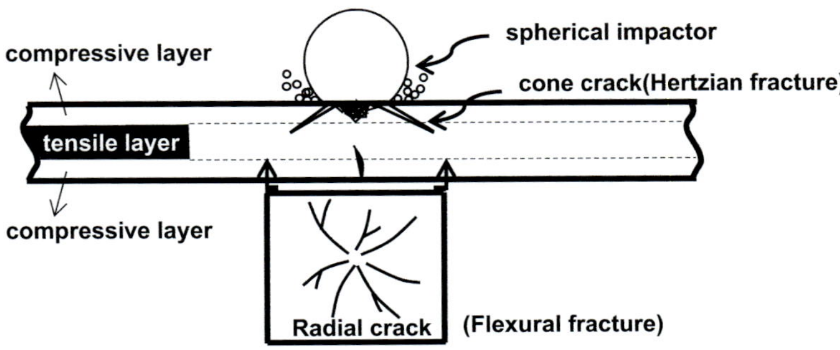

FIGURE 11.1 A schematic diagram of typical failure modes of glass panel when subjected to impact by a spherical impactor.

Hertzian fracture originates on the frontal (impacted) surface of the glass, where conical cracks initiate due to high localised contact pressure. These cracks typically propagate into the glass at an angle of approximately 30° to the surface of the glass [6–9]. The failure is driven by high tensile stresses close to the edge of the contact zone. The presence of pre-existing cracks on the frontal surface significantly increases the risk of Hertzian fracture because of stress concentration at the tip of these cracks. Consequently, deeper surface cracks are associated with lower failure velocities.

In contrast, flexural failure occurs on the rear surface of the glass pane (opposite to the side where contact is made). Crack propagation is triggered when the intensity of the flexural tensile stress exceeds the material strength [10,11]. Circumferential (hoop) stresses associated with flexural deformation are responsible for triggering this type of fracture. As with Hertzian failure, the presence of a critical flaw on the rear surface would lower the threshold velocity of impact to initiate failure.

Experimental observations indicate that flexural failure is the dominant failure mode in glazing panels when subjected to impact by ice, as opposed to Hertzian fracture [12]. This phenomenon is mainly attributed to the relatively low transient contact pressure experienced on the glass surface when impacted by an ice sphere (as the compressive yield strength of ice is relatively low). The yield strength of ice typically ranges from 5 to 35 MPa [13–18]. As a result, the intensity of tensile stresses induced on the glass surface in the vicinity of the contact zone is generally well below the threshold to offset the pre-stored compressive stresses in toughened *D*glass. Hertzian style of fracture is therefore suppressed.

11.2.3 Influences of Crack Depth and Length

Microcracks within the glass panel are known to compromise the mechanical strength and impact resistance of glass. Figure 11.2 illustrates the different types of cracks, their typical cause, and the crack depths. Having knowledge of the influence of these defects on the strength of glass is essential for assessing the long-term performance of glazing products, particularly under service conditions.

Observations from experimental studies involving controlled impact by ice spheres on the glass panes of varying thicknesses demonstrate a clear relationship between crack depth and the likelihood of failure. Results show a gradual decrease in the failure velocity with increasing crack depth, implying reduced resistance to impact [12].

In contrast, results from sensitivity studies on glass specimens with artificially introduced surface scratches of varying lengths reveal no significant correlation between crack length and failure velocity. These findings imply that, for toughened glass subject to impact by hail, the residual resistance is governed primarily by the depth of the critical crack, as opposed to its length [12].

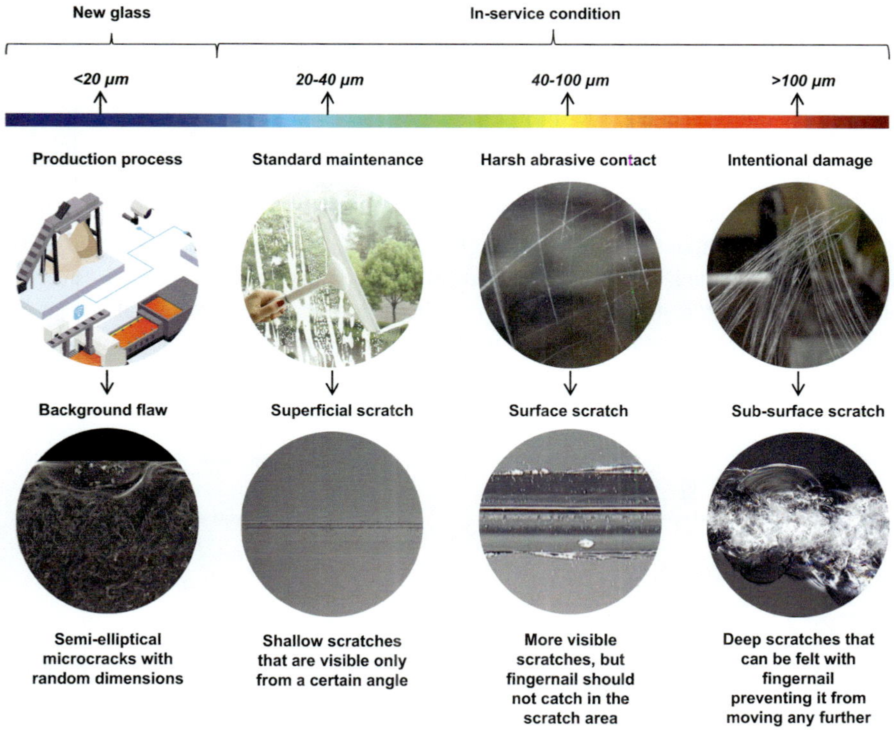

FIGURE 11.2 Representative types of microcrack and its depth observed in glazing panels during service life

11.2.4 Influences by Offset Distance of Impact Point to Critical Crack

The distribution of microcracks within a glass panel is inherently random, as is the disposition of the points of contact with hail. As such, accurate information on the location of the point of contact relative to that of a critical flaw on the glass surface is critical for a reliable assessment of the risk of failure. Experimental findings show that the failure velocity associated with a dominant crack is highly sensitive to the offset distance of the point of impact from the location of the nearest critical flaw [19]. A direct impact on the critical crack (i.e., offset distance $d = 0$) would result in the lowest failure velocity because of the very high stress concentration occurring at the flaw. With increasing offset distance, the failure velocity increases, implying a reduced likelihood of crack activation leading to fracture. This trend continues until the offset reaches a threshold known as the influential radius, beyond which further increases in distance would no longer have any significant influence on the failure velocity. In this regime, failure is instead governed by the random distribution of background flaws which is inherent in the glass by the very nature of the manufacturing process. The behaviour as described is illustrated in Figures 11.3 and 11.4.

11.2.5 Insensitivity of Failure Velocity to Impact History

Glass specimens containing pre-existing (artificial) cracks were subjected to repeated impact at progressively increasing velocity to investigate the influence of the history of exposure to hail hazard on the risks of failure for a given impact scenario [19]. Experimental results show no observable correlation between the number of prior occurrences of impact and the failure velocity.

Digital microscopy was employed to observe changes to a monitored crack following each impact occurrence. No visible growth, nor change in appearance, of the crack was detected across successive impacts within the limit of failure. Repetitive subcritical hail impact on the same toughened glass specimen is shown not to cause any cumulative crack growth, thereby lowering its impact resistance. Consequently, the history of prior strikes by hail on a critical crack in a glazing panel may be assumed to have no bearing on the resistance of the panel to further impact. The

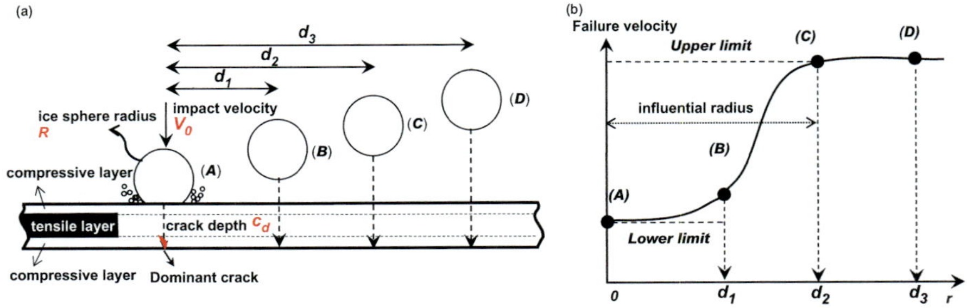

FIGURE 11.3 Variation of failure velocity with offset distance from impact position to critical crack (a) schematics of different impact locations and (b) indication of the corresponding failure velocity (r: radial distance from the dominant crack).

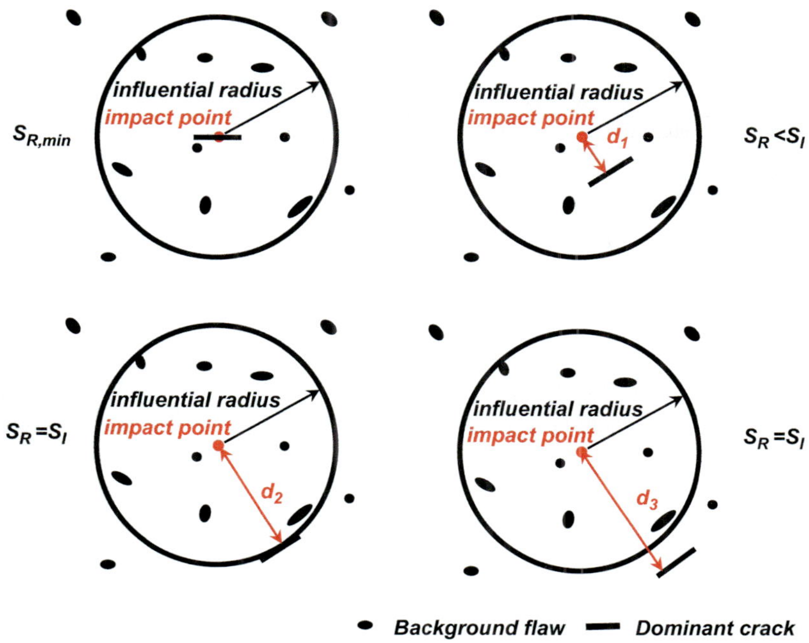

FIGURE 11.4 Schematics of the relative positions of a crack to impact position (S_I: Initial capacity; S_R: Residual capacity; $S_{R,\min}$: Minimum residual capacity).

risk of hail-inflicted failure of a toughened glass panel is controlled primarily by its thickness and pre-existing cracks in the panel.

11.3 DETERMINISTIC APPROACH OF PREDICTING GLASS FAILURE UNDER SINGLE HAIL IMPACT

The damage factor η_D, defined as the ratio of the dynamic stress intensity factor K_{Id} to the dynamic fracture toughness K_{ID}, provides a measure of the margin of safety when subject to impact actions [12,19], as expressed in Eq. (11.2):

$$\eta_D\left(V_f\right)=\frac{K_{\text{Id}}}{K_{\text{ID}}}-1\geq 0 \tag{11.2}$$

where $n_D = 1$ corresponds to failure initiation. The calculation of the dynamic stress intensity factor, K_{Id}, requires knowledge of the three-dimensional stress field in the vicinity of a dominant crack, combined with the transient contact pressure generated by impact of a hailstone.

Closed form expressions for the calculation of K_{Id} and K_{ID} are listed in the subsequent subsection based on the work of the authors [12]. The input parameters and procedure for computing $\eta_D\left(V_f\right)$ are summarised in Figure 11.5.

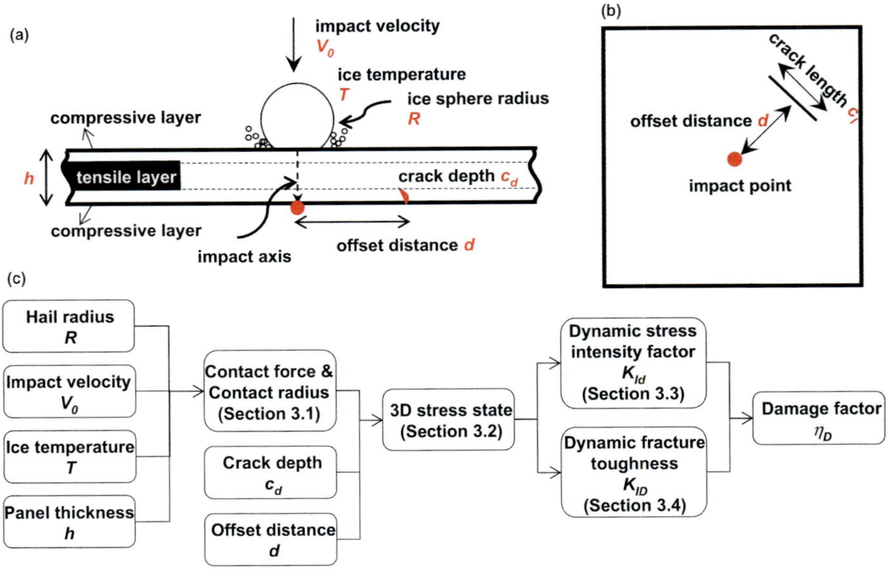

FIGURE 11.5 Diagrams of (a) side view, (b) plan, and (c) flow chart of deterministic model.

11.3.1 Contact Force and Contact Radius

The key relationships required for predicting flexural stresses generated by the impact of hail, as presented in Refs. [12,19] are summarised in Table 11.2. The contact force risetime profile (Figure 11.6a) was derived from an analysis of 149 recorded time histories of the impact force which were captured from experiments which involved accelerating ice specimens onto a modified Hopkinson bar [20]. This baseline forcing function was then adapted in the present study to account for the effects of plate flexibility. The model so developed was then cross validated against flexural strain measurements taken from dynamic tests which involved impacting ice spheres on glass specimens. Parameters governing contact force behaviour include the radius of the ice sphere (R), the velocity of impact (V_0), the tensile strength of ice (σ_T), and a reduction factor β that accounts for the deformability of both the impactor and the targeted panel. The change in contact radius with time, $a(t)$, was also studied through analysing footage captured by a high-speed camera at 8.32 μs time intervals, as illustrated in Figure 11.6b.

11.3.2 Closed-Form Expressions for Fast-Tracking the 3D Stress State Surrounding a Dominant Crack

Circumferential stress $\sigma_{\theta\theta}(r,z,t)$, radial stress $\sigma_{rr}(r,z,t)$, contact stress $\sigma_{zz}(r,z,t)$, and radial shear stress $\sigma_{rz}(r,z,t)$ are the four stress components subject to modelling. Note that neither circumferential shear stresses $\sigma_{\theta z}(r,z,t)$ nor torsional stresses $\sigma_{r\theta}(r,z,t)$ can be generated by a point contact because of axisymmetric loading. The considered stress components can be represented collectively by the generic notation: $\sigma_{ij}(r,z,t)$

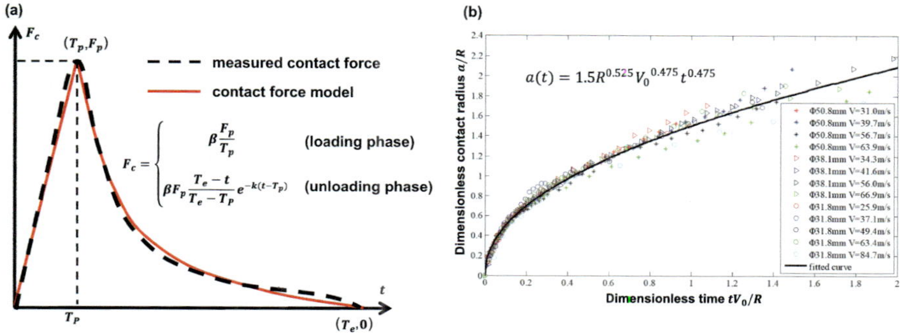

FIGURE 11.6 Contact force and contact radius: (a) a schematic diagram of the contact force model and (b) time-dependent contact radius function.

TABLE 11.2
Determination of Contact Force and Contact Radius

Contact forcing function

$$F_c(t) = \frac{\beta F_p}{T_p} t \quad (0 \le t \le T_p)$$

$$F_c(t) = \beta F_p \frac{T_e - t}{T_e - T_p} e^{-k(t-T_p)} \quad (T_p \le t \le T_e)$$

(1) **Determination of peak contact force** F_p

$$F_p = 3969 R^2 \sigma_T c^{-1.31} V_0^{1.31}$$

(2) **Determination of time at peak force** T_p

$$T_p = 0.07 R \rho^{0.68} \sigma_T{}^{-0.68} V_0^{0.36}$$

$$\sigma_T = -0.0065T + 0.495$$

$$c = \sqrt{E_i / \rho_i}$$

(3) **Determination of total duration** T_e

$$\int_0^{T_e} F_c(t)dt \ge 95\% \left(T_e = 0.8 \text{ ms}\right)$$

(4) **Determination of** k

$$f(k) = \frac{\Delta I \Delta T}{F_p} k^2 - F_p k + 1 - e^{-k\Delta T} = 0$$

$$a(t) = 1.5 R^{0.525} V_0{}^{0.475} t^{0.475}$$

$$\Delta T = T_e - T_p$$

(5) **Determination of β**

$$\beta = \frac{1}{1 + 0.8\lambda}$$

$$\lambda = 10.6 \frac{V_0^{0.309} R^2 \rho^{0.253} \sigma_T^{0.748} c^{-0.583}}{\sqrt{mD}}$$

Contact radius function $a(t)$

$$a(t) = 1.5 R^{0.525} V_0{}^{0.475} t^{0.475}$$

which is a function of time (t), offset from the impact position to the critical crack in the radial direction (r), and depth into the thickness of the glass panel (z). The key original contribution of this study was to approximate $\sigma_{ij}(r,z,t)$ as a triple product of three component functions: (1) time-dependent function: $s_{ij}(t)$, (2) radial distribution function: $f_{ij}(r)$, and depth-dependent function: $g_{ij}(z)$, as shown by Eq. (11.3).

$$\sigma_{ij}(r,z,t) = s_{ij}(t) f_{ij}(r) g_{ij}(z) \quad i,j = r,z,\theta \tag{11.3}$$

Component functions: $s_{ij}(t)$, $f_{ij}(r)$, and$g_{ij}(z)$ for each of the four stress components: $\sigma_{\theta\theta}(r,z,t)$, $\sigma_{rr}(r,z,t)$, $\sigma_{zz}(r,z,t)$, and $\sigma_{rz}(r,z,t)$ are listed in Table 11.3 and displayed in Figure 11.7 based on the earlier work of the authors [19].

The first principal stress $(\sigma_1(r,z,t))$ along the crack path can be determined by use of Eq. (11.4).

$$\sigma_1(r,z,t) = \max\left\{ \begin{array}{l} \sigma_{\theta\theta,\text{net}}(r,z,t), \dfrac{\sigma_{rr,\text{net}}(r,z,t) + \sigma_{zz}(r,z,t)}{2} \pm \\ \sqrt{\left(\dfrac{\sigma_{rr,\text{net}}(r,z,t) - \sigma_{zz}(r,z,t)}{2}\right)^2 + \tau_{rz}^2(r,z,t)} \end{array} \right\} \tag{11.4}$$

11.3.3 Closed-Form Expressions for Calculating Stress Intensity Factor at a Crack

The dynamic stress intensity factor K_{Id} is defined as a product of (1) the quasi-static stress intensity factor (K_{Is}) which is a function of the depth of crack (c_d) and profile of the first principal stress $(\sigma_1(r,z,t))$ along the crack propagation path;

TABLE 11.3
Summary of Expressions for Calculating Stresses Caused by Ice Impact

Stress Type $\sigma_{ij}(z,r,t)$	Time Dependence $s_{ij}(t)$	Depth Dependence $g_{ij}(z)$	Radial Dependence $f_{ij}(r)$	Location of Peak
$\sigma_{\theta\theta}$	$-\dfrac{3(1+v)}{4\pi h^2}F_c(t)\left[\ln\left(\dfrac{a(t)^2}{4\lambda_b t}\right)-1+\gamma\right]$	$2\dfrac{z}{h}-1$	$e^{-\left(\frac{r}{0.4r_f(t)}\right)^2}$	$r=0$ $z=h$
σ_{rr}	$-\dfrac{3(1+v)}{4\pi h^2}F_c(t)\left[\ln\left(\dfrac{a(t)^2}{4\lambda_b t}\right)-1+\gamma\right]$	$2\dfrac{z}{h}-1$	$\left(1-2.5\dfrac{r}{r_f(t)}\right)e^{-\left(\frac{r}{0.64r_f(t)}\right)^2}$	$r=0$ $z=h$
σ_{zz}	$-2\dfrac{F_c(t)}{\pi a^2(t)}$	$e^{-\left(\frac{z}{0.6h}\right)^2}$	$1-\dfrac{r^2}{a^2(t)}$	$r=0$ $z=0$
σ_{rz}	$1.5\dfrac{F_c(t)}{2\pi a(t)h}$	$1-\left(\dfrac{2z}{h}-1\right)^2$	$1-\left(\dfrac{r}{0.75a(t)}-1\right)^2$	$r=0.75a(t)$ $z=h/2$

(2) modification factor $k(v_c)$ which is a function of the crack propagation velocity (v_c); and (3) time-dependent factor $\phi(t)$ to account for the evolution of stress and its reflections from the glass surfaces. A listing of all the formulae for determining the parameter values is shown in Table 11.4.

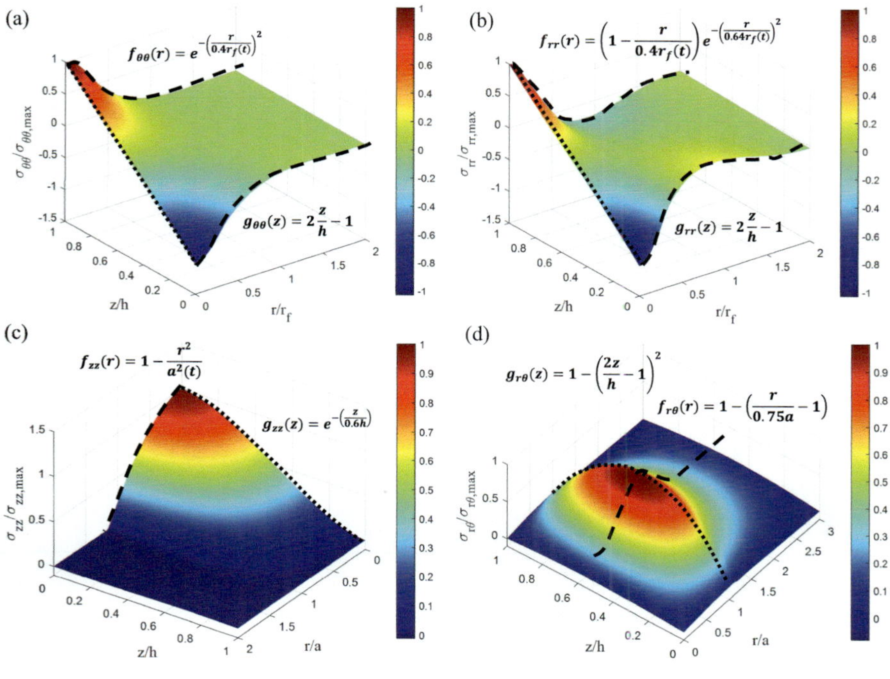

FIGURE 11.7 Distributions of stresses in 3D: (a) $\sigma_{\theta\theta}(r,z,t)$, (b) $\sigma_{rr}(r,z,t)$, (c) $\sigma_{zz}(r,z,t)$, (d) $\sigma_{rz}(r,z,t)$.

TABLE 11.4

Summary of Expressions for Calculating Damage Factor

Dynamic stress intensity factor $K_{\mathrm{Id}}(c_d, t, v_c, \sigma_1(r,z,t))$

$$K_{\mathrm{Id}}(c_d, t, v_c, \sigma_1(r,z,t)) = k(v_c)\phi(t)K_{\mathrm{Is}}(c_d, \sigma_1(r,z,t))$$

Time dependent factor $\phi(t)$

$$\phi(t) = \int_0^t \phi^H\left(\frac{c_S}{c_d}(t-\tau)\right)\frac{\sigma_1'(\tau)}{\sigma_1(t)}d\tau$$

When $c_d \sim 10^2$ µm, $t > 0.15$ µs, $\phi(t) \approx 1$.

Crack velocity factor $k(v_c)$

$$k(v_c) = \left(1 + v_c/(c_R - v_c)\right)^{-1}\left(1 - (v_c/c_L)^{0.5}\right)^{-1}$$

When $K_{\mathrm{Id}} < K_{\mathrm{ID}}$, $v_c \approx 0$, $k(v_c \approx 0) = 1$

Notation

c_d -crack depth

v_c -crack velocity

σ_1-maximum principal stress along crack path

r -radial distance from the impact position

z -depth into the thickness of glass pane

c_S -shear wave speed

c_R -Rayleigh wave speed

c_L -longitudinal wave speed

(Continued)

TABLE 11.4 (*Continued*)
Summary of Expressions for Calculating Damage Factor

Quasi-static stress intensity factor $K_{\mathrm{Is}}\left(c_d,\sigma_1(r,z,t)\right)$ (1) For critical crack generated by wear and tear: $K_{\mathrm{Is}}\left(C_d,\sigma_1(r,z,t)\right)=\frac{2}{\pi}c_d^{1/2}\int_0^{c_d}\frac{\sigma_1(r,z,t)\left(1+F(z/c_d)\right)}{\left(c_d^2-z^2\right)^{1/2}}dz$ $F(z/c_d)=1+\left(1-(z/c_d)^2\right)\left[\begin{array}{c}C_1+C_2(z/c_d)^2+C_3(z/c_d)^4\\+C_4(z/c_d)^6+C_5(z/c_d)^8\end{array}\right]$ (2) For background flaw: $K_{\mathrm{Is,bf}}=\sigma_1(r,z,t)\sqrt{\pi\frac{c_d}{Q}}F\left(\frac{c_d}{c_l},\varphi\right)$ $Q=1+1.464\left(\frac{c_d}{c_l}\right)^{1.65}$ $F\left(\frac{c_d}{c_l},\varphi\right)=\left(1.13-0.09\frac{c_d}{c_l}\right)f_\varphi$ $f_\varphi=\left[\left(\frac{c_d}{c_l}\right)^2\cos^2\varphi+\sin^2\varphi\right]^{0.25}\left[1+0.1(1-\sin\varphi)^2\right]$ When $c_d/c_l<1$, $\max\left(K_{\mathrm{Is,bf}}\right)$ occurs at $\varphi=\pi/2$, thus $f_\varphi=1$.	T_p-time of occurrence of peak strain c_l -crack length c_d -crack depth ϕ -parametric angle of the ellipse

11.3.4 Dynamic Fracture Toughness

Dynamic fracture toughness is also dependent on strain rate [21–26], the effects of which can be estimated using Eq. (11.5).

where K_{IC} is quasi-static fracture toughness, which may be taken as 0.72 ~ 0.78 $0.72\sim0.78\ \mathrm{MPa}\cdot\sqrt{\mathrm{m}}$ [27], while $\gamma(\dot{\varepsilon})$ is the dynamic increment factor which has been determined experimentally by use of the tensile Split Hopkinson Pressure Bar apparatus [22]. For a strain rate $\dot{\varepsilon}\le 350$, the dynamic increment factor has been found to be consistent with estimates from the empirical expression: $\gamma(\dot{\varepsilon})=1.137+0.015\log(\dot{\varepsilon})$. The average strain rate $\bar{\dot{\varepsilon}}$ as defined by Eq. (11.6) can be used to quantify strain rate effects for the purpose of calculating fracture toughness.

$$\bar{\dot{\varepsilon}}=\max\left(\Delta\varepsilon_{\mathrm{net}}(r,z)\right)/t_p \tag{11.6}$$

where t_p is the time of occurrence of maximum strain, $\max\left(\Delta\varepsilon_{\mathrm{net}}\right)$ is the strain increment during the period t_p at the specific location (r, z), which is calculated as $\max\left(\Delta\varepsilon_{\mathrm{net}}\right)=\varepsilon_{\mathrm{net}}\left(r,z,t=t_p\right)-\varepsilon_{\mathrm{net}}\left(r,z,t=0\right)$.

11.3.5 Worked Example

The application of the expressions presented in the earlier sections for predicting the outcome of an impact by an ice sphere is demonstrated here using a single-impact scenario (Tables 11.5a & b and 11.6). The test specimen is a 4 mm-thick glass pane which has been prestressed to a maximum surface compressive stress of 85 MPa. A microcrack with a depth of 64.6 μm was introduced prior to the impact. The impactor is a spherical ice particle with a diameter of 50.8 mm (radius 25.4 mm), striking the glass pane at a location which is offset by 15 mm from the crack tip. The objective is to calculate the stress intensity factor at the nearest crack tip, corresponding to a radial distance of $r = 15$ cm.

TABLE 11.5A
Material Properties of the Ice Sphere

Radius of Sphere R (mm)	Young's Modulus E (GPa)	Density ρ(kg/m³)	Temperature T (°C)	Impact Velocity V_0 (m/s)
25.4	5	870	−30	53.5

TABLE 11.5B
Material Properties of the Glass Panel

Prestress σ_{pc} (MPa)	Thickness h (mm)	Density ρ_g (kg/m³)	Young's Modulus E_g (GPa)	Poisson's Ratio ν_g	Fracture Toughness K_{IC} (MPa·m$^{0.5}$)
85	4	2500	69	0.23	0.72

TABLE 11.6
Calculations of Key Properties

1. Properties of ice

1.1 Tensile strength of ice

$\sigma_T = -0.0065 + 0.495T = 0.69$ MPa

1.2 Wave velocity in ice

$c = \sqrt{E/\rho} = 2{,}397$ m/s

2. Properties of panel

2.1 Unit mass per area

$m = \rho_g h = 10$ kg/m²

2.2 Flexural rigidity

$D = E_g h^3 / 12\left(1 - \nu_g^2\right) = 394\ \mathrm{N \cdot m}$

2.3 Flexural parameter

$\lambda_b = \sqrt{D/m} = 6.28\ \mathrm{m^2/s}$

3. Parameters of contact force

3.1 Peak contact force

$F_p = 3969 R^2 \sigma_T c^{-1.31} V_0^{1.31} = 12.1$ kN

3.2 Flexural rigidity

$T_p = 0.07 R \rho^{0.68} \sigma_T^{-0.68} V_0^{0.36} = 79.5\ \mu s$

3.3 Flexural parameter

$T_e = 0.8$ ms

3.4 Decay rate

$k = 2{,}291\ \mathrm{s^{-1}}$

3.5 Reduction factor

$$\beta = \frac{1}{1 + 0.8\lambda} = 0.708$$

$$\lambda = \frac{10.6 V_0^{0.309} R^2 \rho^{0.253} \sigma_T^{0.748} c^{-0.583}}{\sqrt{mD}} = 0.515$$

11.3.5.1 Calculation of Contact Forcing Function and Contact Radius

Loading phase: $F_c(t) = \dfrac{F_p}{T_p} t = 1.525 \times 10^8 t$

Unloading phase: $F_c(t) = F_p \dfrac{T_e - t}{T_e - T_p} e^{-k(t - T_p)} =$

$$1.684 \times 10^7 (0.0008 - t) e^{-2,291(t - 7.95 \times 10^{-5})}$$

Time-dependent contact radius: $a(t) = 1.5 R^{0.525} V_0^{0.475} t^{0.475} = 1.444 t^{0.475}$

11.3.5.2 Calculation of Stress State

Prestress of toughened glass: $\sigma_p(z) = \sigma_{\text{pc}} \left(1 - 6\dfrac{z}{h} + 6\dfrac{z^2}{h^2}\right) =$

$$85 \times 10^6 \left(1 - 1,500z + 375,000z^2\right)$$

The expressions for the stress state have been determined by the equations in Table 11.3.

To ascertain the net stresses, subtract the prestress from the impact-induced stresses. The amount of net stress so calculated can be used for determining the intensity of the principal stress, as shown in Figure 11.8a.

11.3.5.3 Calculation of Damage Factor

The estimated value of the first principal stress was used for the calculation of the maximum dynamic stress intensity factor ($K_{\text{Id,max}}$) by use of equations listed in Table 11.4 involving numerical integration as shown in Figure 11.8b. Calculation of the dynamic fracture toughness involves evaluation of the average strain rate. With the example shown in Figure 11.8a: at $t = 0$, $\sigma_1 = \sigma_2 = -85$ MPa, and at $t_p = 121\ \mu s$, $\sigma_1 = 134.5$ MPa, $\sigma_2 = 15.3$ MPa. The flexural strain can be inferred from the estimated stress values using the elementary relationship: $\varepsilon_{\text{net}} = \dfrac{1}{E_g}(\sigma_1 - \nu\sigma_2)$.

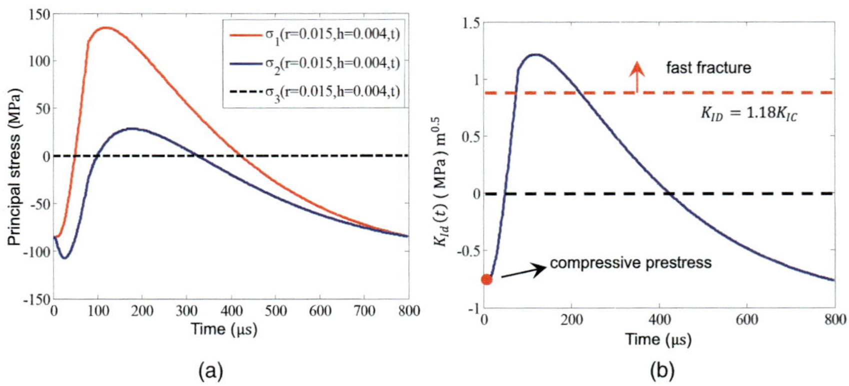

FIGURE 11.8 Simulation results: (a) Time history of principal stresses, (b) dynamic stress intensity factor.

The maximum strain increment during the course of impact can be calculated as $\max(\Delta\varepsilon_{\text{net}}) = 2.863 \times 10^{-3}$. Average strain rate: $\bar{\dot{\varepsilon}} = \max(\Delta\varepsilon_{\text{net}})/t_p = 26.76\ \text{s}^{-1}$.

Dynamic increment factor: $\gamma(\dot{\varepsilon}) = 1.137 + 0.015\log(\dot{\varepsilon}) = 1.186$

Dynamic fracture toughness: $K_{\text{ID}} = \gamma(\dot{\varepsilon})K_{\text{IC}} = 0.854\ \text{MPa}\cdot\text{m}^{0.5}$

Damage factor: $\eta_D = \dfrac{K_{\text{Id,max}}}{K_{\text{ID}}} = \dfrac{1.211}{0.854} > 1$

Damage factor exceeding unity infers triggering of fast fracture as illustrated in Figure 11.8b.

11.4 FRAGILITY CURVES FOR HAIL RESISTANCE OF TOUGHENED GLASS

Fragility curves provide a probabilistic measure of failure likelihood under varying impact intensities. This section presents the development and interpretation of fragility curves for toughened glass subjected to hail impact, enabling quantitative assessment of hail resistance.

11.4.1 Fragility Curves for Ultimate Failure Caused by a Single Impact

Fragility curves for ultimate failure of toughened glass involve determining its initial capacity (S_I) and residual capacity (S_R), the meaning of which is explained diagrammatically in Figure 11.9.

The residual strength of toughened glass is primarily governed by dominant cracks, which can be evaluated using the predictive model presented earlier. In contrast, the initial hail resistance of toughened glass is controlled by background flaws, small surface-adjacent imperfections located on the tensile side of the panel. These flaws typically have depths on the order of 10 μm, with a maximum depth rarely exceeding 20 μm.

A generic semi-elliptical flaw model, defined by its half-length c_l and crack depth c_d, is depicted in Figure 11.10. The procedure for calculating the stress intensity factor

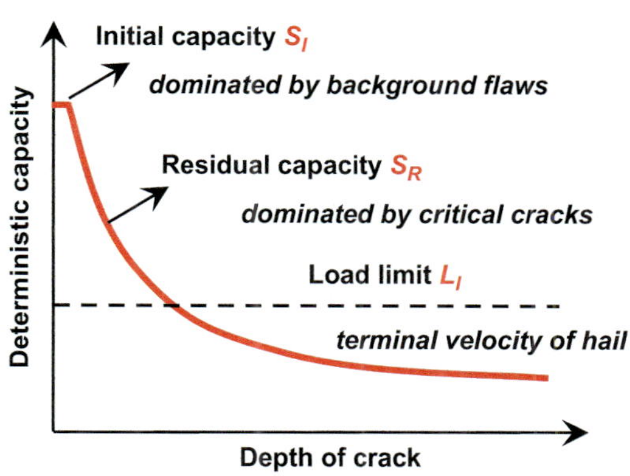

FIGURE 11.9 Capacity curve of glass pane when impacted by hail.

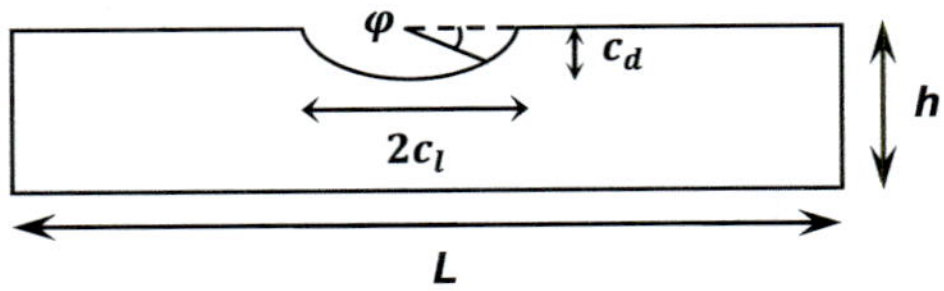

FIGURE 11.10 The geometry of a typical background flaw.

at such flaws is summarised in Section 11.3.3. The probabilistic distribution of c_d for all flaws on the glass pane was reported to be of the *Log-Normal* form, as defined by Eq. (11.7) and is characterised by the dual parameters (μ_c, ε_c) [8,28]. The value of the dual parameters can be found by calibrating against results from quasi-static tests using the procedure introduced in [29].

$$f(c_d) = \frac{1}{\sqrt{2\pi} c_d \varepsilon_c} \exp\left[-\frac{1}{2} \left(\frac{\ln(c_d) - \mu_c}{\varepsilon_c} \right)^2 \right] \tag{11.7}$$

The *Cumulative Probability Distribution* (CPD) of failure velocity in a hail impact can be determined by the procedure summarised in the flowchart of Figure 11.11 and the following list of steps:

1. Randomly dispose micro-flaws (of size distribution conforming to the calibrated probabilistic distribution) on the glass pane,
2. Identify possible critical flaws and their respective depths surrounding the point of impact,
3. For each identified flaw, determine the failure velocity to trigger fracture,
4. The critical flaw is the flaw that has been identified to have the lowest failure velocity.

Repeat steps (1)–(4) N times each with a different set of randomly generated background flaws sizes to derive a list of failure velocity values which are then ranked to construct the CPD function, as shown in Figure 11.11b. The nominal initial hail resistant capacity is taken as the median (50% exceedance) value from the CPD function.

Figure 11.12 shows an illustrative example of determining a *fragility surface* (consisting of a number of *fragility curves*). The failure velocity is expressed as a function of the crack depth and offset distance from the dominant flaw (c_d and d). The failure velocity converges to the initial capacity controlled by the background flaws as the offset distance increases. The determination of the critical condition to trigger fracture is illustrated in the example of Figure 11.13a wherein the size (diameter) of hail D is 50.8 mm, glass pane thickness, h is 4 mm, and prestress near the glass surface σ_{pc} is 85 MPa. The velocity of impact of a hailstone may be taken as the terminal velocity, which can be found using Eq. (11.8).

$$V_T = \sqrt{\frac{4\rho_{\text{hail}} D_h g}{3\rho_{\text{air}} C_D}} \tag{11.8}$$

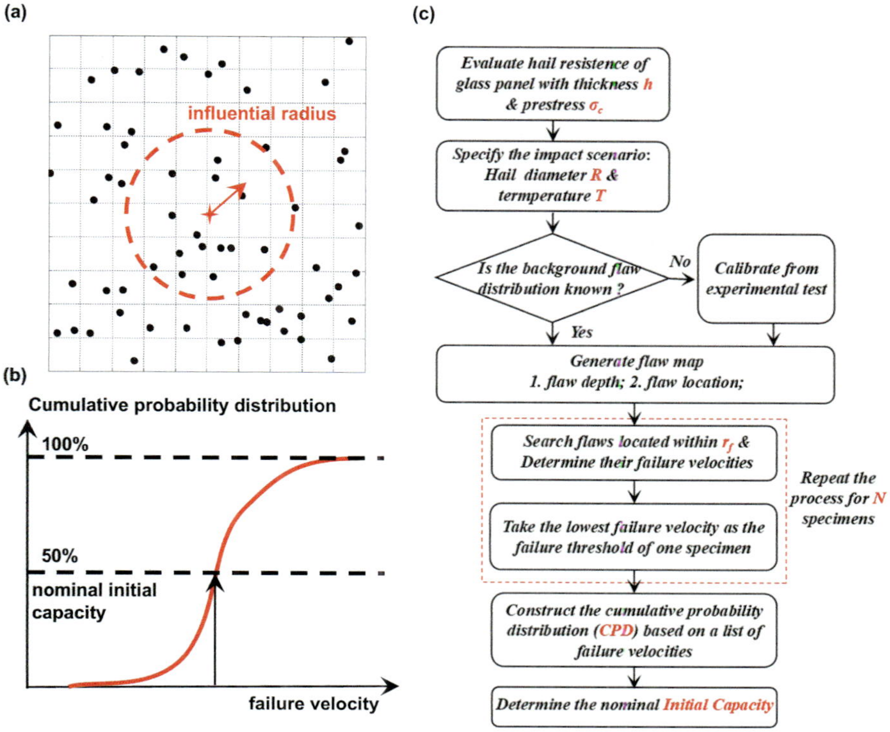

FIGURE 11.11 Determination of initial hail resistant capacity of new glass (a) stochastic simulation of the background micro-flaws, (b) cumulative probability distribution of failure velocity, and (c) a flow chart.

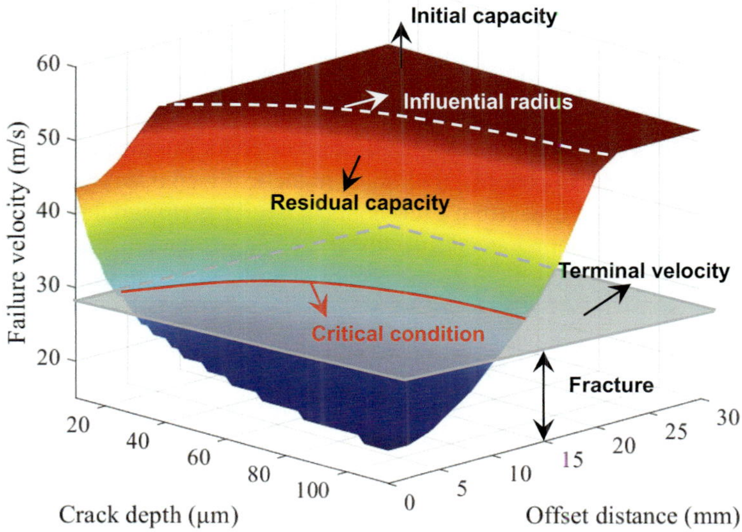

FIGURE 11.12 Variation of failure velocity with depth and location of crack.

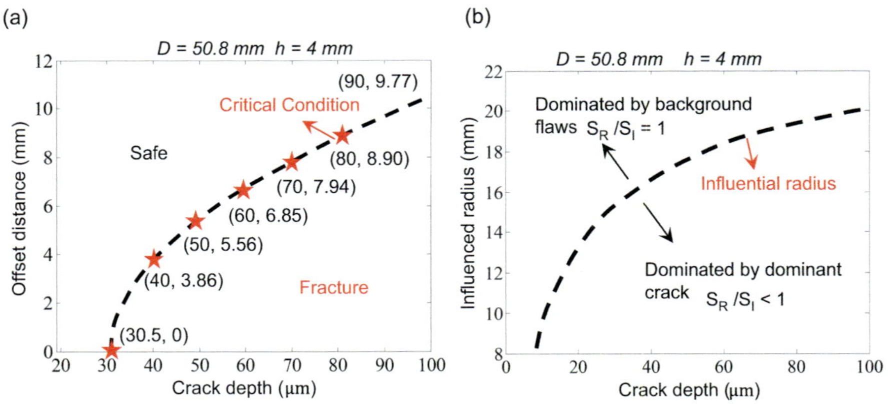

FIGURE 11.13 Diagrams of determination of (a) outcome of a hail impact and (b) influential radius.

where, D_h is the diameter of the hailstone (=2R), ρ_{hail} is density of hail, ρ_{air} is density of air (= 1.23 kg/m^3), C_D is drag coefficient (= 0.6) [30,31], and g is gravitational acceleration (= 9.81 m/s^2).

The outcome of the impact can be predicted reliably using Figure 11.12 provided that the depth of the critical crack, offset distance from the crack, and velocity of impact are all known. The influential radius (centred at the point of impact) can also be found by equating the residual capacity (S_R) to the initial capacity (S_I) of the same glass pane which is without a dominant flaw (Figure 11.13b). Knowledge of the influential radius can be used to expedite the prediction of the outcome of a hailstorm where thousands of repeated impacts need to be factored into the evaluation. In assessing individual strikes by hail during a storm, only dominant cracks that are within the influential radius of the considered strike warrant consideration. A significant amount of computation time would be saved as a result. The proposed probabilistic modelling method is to simulate a hailstorm by resolving it into a series of stochastic individual strikes. The outcome of a strike in isolation is determined using closed-form expressions derived from the deterministic predictive model as presented in Ref. [19]. The computational time required for solving these closed-form solutions is negligible. In contrast, traditional numerical methods typically require a considerable amount of time (from a few hours to a full day) to simulate the outcome of a single strike. Simulating an entire hailstorm using conventional numerical modelling is therefore impractical. With the proposed probabilistic modelling approach, the outcome of a 1 m × 1 m glass panel experiencing a simulated hailstorm can be predicted in less than a minute.

11.4.2 Fragility Curves in a Hailstorm

Probabilistic modelling of the performance of toughened glass involves tracking the probability of ultimate failure (P_f) in a hailstorm of increasing severity. Results can be presented in the form of a fragility curve as presented in the schematic diagram of Figure 11.14. The rationale of the modelling methodology is explained herein (Figure 11.15). As revealed by recent experimental investigations conducted by the authors, repetitive impact by hail within the failure threshold would not result in any cumulative crack growth [12,19]. Thus, the history of strikes sustained by a critical crack need not be factored into the evaluation. The two random variables involved in the modelling are: (1) amount of offset from the position of impact to critical flaws in its vicinity, and (2) distribution of hailstones size in a hailstorm.

The probabilistic simulation can be accomplished by combining these two random variables: impact position and size of hailstones (refer Figure 11.16). The probability of failure is then solved by employing *Monte Carlo* simulations. The probabilistic modelling procedure is listed as follows (refer flow chart of Figure 11.17).

(1) Set up a Cartesian coordinate for identifying positions of the critical flaws and points of impact (Figure 11.16), (2) Model distribution of size of hail and their

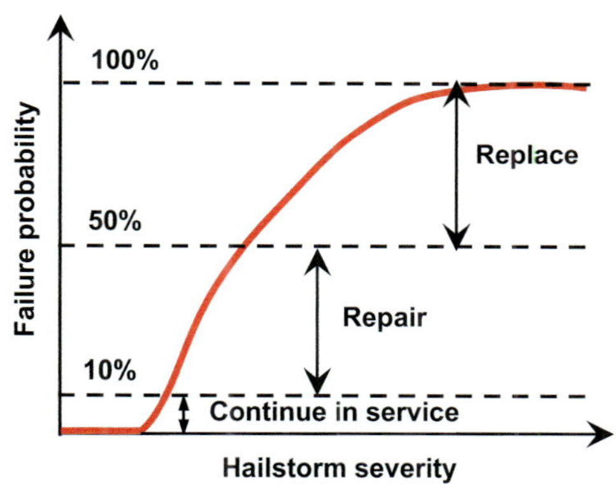

FIGURE 11.14 Probabilistic modelling of glass failure when impacted by hail.

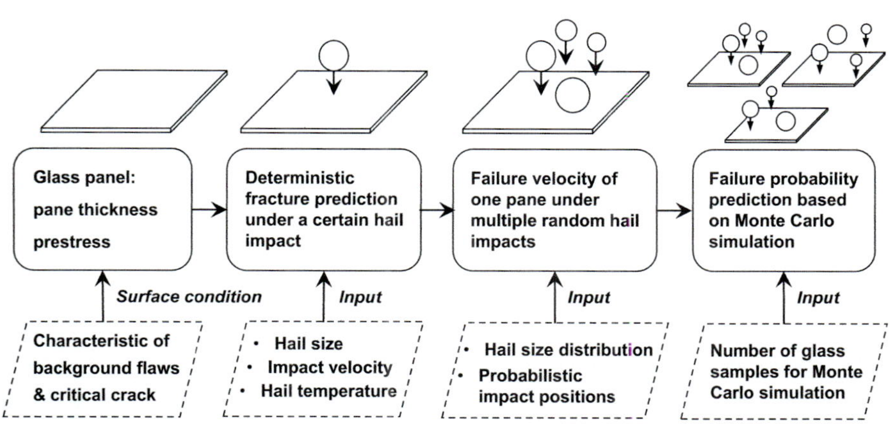

FIGURE 11.15 Schematic of the simulation of failure probabilistic in a hailstorm.

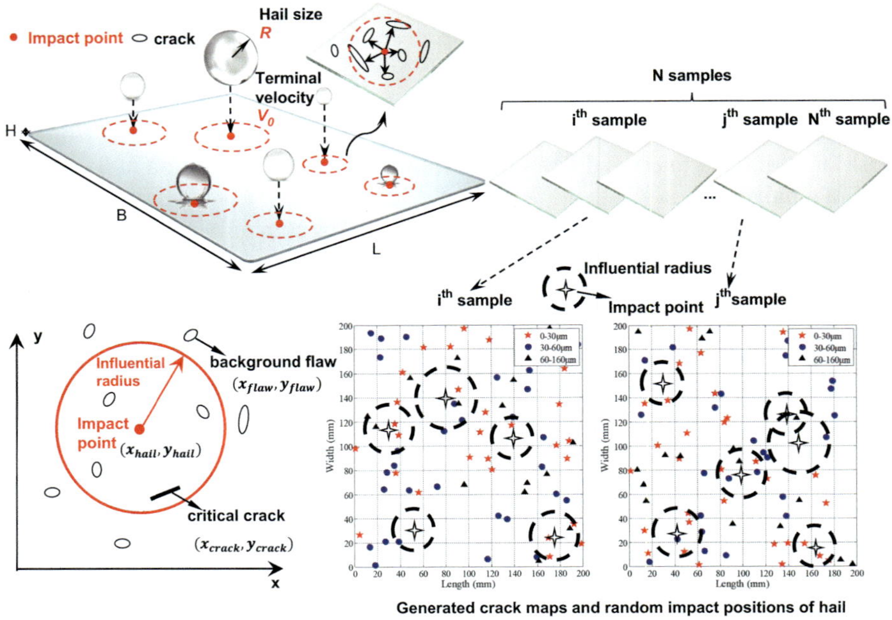

FIGURE 11.16 Schematic process of the probabilistic simulation.

impact position $\left(x_{\text{hail}}, y_{\text{hail}}\right)$ for given storm severity D_L, (3) Model background flaws as per the calibrated distribution function of the flaw size $f\left(c_d^{\text{flaw}}\right)$, followed by random generation of their disposition $\left(x_{\text{flaw}}, y_{\text{flaw}}\right)$, (4) Input the depth of dominant cracks c_d^{crack} and their position $\left(x_{\text{crack}}, y_{\text{crack}}\right)$ as identified from field surveys, (5) Determine the influential radius of each hailstone (each of which is identified with subscript "i") and the depth of flaws $c_{d,1} \sim c_{d,j}$ where subscript "j" identifies the flaw; and then calculate the offset distance from the impact position $d_1 \sim d_j$, (6) Determine the velocity of impact to trigger failure of each identified crack $V_{f,1} \sim V_{f,j}$ for the ith hailstone, (7) Check if the failure velocity is exceeded by the terminal velocity for that size of hailstone to determine if fracture would occur, (8) Repeat Steps 5–7 for all strikes occurring in the hailstorm to determine if the glass would experience fracture, and (9) Repeat these steps covering all the glazing specimens to determine the total number of failure counts $\left(N_{\text{total}}\right)$ for calculating the probability of failure.

11.5 APPLICATIONS

This predictive methodology developed in this paper enables quantitative assessment of the risk of failure, either deterministically or probabilistically, for either blend new or in-service glass when subject to impact by hail. Multi-level quantitative criteria are thus proposed and presented in Table 11.7. Worked examples are presented in this section to illustrate the methodology [29].

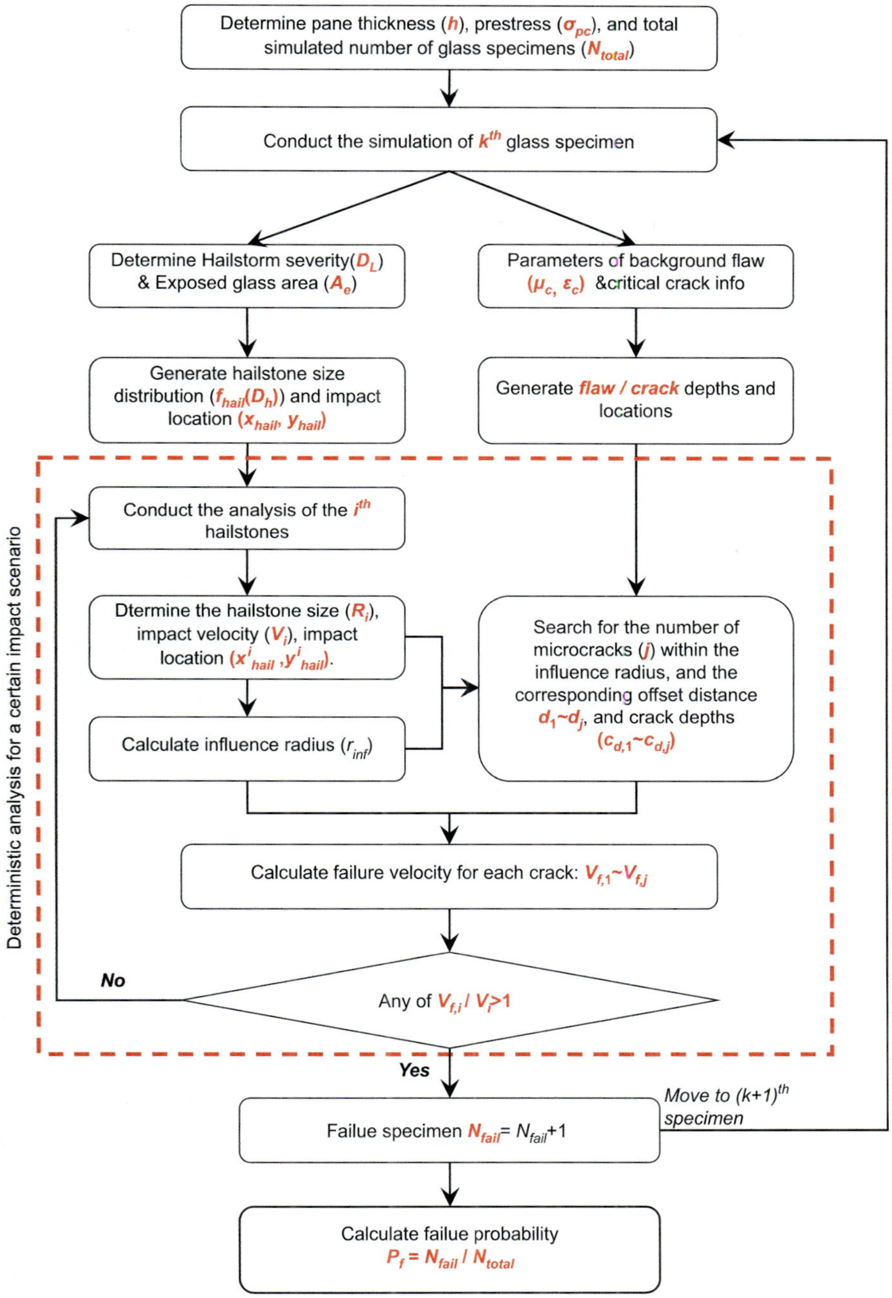

FIGURE 11.17 A flow chart illustrating the procedure of probabilistic simulation.

TABLE 11.7
Multi-Level Criteria for Evaluating Hail Resistance of Glass

Method	N	Criterion	Glass Type	Measures to Be Taken	
Deterministic (isolated impact)	1	$S_I < L_I$	New glass	Optimise thickness or prestress	
	2	$S_I > L_I$	New glass	Pass	
	3	$S_I > L_I$ & $S_R < L_I$	In-service glass	Maintain or replace glass	
	4	$S_I > L_I$ & $S_R > L_I$	In-service glass	Allow to remain in service	
Probabilistic (hailstorm)	1	$P_f < 10\%$	New glass	Pass	
	2	$P_f < 10\%$	In-service glass	Slight	Allow to remain in service
	3	$10\% < P_f < 50\%$	In-service glass	Moderate	Maintain the glass
	4	$P_f > 50\%$	In-service glass	Severe	Replace the glass

11.5.1 Deterministic Failure Prediction

The lowest residual capacity (based on the worst scenario of a direct strike on a dominant crack) may be calculated to provide a conservative deterministic prediction of the capacity of an in-service glass installation in which the depth of the critical crack is known from a field survey. A closed-form expression for the damage factor η_D, as defined by Eq. (11.2), was derived for the case where the crack is subject to a direct strike by hail. The derivation follows the procedures outlined in Section 11.5, with further details available in Ref. [12,19]. The failure velocity V_f can be determined by solving $\eta_D\left(V_f\right)-1=0$.

$$\eta_D\left(V_f\right)=\frac{c_d^{0.5}}{K_{\mathrm{Ic}}}\left\{\frac{0.042R^2\sigma_T V_f^{1.31}h^{-2}}{(1+0.8\lambda)}\left[\ln\left(3.233\times10^{-4}RV_f^{0.932}\sigma_T^{0.034}h^{-1}\right)-0.423\right]-0.950\sigma_c\right\} \tag{11.9}$$

where R (radius) is half the size of the hailstone, σ_T is the tensile strength of ice, h is the thickness of the glass pane, σ_c is the surface compressive stress in the toughened glass, c_d is the depth of the crack, K_{IC} is the quasi-static fracture toughness, λ is the reduction factor of plate flexibility (refer Table 11.2). The example glass pane for illustration (which has been prestressed to a surface compression stress of 85 MPa) features a dominant crack of depth equal to 100 microns on the glass surface. The predicted lowest residual capacity is shown to be dependent on the thickness of the glass pane and size of hail (Figure 11.18a). If the velocity of impact is taken to be equal to the terminal velocity of the hailstone, the failure velocity threshold can be correlated against the size (diameter) of the hail for different glass pane thickness, as shown in Figure 11.18b.

For a glass pane of known thickness, surface prestress, and maximum scratch depth (from field surveys) the failure velocity value that is calculated from the methodology (illustrated above) can be used to infer the remaining lifespan of the glazing installation through reference to regional hailstorm statistics. An example of such

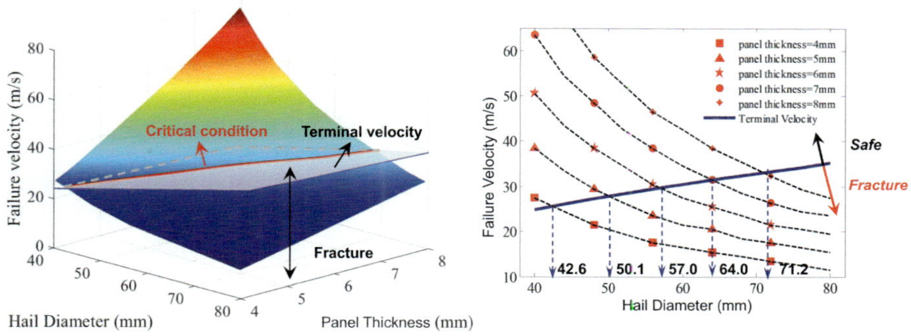

(a) (b)

FIGURE 11.18 Failure velocity versus hail diameter and glass pane thickness (a) surface plot and (b) line.

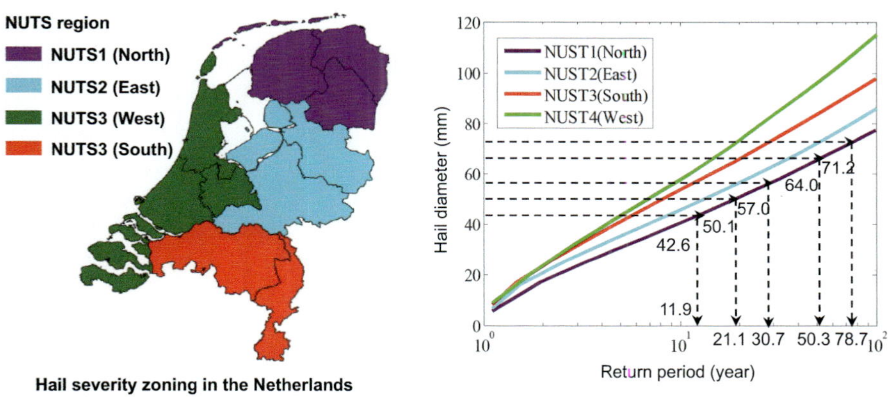

(a) (b)

FIGURE 11.19 Hailstorms in the Netherlands: (a) hail severity zoning and (b) hail diameter versus return period.

statistics is presented in Figure 11.19 which shows the predicted diameter of hail for different return periods in different parts of the Netherlands [32].

11.5.2 Probabilistic Failure Prediction

The toughened glass pane in the illustrative example presented herein has an exposed surface area of 0.5 m × 0.5 m. Value of D_L representing hailstorm severity in Australia is taken as 40 ~ 150 cm, given that the latest record of the largest hailstone across the continent is 150 cm. The probabilistic distribution of the hailstone size in a storm can be calculated according to the function introduced in Section 11.3 from the previous chapter. The size distribution of the background micro-flaws can be generated by use of Eq. (11.7) based on the statistical parameters: $\mu_c = 2.408$ and $\varepsilon_c = 0.642$ 29].

Uniform disposition of the background flaws across the surface of the glass pane is assumed.

11.5.2.1 Probabilistic Analyses Generating Fragility Curves for New Toughened Glass

Parametric studies were conducted initially on new glazing installations. The failure probability of a 6-mm-thick toughened glass pane is presented as a function of hailstorm severity in Figure 11.20, and the influence of the glass pane thickness is shown in Figure 11.21a. Charts of this form would enable the rational selection of skylights to be tailored for different areas around the globe. The influence of the surface condition of the newly manufactured glass (which controls the distribution of the size of the background micro-flaws) is also presented in Figure 11.21b. This type of information would be of use to stakeholders, architects, insurers, and manufacturers.

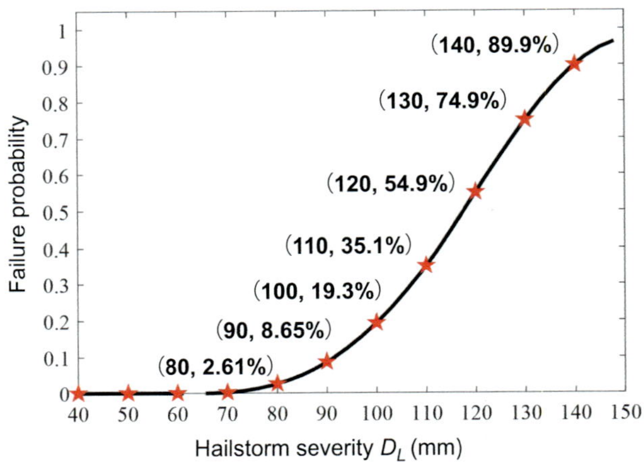

FIGURE 11.20 Influence of hailstorm severity on the failure probability.

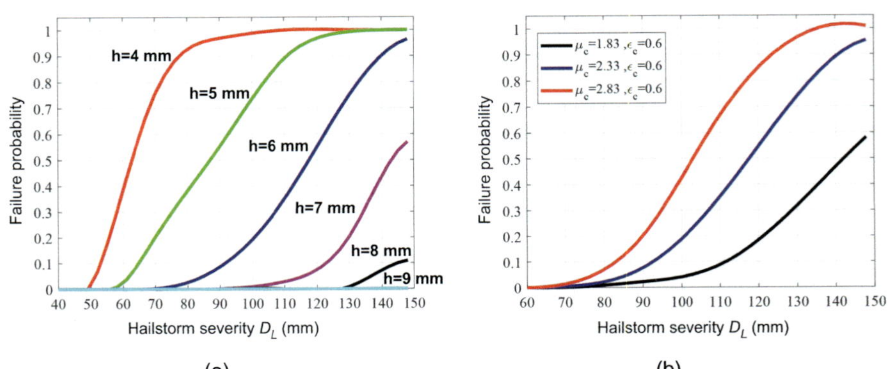

FIGURE 11.21 Influence by (a) pane thickness and (b) surface condition on the failure probability.

11.5.2.2 Probabilistic Analyses Generating Fragility Curves for In-Service Toughened Glass

In another illustrated example, an in-service installation which was found to have a scratch which was offset by some 5 cm from the centre of the glass pane. The influence of the depth and length of a critical crack on the fragility curves showing their respective probability of failure is presented in Figures 11.22a and b. This type of information would greatly facilitate the assessment of the residual resistance of in-service glass panes in which relevant details of the dominant cracks have been found from field surveys. Notably, in addition to crack depths and lengths, the failure probability was found to be particularly sensitive to the disposition of the cracks. Figure 11.23a shows the failure probability of a 10 cm crack which has been divided into 10, 20, 100, and 200 segments dispersed across the plate. Multiple distributed cracks are found to pose a higher risk of failure compared to a single crack of the same equivalent length. This observation can be explained by referring to the larger influential area of multiple distributed cracks compared to a single (long) crack, as illustrated in Figure 11.23b.

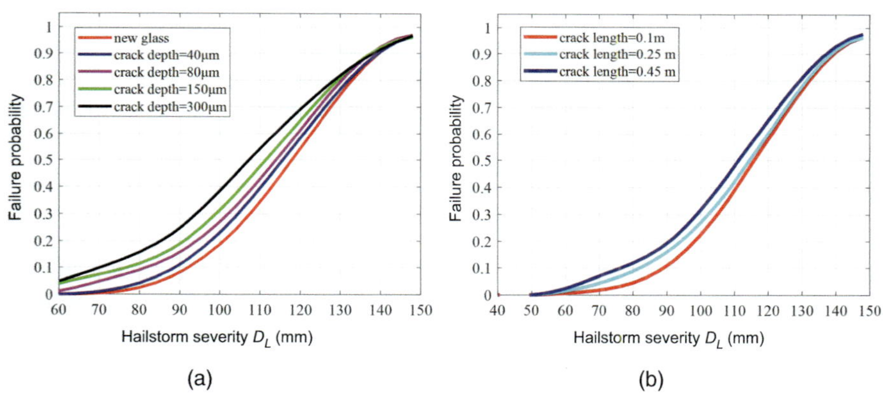

FIGURE 11.22 Influence by (a) crack depth and (b) crack length on the failure probability.

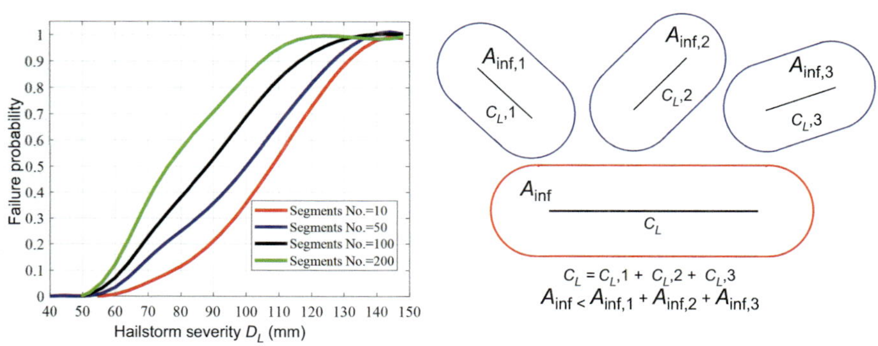

FIGURE 11.23 Influence of crack distribution on the failure probability.

11.6 CONCLUSION

A detailed framework for evaluating the resistance of toughened glass to hail hazards is presented in this chapter, combining experimental observations with both deterministic and probabilistic modelling approaches. The fundamental failure mechanisms of glass were treated in detail with particular reference to flexural cracking, Hertzian fracture, and the significance of the offset distance of the point of impact from the pre-existing critical flaws. The influence of key parameters, namely depth of crack, offset distance, and disposition of the background flaw, on the probability of failure was studied in support of the development of the accurate risk assessment methodology.

To quantify the margin of safety margin in an impact scenario, a deterministic approach utilising a damage factor was introduced, and supported by analytical expressions for estimating the stress intensity factors. These calculations were further extended using probabilistic techniques to develop fragility curves that capture the inherent variability in flaw characteristics and impact conditions. This integrated approach enables the prediction of failure likelihood under realistic hail scenarios, providing practical guidance for engineering design and the performance evaluation of glazing systems.

The methods and insights outlined in this chapter aim to support the development of more resilient, reliable, and climate-adaptive building envelopes, contributing to improved safety, material efficiency, and long-term durability in the face of increasingly extreme weather patterns.

ACKNOWLEDGEMENT

The authors gratefully acknowledge the financial support provided by the Australian Research Council (ARC) through projects LP190100208, IC230100015, IH240100006, and LP240100097. The authors also thank Mr. Ian Bennie (General Manager, Ian Bennie and Associates Pty Ltd) and Mr. Suresh Sutrave (Managing Director, Atlite Skylights Pty Ltd) for their valuable cash and in-kind contributions as Partner Investigators.

REFERENCES

1. P. R. Sparks. Wind speeds in tropical cyclones and associated insurance losses. *Journal of Wind Engineering and Industrial Aerodynamics* 2003; 91:1731–1751. https://doi.org/10.1016/j.jweia.2003.09.018
2. P. R. Sparks, S. D. Schiff, and T. A. Reinhold. Wind damage to envelopes of houses and consequent insurance losses. *Journal of Wind Engineering and Industrial Aerodynamics* 1994; 53.1–2:145–155. https://doi.org/10.1016/0167-6105(94)90023-X
3. J. E. Minor. Windborne debris and the building envelope. *Journal of Wind Engineering and Industrial Aerodynamics* 1994; 53.1–2:207–227. https://doi.org/10.1016/0167-6105(94)90027-2
4. S. A. Changnon. Temporal and spatial distributions of damaging hail in the continental United States. *Physical Geography* 2008; 29.4:341–350. https://doi.org/10.2747/0272-3646.29.4.341

5. Y. Cui, N. Lam, S. Shi, et al. Flexural strains in a toughened glass panel generated by impact of an ice sphere. *Int J Solids Struct* 2023; 281. https://doi.org/10.1016/j.ijsolstr.2023.112438
6. R. Danzer, P. Supancic, J. Pascual, et al. Fracture statistics of ceramics – Weibull statistics and deviations from Weibull statistics. 2007; 74:2919–2932. https://doi.org/10.1016/j.engfracmech.2006.05.028
7. C. Lu, R. Danzer, and F. D. Fischer. Scaling of fracture strength in ZnO: Effects of pore/grain-size interaction and porosity. 2004; 24:3643–3651. https://doi.org/10.1016/j.jeurceramsoc.2003.12.001
8. M. Pathirana, N. Lam, S. Perera, et al. Risks of failure of annealed glass panels subject to point contact actions. *Int J Solids Struct* 2017; 129:177–194. https://doi.org/10.1016/j.ijsolstr.2017.09.001
9. I. Nurhuda, N. T. K. Lam, E. F. Gad, et al. Estimation of strengths in large annealed glass panels. 2010; 47:2591–2599. https://doi.org/10.1016/j.ijsolstr.2010.05.015
10. Z. Wang, T. Guan, T. Ren, et al. Effect of normal scratch load and HF etching on the mechanical behavior of annealed and chemically strengthened aluminosilicate glass. 2020; 46:4813–4823. https://doi.org/10.1016/j.ceramint.2019.10.214
11. A. Ball and H. W. McKenzie. On the low velocity impact behaviour of glass plates. 1994; 04:C8-783–C8-788. https://doi.org/10.1051/jp4:19948121
12. Y. Cui, N. Lam, S. Shi, et al. Deterministic failure prediction of toughened glass when impacted by ice. *Int J Impact Eng* 2024; 185:https://doi.org/10.1016/j.ijimpeng.2023.104833
13. I. V. Roisman. Hydrodynamic model of a collision of a spherical plastic ice particle with a perfectly rigid substrate. 2022; 159:https://doi.org/10.1016/j.ijimpeng.2021.104019
14. H. Kim and J. N. Keune. Compressive strength of ice at impact strain rates. 2007; 42:2802–2806. https://doi.org/10.1007/s10853-006-1376-x
15. S. J. Jones. High strain-rate compression tests on ice. *The Journal of Physical Chemistry B* 1997; 101.32:6099–6101. https://doi.org/10.1021/jp963162j
16. M. Shazly, V. Prakash, and B. A. Lerch. High-strain-rate compression testing of ice (No. NASA/TM-2006-213966). *Cold Regions Science and Technology* 2006. https://doi.org/10.1016/0165-232X(82)90015-5
17. M. Mellor and D. M. Cole. Deformation and failure of ice under constant stress or constant strain-rate. *Journal of Offshore Mechanics and Arctic Engineering* 1982; 5.3: 201–219. https://doi.org/10.1115/1.2920104
18. G. A. Kuehn, Schulson, E. M., Jones, D. E., and Zhang, J. The compressive strength of ice cubes of different sizes. 1993; 115(2):142–148.
19. Y. Cui, N. Lam, S. Shi, et al. Fast fracture in toughened glass when impacted randomly by ice. *Int J Impact Eng* 2024; 197:105091.
20. S. Chen, E. Gad, L. Zhang, et al. Experiments on an ice ball impacting onto a rigid target. 2022; 167: https://doi.org/10.1016/j.ijimpeng.2022.104281
21. K. Osnes, O. S. Hopperstad, and T. Børvik. Rate dependent fracture of monolithic and laminated glass: Experiments and simulations. 2020; 212: https://doi.org/10.1016/j.engstruct.2020.110516
22. X. Zhang, H. Hao, and G. Ma. Dynamic material model of annealed soda-lime glass. 2015; 77:108–119. https://doi.org/10.1016/j.ijimpeng.2014.11.016
23. B. Jiang, M. Atif, Y. Ding, et al. An experimental study on the dynamic flexural tensile behavior of glass. 2022; 266: https://doi.org/10.1016/j.engfracmech.2022.108417
24. R. J. Charles. Dynamic fatigue of glass. *Journal of Applied Physics* 1958; 29.12: 1657–1662. https://doi.org/10.1063/1.1723019
25. J. E. Ritter. Dynamic fatigue of soda-lime-silica glass. 1969; 40:340–344. https://doi.org/10.1063/1.1657056

26. X. Nie, W. W. Chen, A. A. Wereszczak, et al. Effect of loading rate and surface conditions on the flexural strength of borosilicate glass. 2009; 92:1287–1295. https://doi.org/10.1111/j.1551-2916.2009.03019.x
27. M. Haldimann. Fracture strength of structural glass elements: analytical and numerical modelling, testing and design. 2006; https://doi.org/10.5075/epfl-thesis-3671
28. M. Pathirana, N. Lam, S. Perera, et al. Probabilistic modelling of Hertzian fracture of glass by flying objects impact in bad weather. *Int J Impact Eng* 2018; 118:11–23. https://doi.org/10.1016/j.ijimpeng.2018.03.010
29. Y. Cui, N. Lam, S. Shi, et al. Fragility curves for hail resistance of toughened glass. 2025; 104: https://doi.org/10.1016/j.jobe.2025.112249
30. J. Sun, N. Lam, L. Zhang, et al. Contact forces generated by hailstone impact. *Int J Impact Eng* 2015; 84:145–158. https://doi.org/10.1016/j.ijimpeng.2015.05.015
31. S. Shi, N. Lam, S. Chen, et al. An analytical approach for modelling contact forcing function of hailstone impact. *Int J Solids Struct* 2023; 112214. https://doi.org/10.1016/j.ijsolstr.2023.112214
32. L. Wouters, M. Boon, D. van Putten, et al. A Hail Climatology of the Netherlands. 2019. https://hdl.handle.net/1871.1/6ff0b1d2-863b-42f6-9f13-b88127569027

12 Performance Modelling of Aluminium Building Cladding under Hail Hazards

Shuangmin Shi[a], *Nelson Lam*[a], *Yiwen Cui*[a], *Emad Gad*[b], *and Lihai Zhang*[a]
[a]Department of Infrastructure Engineering, The University of Melbourne, Parkville, VIC, Australia
[b]School of Engineering, Swinburne University of Technology, Hawthorn, VIC, Australia

12.1 INTRODUCTION

Hailstorms are among the most frequent and damaging forms of localised extreme weather events, particularly in regions with high convective storm activity. Their impacts on the built environment can be severe, affecting both structural and non-structural components. Building envelope systems such as cladding panels, skylights, and roofing elements are especially vulnerable due to their exposed locations and are susceptible to localised damage caused by the high transient pressure occurring at the point of contact with the hailstones. With changing climate patterns contributing to increased variability in storm intensity and frequency, the need to control hail-inflicted damage through rational design is becoming increasingly urgent.

Amid the widespread use of aluminium alloy cladding in modern construction in view of its lightweight, corrosion resistance, and aesthetic qualities, the susceptibility to damage by hail remains a cause for concern. Existing design standards and codes fall short of accounting for repeated and variable impact events, leaving a knowledge gap in performance-based assessments and long-term risk evaluation. Filling in the omission requires not only accurate physical modelling of impact behaviour but also integration with probabilistic hazard representation and system-level performance assessment.

This chapter presents a comprehensive framework for evaluating the performance of aluminium cladding panels that are exposed to hail hazard, from the perspectives of contact mechanics through to life-cycle deterioration and retrofitting. The methodology combines analytical modelling of contact force and damage, probabilistic representation of hailstorm occurrence based on meteorological data, and reliability

DOI: 10.1201/9781003567332-12

and resilience-based metrics for evaluating performance over time. A structured approach for assessing and enhancing the performance of building envelope systems in the face of environmental hazards is presented.

12.2 HAIL IMPACT CONTACT FORCE

The contact forcing function generated by the impact of a hailstone is crucial to the reliable prediction of damage to a range of building components. This section presents an analytical model for accurate prediction of the forcing function of hail impact which is dependent on the size of the hailstone, its temperature, and velocity of impact.

Accurate modelling of impact forces depends heavily on the choice of contact law. The Hertzian contact law as derived from classical elastic impact theory [1] does not take into account energy dissipation. The Kelvin–Voigt model [2] which combines a linear spring and damper in parallel introduces the effects of damping but has notable limitations, including discontinuities at the onset of (nonphysical) negative contact forces following the rebound of the hailstone. The Hunt and Crossley model which was developed to address these limitations [3] integrates a non-linear visco-elastic formulation with hysteresis damping, offering a more realistic representation of energy loss occurring in the course of impact. The effectiveness of the model in representing the impact force occurring at the contact with ice, in particular, has been demonstrated [4,5].

12.2.1 Contact Force Model Overview

The contact force model introduced herein adopts a two-degree-of-freedom (2DoF) spring-connected lumped-mass system, consisting of a frontal mass m_1 representing the hailstone travelling at an initial velocity v_0, and a rear mass m_2 representing the target, as shown in Figure 12.1. The interaction between the hail impactor and the target surface is governed by a non-linear viscoelastic spring following the Hunt and Crossley framework, while the rear spring, with stiffness k_2, is linearly elastic to emulate the flexibility of the target. The user of the model is required to specify the size of hail, its temperature, and the velocity of impact as input parameters. The impact action is categorised into two stages: (1) Stage I is the loading phase in which the contact force is characterised by an instant reduction in the effective mass of the hailstone as an impactor, and (2) Stage II is the unloading phase in which the contact stiffness of hail is gradually reduced to zero at the end of the impact event.

The governing equation describing the non-linear force-indentation relationship of the frontal spring is written as Eq. (12.1).

$$F_c = k_n \delta^p + D_n \delta^p \dot{\delta} \tag{12.1}$$

where, F_c is the contact force generated by the impact, k_n is the non-linear stiffness coefficient, p is the exponent characterising the non-linear behaviour of the spring, D_n is the damping coefficient, δ is the indentation of the impactor into the target

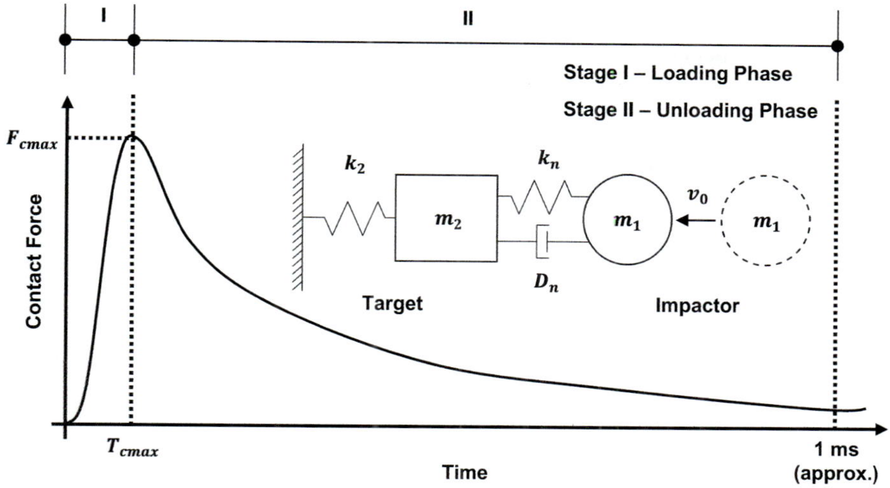

FIGURE 12.1 Schematic representation of the loading and unloading phases during the impact process and the equivalent 2DoF model.

surface defined as the relative displacement by Eq. (12.2a), and $\dot{\delta}$ is the rate of indentation defined as the relative velocity by Eq. (12.2b).

$$\delta = x_1 - x_2 \tag{12.2a}$$

$$\dot{\delta} = \dot{x}_1 - \dot{x}_2 \tag{12.2b}$$

where x_1 and x_2 are the time-dependent displacements, and $\dot{x}_1$ and $\dot{x}_2$ are the time-dependent velocities, of the impactor and target, respectively, in the course of the impact.

The damping coefficient D_n can be expressed in a generalised form as defined by Eq. (12.3).

$$D_n = (0.2p + 1.3) \times \left(\frac{1 - \text{COR}}{\text{COR}}\right) \frac{k_n}{\dot{\delta}_0} \tag{12.3}$$

where COR is the coefficient of restitution and $\dot{\delta}_0$ is the initial indentation velocity ($= v_o$). Expressions for determining the value of k_n, p, and COR are listed as per Eqs. (12.4a), (12.4b) and (12.4c).

$$k_n = (2{,}200 v_0 + 170{,}269)\left(\frac{\sigma_T}{\sigma_{T=-10°\text{C}}}\right)^3 \tag{12.4a}$$

$$p = (0.01 v_0 + 1.263)\sqrt{\frac{\sigma_T}{\sigma_{T=-10°\text{C}}}} \tag{12.4b}$$

TABLE 12.1
Tensile Strength of Ice at Temperature of −10°C and −30°C [6]

Temperature of Ice Ball (°C)	Tensile Strength σ_T (MPa)
−10	0.56
−30	0.69

$$\text{COR} = 0.085 v_0^{-0.97} \times \left(\frac{\sigma_{T=-10^\circ\text{C}}}{\sigma_T} \right)^3 \times \left(\frac{0.0508}{2R_h} \right) \tag{12.4c}$$

where, R_h is radius of the hail impactor, and σ_T is tensile strength of ice to account for temperature effects. Some experimental measurements of σ_T are listed in Table 12.1.

12.2.2 Loading Phase

The loading phase begins at the instance the spherical ice specimen makes the first contact with the surface of the target. In less than half a millisecond, the contact force rises rapidly to its peak. During this extremely short period of time, cracks initiate, and propagate, through the ice sphere at a speed approaching 2,000 m/s. While the external shape and strength of the ice sphere are preserved, its internal structure deteriorates rapidly. Consequently, only a portion of the mass of the ice sphere contributes to the immediate transfer of momentum, thereby reducing the effective inertia force. This behaviour is represented accurately by the Hill function which models the time-dependent effective mass, $m_{1\,\text{eff}}(t)$ as shown in Eq. (12.5).

$$m_{1\,\text{eff}}(t) = m_1 \times \left(\frac{K_1^{n_1}}{K_1^{n_1} + \left(\frac{F_c(t)}{\sigma_T R_h^2} \right)^{n_1}} \right) \tag{12.5}$$

where, K_1 is the threshold contact force to initiate rapid change in the effective mass of the impactor, n_1 is a parameter for characterising the rate of decay of the effective mass, and $F_c(t)$ is the evolving contact force. High-speed camera observations indicate that K_1 is closely dependent on the initial impact velocity v_0, and can be estimated as $K_1 = 0.12 v_0^{1.31}$. The value of n_1 is empirically calibrated to a value of 7.0.

12.2.3 Unloading Phase

The unloading phase begins immediately at the instance when the contact force has risen to its peak. The unloading phase lasts roughly ten times longer than the loading phase, and features the progressive disintegration of the ice specimen into fine debris which disperses radially. During this process, the contact stiffness, $k_n(t)$, decreases significantly over time as modelled by the Hill function of Eq. (12.6).

$$k_n(t) = k_{no} \times \left(\frac{K_2^{n2}}{K_2^{n2} + \left(\frac{t}{T_{c\max}} - 1 \right)^{n2}} \right) \tag{12.6}$$

where, k_{no} is the initial contact stiffness as calculated from Eq. (12.4a), t is time at the commencement of each time step, $T_{c\max}$ is the time corresponding to the occurrence of the peak contact force, K_2 is the threshold stiffness that triggers fragmentation resulting in a marked reduction in stiffness, while n_2 defines the rate of decay in stiffness on unloading. At −10°C, K_2 has the value of 0.84. With falling temperature, the value will decrease. With the value of n_2 calibrated to unity, the dependence of K_2 on temperature can be modelled by the simplified relationship: $K_2 = 0.84\left(\sigma_{T=-10^\circ C} / \sigma_T\right)^2$.

The fragmentation and disintegration of the ice sphere (impactor) on unloading would also reduce the rate of indentation into the target, as represented by Eq. (12.7).

$$\delta(t) = 2\dot{\delta}_{c\max} \times \left(\frac{1}{1 + \left(\frac{t}{T_{c\max}} \right)^{n3}} \right) \tag{12.7}$$

where $\dot{\delta}_{c\max}$ is the rate of indentation at the instance when the contact force has risen to the peak; and parameter n_3 which characterises the rate of decrease is found to be equal to unity.

12.2.4 Computational Simulation of Contact Force

The contact force history is simulated through discrete time-step integration based on the expressions introduced earlier, beginning with the dynamic equilibrium equations given in Eqs. (12.8a) and (12.8b).

$$k_n \delta^p + D_n \delta^p \dot{\delta} = m_1 \ddot{x}_1 \tag{12.8a}$$

$$k_n \delta^p + D_n \delta^p \dot{\delta} = m_2 \ddot{x}_2 + k_2 x_2 \tag{12.8b}$$

Initially, both the impactor and target are assigned zero displacement, velocity, and acceleration, except for the impactor's velocity, which is set to the initial incident velocity. The timestep increment Δt is typically set to the order of 10^{-6} seconds, or less. At each new timestep, the value of the displacement and velocity of both the impactor and target can be determined in sequence by use of parameters calculated from the previous timestep, using Eqs. (12.9a) and (12.9b).

$$x_{1,2}(i+1) = x_{1,2}(i) + \dot{x}_{1,2}(i)\Delta t \tag{12.9a}$$

$$\dot{x}_{1,2}(i+1) = \frac{x_{1,2}(i+1) - x_{1,2}(i)}{\Delta t} + \ddot{x}_{1,2}(i)\Delta t \tag{12.9b}$$

The relative displacement and rate of indentation are hence calculated using Eqs. (12.2a) and (12.2b). The value of the contact force is then obtained via Eq. (12.1). The accelerations of both bodies at the new timestep may then be calculated using Eqs. (12.10a) and (12.10b).

$$\ddot{x}_1(i+1) = -\frac{F_c(i+1)}{m_{1\,\text{eff}}(t)} \tag{12.10a}$$

$$\ddot{x}_2(i+1) = \frac{F_c(i+1) - k_2 x_2(i+1)}{m_2} \tag{12.10b}$$

The calculation proceeds iteratively until the contact force, F_c, stops increasing with time, marking the onset of the unloading phase. At each time step, the indentation rate, $\dot{\delta}(t)$ is evaluated using Eq. (12.7), and the relative displacement, δ, is updated using the trapezoidal approximation, as defined in Eq. (12.11).

$$\delta(i+1) = \delta(i) + \frac{1}{2}\Delta t\left[\dot{\delta}(i+1) + \dot{\delta}(i)\right] \tag{12.11}$$

The time-dependent contact stiffness $k_n(t)$ which is calculated using Eq. (12.6) is then substituted into Eq. (12.1) for estimating the value of the contact force. This calculation loop continues with a time-step size of 10^{-6} seconds until the contact force value is effectively reduced to zero.

12.2.5 Worked Example – Simulated Contact Force Time Histories

Figure 12.2 presents the simulated time histories of the contact force generated from the calculation procedure as described in the foregoing. The influence of various input parameters, namely hail size, temperature, and incident velocity, can be viewed at a glance. An increase in either the impactor size or velocity, both of which increase the total amount of momentum delivered by the impact, would result in a higher value of the peak contact force along with a slight increase in the rise time, as shown in Figure 12.2a and b. In contrast, a lower hail temperature would result in a higher value of the peak contact force which is reached over a marginally shorter duration, as shown in Figure 12.2c and d.

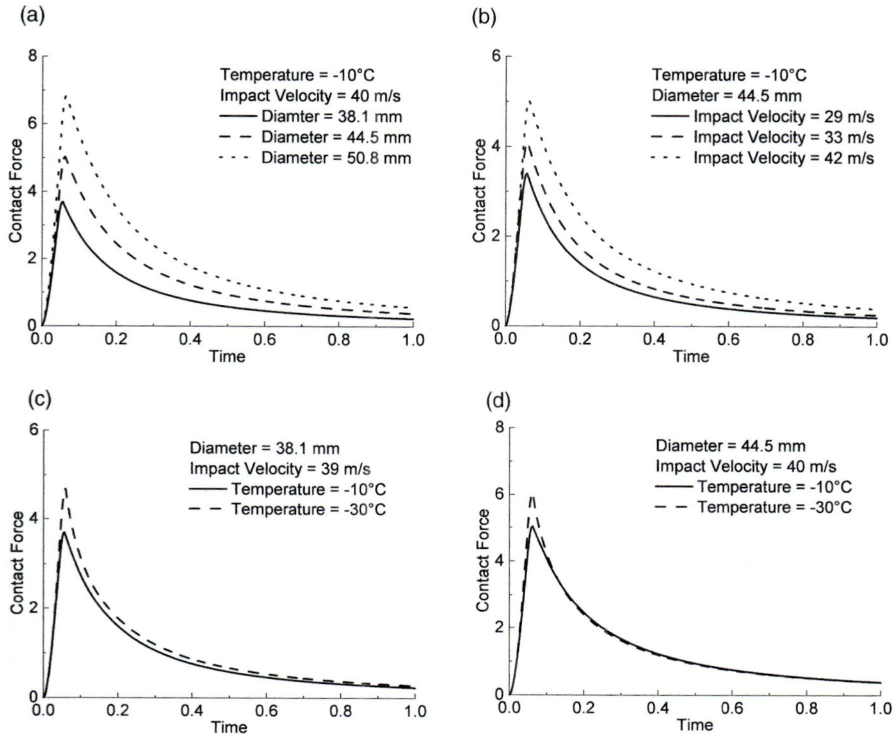

FIGURE 12.2 Comparisons of the model simulated contact force time histories: (a) hail impactor at −10°C of different sizes with common impact velocity of 40 m/s, (b) hail impactor at −10°C of 44.5 mm in diameter with different impact velocities, (c) hail impactor of 38.1 mm in diameter with common impact velocity of 39 m/s under different temperatures, and (d) hail impactor of 44.5 mm in diameter with common impact velocity of 40 m/s under different temperatures.

12.3 HAILSTORM INTENSITY AND HAIL SIZE DISTRIBUTION

Many studies on the damaging potential of hail were based on scenarios involving hailstones of a predefined size. However, to enable a more realistic assessment, having knowledge of how the size of hailstones is distributed in a natural hailstorm would be essential. This section introduces the modelling of the particle size distribution (PSD) of hail to be representative of the behaviour of a real hailstorm.

12.3.1 Hail Particle Size Distribution

The hail PSD presented in this section is a function of the target area, which can be the surface area of a cladding panel, and the maximum diameter of a hailstone, based on field observations from meteorological studies [7]. The distribution is modelled using an exponential function, as given in Eq. (12.12).

$$n(D) = N_0 \exp(-\lambda D) \tag{12.12}$$

where $n(D)$ is the PSD of hail within an air parcel, expressed in terms of the number of hailstones in the diameter interval of $(D, D + dD)$ per unit volume of air. Parameters λ and N_0 have values as defined by Eqs. (12.13a) and (12.13b), respectively.

$$\lambda = \frac{1}{D_A} \frac{\gamma(2.5, \lambda D_S) - \gamma(2.5, \lambda D_L)}{\gamma(1.5, \lambda D_S) - \gamma(1.5, \lambda D_L)} \tag{12.13a}$$

$$N_0 = \frac{\text{Hr}}{v_{TO}} \frac{\lambda_{1.5}}{\gamma(1.5, \lambda D_S) - \gamma(1.5, \lambda D_L)} \tag{12.13b}$$

where, D_A, D_S, and D_L are the average, smallest, and largest diameter of hail, respectively, Hr is the hit rate of hail onto the target, and v_{TO} is the terminal velocity which can be estimated using Eq. (12.14).

$$v_T(D) = \sqrt{\frac{4\rho_h g D}{2\rho_a C_d}} = \sqrt{\frac{4\rho_h g}{2\rho_a C_d}} \sqrt{D} = v_{TO} D^{1/2} \tag{12.14}$$

where, ρ_h is the density of hail, ρ_a is the density of air, and C_d is the drag coefficient.

To improve computational efficiency, Eqs. (12.13a) and (12.13b) can be simplified into a form where D_L is the sole variable. This is justified as D_L is the most widely used parameter for characterising the intensity of a hailstorm. The simplification is achieved by correlating D_L with other parameters, namely D_A, D_S, and Hr, as expressed in Eqs. (12.15a), (12.15b), and (12.15c).

$$D_A = 1.449 D_L^{0.649} \tag{12.15a}$$

$$D_S = 1.004 D_L^{0.542} \tag{12.15b}$$

$$\text{Hr} = 153.403 D_L^{-1.087} \tag{12.15c}$$

where, D_A, D_S, and D_L are in the unit of mm, whereas Hr is in the unit of $\text{min}^{-1}\text{m}^{-2}$.

A list of values for λ and N_0 for a range of common values of D_L is presented in Table 12.2.

The size distribution model described above is essential for accurately estimating the extent of hail damage caused to skylights, roofing panels, claddings, and automotive components. Equation (12.16) provides a means of calculating the total number of hailstones, N_t, which is expected to strike a panel of one square metre in size in a hailstorm event.

$$N_t = 73{,}616 D_L^{-0.917} \tag{12.16}$$

where N_t is in the unit of m^{-2}. For a given cladding surface exposed to hail, the expected number of strikes by hailstones within each size range can be calculated by taking the product of the value of N_t calculated using Eq. (12.16) and the exposed area.

TABLE 12.2
Calculated Values of N_0 and λ for Some Predefined Values of D_L

D_L (mm)	λ	N_0
20	0.1873	0.1683
30	0.1562	0.0869
40	0.1328	0.0481
50	0.1158	0.0292
60	0.1030	0.0191

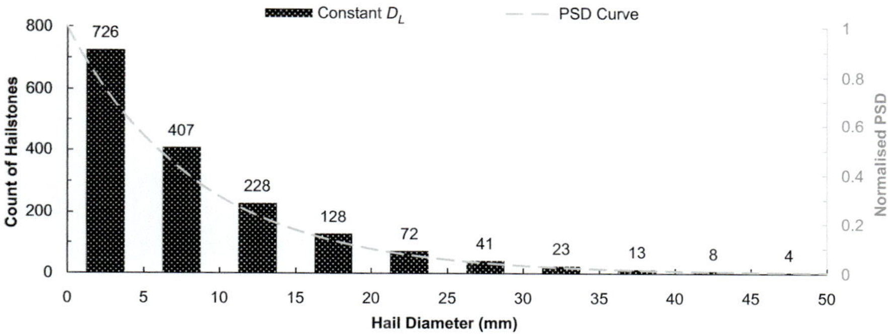

FIGURE 12.3 Distribution of hailstone impacting a cladding panel with planar dimensions of 900 mm × 900 mm in a hailstorm event with an expected maximum diameter of 50 mm.

12.3.2 Worked Example – Hail Size Distribution

For a well-defined target such as the surface of a cladding panel exposed to hail hazard, the expected number of strikes by hailstones in each size range can be determined by taking the product of N_t as found from Eq. (12.16) and the exposed area. For example, consider a hailstorm characterised by a D_L of 50 mm. The modelled distribution of hailstones impacting on a panel with planar dimensions of 900 mm by 900 mm is shown in Figure 12.3.

12.4 DETERMINISTIC DAMAGE MODELLING INDENTATION

There is currently no rational method of design for claddings, or rooftops, to counter hail hazard despite the scale of damage that was inflicted on aluminium panels in hailstorm events of the recent past. This section presents an analytical model for predicting permanent indentation on the surface of an aluminium cladding panel when impacted by a hailstone.

The hailstone-panel contact process can be divided into four stages. Stage I marks the initial contact between the hailstone and the surface of the panel. In less than half a millisecond, both the contact force and the resulted in-plane strain rise rapidly until reaching the peak (Stage II), during which time cracks fully propagate through the hailstone while local indentation begins to form on the panel. As the hailstone disintegrates, both the contact force and the in-plane strain decrease to zero. The global displacement of the panel, which evolves at a much slower pace, continues to increase until the maximum deflection is reached (Stage III). This global response, which is dominated by elastic deflection and vibration, eventually subsides through energy dissipation by damping. Finally, a permanent indentation remains at the location of the "impact site" where contact is made with the hailstone (Stage IV).

12.4.1 Effect of Strain Rate

The effect of strain rate must not be overlooked in hailstorm scenarios. Unlike static or quasi-static loading, the dynamic behaviour of a panel in response to impact by hail requires consideration of rate-dependent material behaviour. In modelling the dynamic response behaviour, the mean dynamic flow stress σ_d based on the Tresca yield criterion for rigid plastic material is introduced as a parameter [8]. The expression of σ_d, which is commonly known as the Cowper-Symonds equation, is given in Eq. (12.17).

$$\sigma_d = \sigma_y \left[1 + \left(\frac{\dot{\varepsilon}_m}{C} \right)^{1/q} \right] \tag{12.17}$$

where, σ_y is the yield stress of the aluminium alloy material, $\dot{\varepsilon}_m$ is the mean strain rate, C and q are material constants which may be taken to be equal to 6,700 and 4, respectively, for aluminium alloy [9].

The mean strain rate is defined as the average rate of increase (the slope) within the rise time, which corresponds to the period when the in-plane surface strain increases steadily from zero to its peak. Based on a series of impact experiments, a power-law correlation has been established between the mean strain rate and the ratio of the impact velocity v_0 (m/s) and plate thickness h (m), as shown in Eq. (12.18).

$$\dot{\varepsilon}_m = 2.09 \left(\frac{v_0}{h} \right)^{0.45} \tag{12.18}$$

Consider a hypothetical scenario where a hailstone impacts a 3 mm thick aluminium panel of grade ASTM 5052-H32 at about 40 m/s. The mean strain rate is estimated to be 150.1 s^{-1} using Eq. (12.18). Substituting the calculated mean strain rate into Eq. (12.17) gives the estimated dynamic flow stress of 222 MPa approximately which is almost 1.4 times higher than the yield (proof) strength of the aluminium material (160 MPa). The significance of strain rate effects is evident.

12.4.2 Indentation Profile

A generalised algebraic expression for defining the shape of indentation at the contact region is crucial for accurately estimating the amount of energy absorption.

Table 12.3 lists several indentation functions as reported in the literature. The reported shape function which is presented in the generalised form of Eq. (12.19) as shown in Figure 12.4 is compared against experimental measurements along with the other published shape functions (as summarised in Table 12.3).

$$w(r) = w_0 e^{a(r/h)^b} \tag{12.19}$$

where, $w(r)$ is the normalised profile of the permanent indentation, w_0 is the maximum amount of permanent indentation occurring at the impact point, r is the radial distance measured from the point of impact, h is the thickness of the aluminium panel. Values of the coefficients a and b, which can be determined through calibration against experimental data, are −0.09 and 1.4, respectively [10]. As shown in Figure 12.4, the indentation profile as defined by Eq. (12.19) matches with experimental measurements much better than the three published shape functions (as listed in Table 12.3).

TABLE 12.3
Empirical Indentation Shape Functions Reported in the Literature

No.	Shape Function	Reference
1.	$w(r) = w_0 \cos\left(\frac{\pi r}{2R}\right)$	[11]
2.	$w(r) = w_0\left[1 - \left(\frac{r}{R}\right)^2\right]$	[11]
3.	$w(r) = w_0 J_0\left(\frac{ar}{R}\right)$	[8]

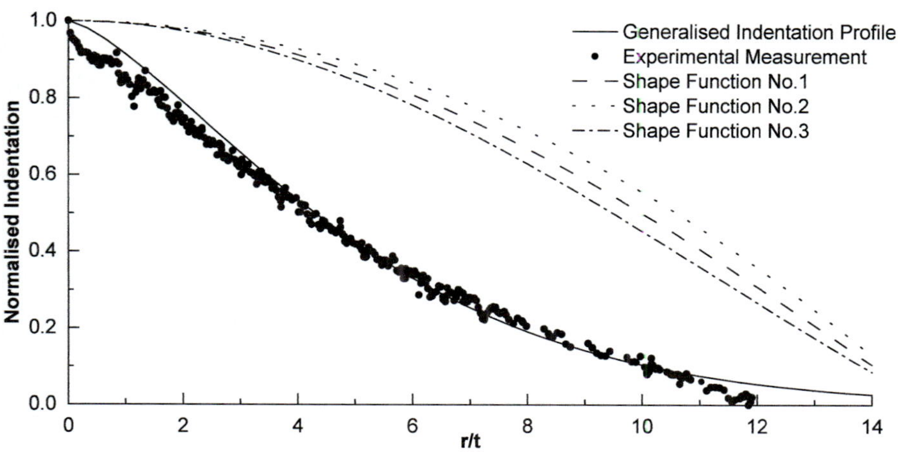

FIGURE 12.4 Comparison of shape functions reported in the literature, the generalised indentation profile proposed in Eq. (12.19), and experimental measurements.

12.4.3 Predictive Model

The total energy dissipated through plastic deformation leading to permanent indentation, referred to herein as plastic work, is given by the volume integral of the sum of the products of radial and circumferential stresses with their corresponding strains. This relationship is expressed in Eq. (12.20).

$$U_p = \int_{\text{vol}} \left[\int (\sigma_r d\varepsilon_r + \sigma_\theta d\varepsilon_\theta) \right] dV \tag{12.20}$$

where, U_p is the plastic work, and σ_r, σ_θ, ε_r, and ε_θ are the stresses and strains in the radial and circumferential directions, respectively. It is important to note that the equality in Eq. (12.20) is based on thin plate theory assuming plane stress conditions and symmetry, consistent with the Kirchhoff hypothesis. These assumptions are justified by experimental observations showing nearly identical strain measurements on both the front surface (which is exposed to the impact) and rear surface of the plate, indicating that the neutral axis is little affected by the impact.

The relevant strains can be further expressed in Eqs. (12.21a) and (12.21b).

$$\varepsilon_r = \varepsilon_m + \varepsilon_{rb} = \frac{1}{2}\left(\frac{\partial w}{\partial r}\right)^2 + z\kappa_r \tag{12.21a}$$

$$\varepsilon_\theta = \varepsilon_{\theta b} = z\kappa_\theta \tag{12.21b}$$

where, ε_m, ε_{rb}, and $\varepsilon_{\theta b}$ are the radial membrane strain, radial bending strain, and circumferential bending strain, respectively, z represents the direction of indentation (transverse) in a cylindrical coordinate system as depicted in Figure 12.5, κ_r and κ_θ are the radial and circumferential curvatures, respectively, as expressed in Eqs. (12.22a) and (12.22b).

$$\kappa_r = -\frac{\partial^2 w}{\partial r^2} \tag{12.22a}$$

$$\kappa_\theta = -\frac{1}{r}\frac{\partial w}{\partial r} \tag{12.22b}$$

The radial and circumferential stresses are evaluated using the mean dynamic flow stress σ_d as defined by Eq. (12.17). The total plastic work is then determined by substituting Eqs. (12.21) and (12.22) into Eq. (12.20), resulting in Eq. (12.23).

$$U_p = U_m + U_{rb} + U_{\theta b} = \int_{\text{vol}} \sigma_d (\varepsilon_m + \varepsilon_{rb} + \varepsilon_{\theta b}) dV \tag{12.23}$$

where, U_m, U_{rb}, and $U_{\theta b}$ are the plastic work components attributed to the radial membrane force (stress), radial bending moments, and circumferential bending moments,

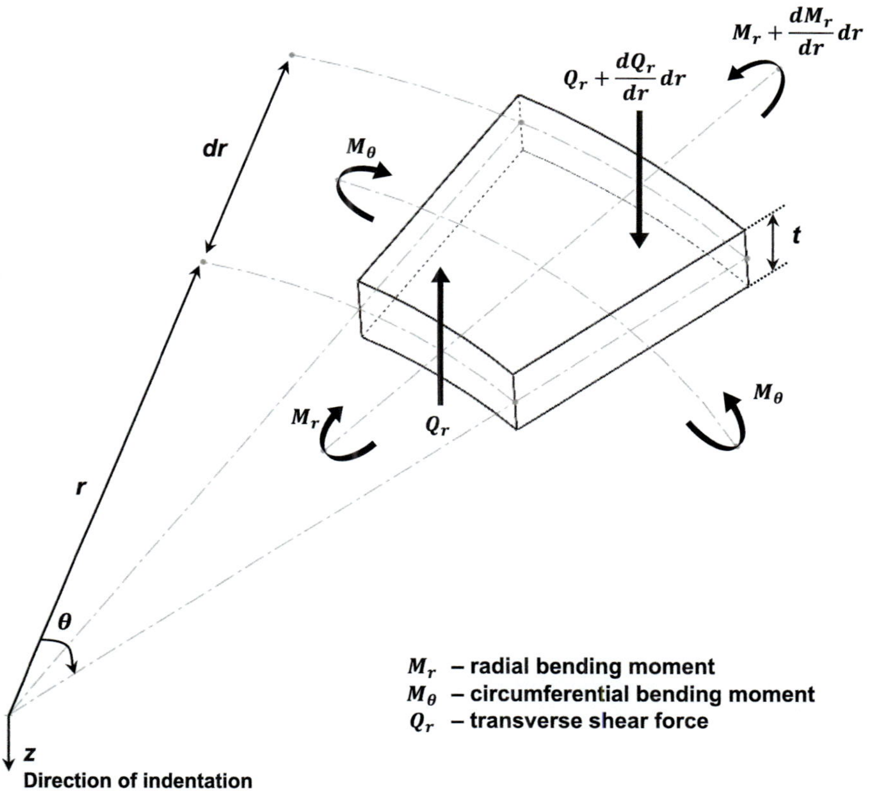

FIGURE 12.5 Cylindrical coordinate system and notation used in this study.

respectively. The full expressions for all three plastic work terms are shown in Eqs (12.24a), (12.24b), and (12.24c).

$$U_m = \pi \sigma_d h \int_0^R \left(\frac{\partial w}{\partial r} \right)^2 r\, dr \tag{12.24a}$$

$$U_{rb} = \frac{1}{2} \pi \sigma_d h^2 \int_0^R \left(-\frac{\partial^2 w}{\partial r^2} r \right) dr \tag{12.24b}$$

$$U_{\theta b} = \frac{1}{2} \pi \sigma_d h^2 \int_0^R \left(-\frac{\partial w}{\partial r} \right) dr \tag{12.24c}$$

where R is the radial distance measured from the point of contact to the edge of the dented area, and w is the indentation profile as defined by Eq. (12.19).

Substituting Eqs. (12.19) and (12.24) into Eq. (12.23) which can be rearranged into Eq. (12.25) and can be used for predicting the amount of permanent indentation w_0.

$$w_0 = \frac{1}{2}\left(-h\frac{\mathrm{B}}{\mathrm{A}} + \sqrt{t^2\frac{\mathrm{B}^2}{\mathrm{A}^2} + \frac{4U_p}{\pi\sigma_d h\mathrm{A}}}\right) \tag{12.25}$$

where A and B are coefficients (having constant values) for a given thickness h of the panel. The values of these coefficients for two common panel thicknesses are listed in Table 12.4.

Equation (12.25) can be used for fast tracking the estimation of w_0 for given projected impact scenarios, as correlations of experimentally measured results have been developed between the amount of plastic work U_p and the initial kinetic energy KE_0 delivered by the projectile. These correlations can be summarised into Eqs. (12.26a) and (12.26b).

$$\left.\frac{U_p}{\mathrm{KE}_0}\right|_{t=3\ \mathrm{mm}} \approx 6.55\% \text{ for a plate with thickness of 3 mm} \tag{12.26a}$$

$$\left.\frac{U_p}{\mathrm{KE}_0}\right|_{t=4\ \mathrm{mm}} \approx 2.09\% \text{ for a plate with thickness of 4 mm} \tag{12.26b}$$

where $\mathrm{KE}_0 = \frac{1}{2}m_h v_0^2$ with m_h being the mass of the hailstone.

12.4.4 Worked Example

Consider an impact scenario where a hailstone of 50.8 mm in diameter and weighing 59.5 g, impacting an aluminium cladding panel with an incident velocity of 50 m/s. The targeted panel, fully clamped on all four edges, has dimensions of 900 mm × 900 mm × 3 mm thick and is fabricated from alloy grade ASTM 5052-H32. The yield strength of this grade is taken as 160 MPa.

Solution:
The mean strain rate ($\dot{\varepsilon}_m$) which is estimated using Eq. (12.18) is substituted into Eq. (12.17) for calculation of the dynamic flow stress (σ_d).

$$\dot{\varepsilon}_m = 2.09\left(\frac{v_0}{t}\right)^{0.45} = 2.09\left(\frac{50}{0.003}\right)^{0.45} = 166\ \mathrm{s}^{-1}$$

TABLE 12.4
Numerical Values of A and B for Panels with a Thickness of 3 and 4 mm

Aluminium Panel Thickness (mm)	A	B
3	0.3499	0.9422
4	0.3454	0.7743

$$\sigma_d = \sigma_y \left[1 + \left(\frac{\dot{\varepsilon}_m}{D} \right)^{\frac{1}{q}} \right] = 160 \left[1 + \left(\frac{166}{6{,}700} \right)^{\frac{1}{4}} \right] = 223.5 \text{ MPa}$$

Equation (12.26a) is then made use of for predicting the amount of energy (U_p) expended in causing permanent indentation. Results so obtained are then substituted into Eq. (12.25) for determination of the maximum indentation (w_0).

$$U_p = 6.55\% \times \text{KE}_0 = 6.55\% \times \frac{1}{2}(0.0595)(50)^2 = 4.87 \text{ J}$$

$$w_0 = \frac{1}{2}\left(-h\frac{\text{B}}{\text{A}} + \sqrt{t^2 \frac{\text{B}^2}{\text{A}^2} + \frac{4U_p}{\pi \sigma_d h \text{A}}} \right)$$

$$= \frac{1}{2}\left(-0.003\frac{0.9422}{0.3499} + \sqrt{0.003^2 \frac{0.9422^2}{0.3499^2} + \frac{4 \times 4.87}{\pi\left(223.5 \times 10^6\right)(0.003)(0.3499)}} \right)$$

$$= 0.75 \text{ mm}$$

12.5 RELIABILITY-BASED PERFORMANCE ASSESSMENT

Cladding panels in buildings are expected to perform satisfactorily and reliably, in severe weather conditions such as hailstorms. Once installed in exposed locations, these panels become susceptible to cumulative structural damage from repeated impacts by debris and hail across multiple storm events over their service life. This section presents a reliability-based framework for predicting hail-inflicted damage to cladding panels, and for estimating the probability of required interventions throughout the lifespan of the installations.

12.5.1 Framework Overview

Reliability-based models provide insights into how a system or component performs and deteriorates over its service life. A schematic of the proposed framework is shown in Figure 12.6. The performance metric used is permanent indentation, $W(t)$. In the life-cycle analysis, it is assumed that damage begins to accumulate from the time of installation at an exposed location, progressing towards failure or severe deterioration requiring intervention. Given the random nature of hailstorm events, damage accumulation is modelled as a jump process incorporating stochastic interevent time intervals. As the amount of indentation increases with each event, the extent of damage would continue to increase over time, eventually reaching the failure threshold requiring replacement.

12.5.2 Intervention Criteria

In the context of hailstorm hazards, damage is assessed based on the extent of permanent indentation on the panel surface. When indentation exceeds a defined threshold,

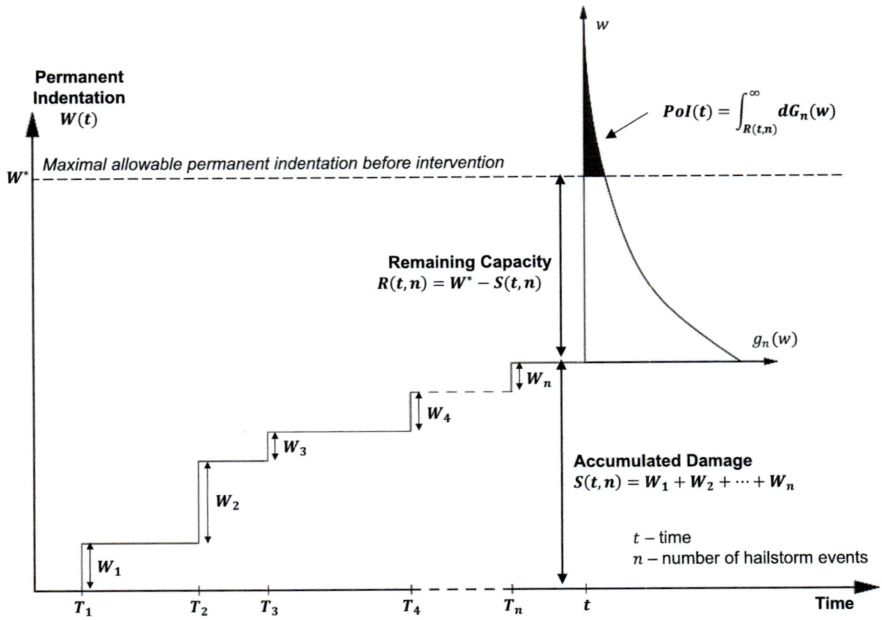

FIGURE 12.6 Schematic representation of life-cycle reliability calculation.

intervention becomes necessary. This threshold may be interpreted in different ways: a lower threshold relates to aesthetic concerns, while an upper threshold corresponds to compromised functionality or structural integrity. For aluminium alloy cladding panels, hail-induced indentations are typically shallow relative to the panel thickness. Therefore, this framework adopts the aesthetic threshold, which is defined as indentation visible to the naked eye (as a trigger for intervention). This lower threshold, denoted as W^* in Figure 12.6, is not specified in design codes but inferred from practical experience. Visibility of indentation depends on factors such as lighting, panel colour, viewing angle, and observation distance. Experimental findings [10,12] suggest that a threshold in the range of 0.3 ~ 0.5 mm is appropriate.

12.5.3 Hazard Function

An effective method for modelling the rate of occurrence of hailstorms and the associated damaging potential is through the use of a marked point process, as described in this section. In the context of life-cycle analysis, the probability that indentation on the panel exceeds a predefined threshold within a given time span [0, t] with notation: $P(T \le t)$ is expressed by Eq. (12.27).

$$P(T \le t) = F_T(t) = 1 - \overline{F_T}(t) \tag{12.27}$$

where, $P(T \le t)$ is the cumulative probability that failure (i.e., damage exceeding the threshold) occurs by time t, $F_T(t)$ is the cumulative distribution function of the damage occurrence time, and $\overline{F_T}(t)$ is the complementary cumulative distribution function, also referred to as the reliability function. The reliability function represents the probability that the system survives without exceeding the threshold up to time t.

To further quantify the risk of failure at a specific moment, the concept of the hazard function is introduced, also known as the hazard rate. This function provides the instantaneous rate at which the event (e.g., threshold exceedance) is expected to occur at time t, given that it has not occurred before that time. Mathematically, it is defined as the conditional probability that failure occurs in the small interval $[t, t+dt]$, given survival up to time t, as shown in Eq. (12.28).

$$P(\text{exceeding threshold}, t \le T \le t+dt) = \frac{P(t \le T \le t+dt)}{1-P(T \le t)} \tag{12.28}$$

This definition captures how the likelihood of damage evolves over time, accounting for the history of system performance. The numerator reflects the probability that failure occurs in the infinitesimal interval $[t, t+dt]$, while the denominator adjusts for the fact that the system has not yet failed by time t.

As the time interval dt approaches zero ($dt \to 0$), the numerator in Eq. (12.28) can be interpreted as the time derivative of function $F_T(t)$ as shown by Eq. (12.29).

$$P(t \le T \le t+dt) \approx \frac{d}{dt}\left[F_T(t)\right] = f_T(t) \tag{12.29}$$

where $f_T(t)$ is the probability density function (PDF) associated with $F_T(t)$. This function characterises the instantaneous probability of failure per unit time at time t.

Substituting Eq. (12.29) into Eq. (12.28) allows us to express the hazard function $h_T(t)$ in a more tractable form, as shown in Eq. (12.30).

$$h_T(t) = \frac{f_T(t)}{1-F_T(t)} = \frac{f_T(t)}{1-\int_0^t f_T(t)} \tag{12.30}$$

This final expression defines the hazard rate as the ratio of the probability density to the reliability function. It characterises the system's time-dependent risk profile and plays a key role in reliability analysis and risk-informed decision-making. In practice, it helps quantify how the failure risk evolves over time, guiding maintenance, and mitigation strategies.

12.5.4 Worked Example – Hazard Function

The relationship between return period and hailstorm severity is essential for defining the hazard function, as given by Eq. (12.30). The return period refers to the average time interval between hailstorm events, whereas severity refers to the storm intensity which is expressed in terms of the maximum recorded hailstone diameter.

In general, more severe hailstorms with larger size of hailstones are associated with longer return periods than milder events. In this example, data from two Australian states, Victoria and New South Wales, are examined. Table 12.5 presents the number of hailstorms recorded in Victoria (1960–2016) and New South Wales (1900–2019) across different ranges of maximum hailstone diameter.

The information presented was extracted from the Australian Bureau of Meteorology (BoM) archive of severe hailstorms. Note, hailstones with diameter of up to 20 mm are likely to melt shortly following impact on the ground, thereby introducing uncertainty in the interpretation of data. This is a source of error posing challenges in the correct interpretation of data. Thus, small hailstones of diameter in the range: [0, 20) may be disregarded given the minor extent of damage caused by hailstones of this range of size. The occurrence of hailstorms across different ranges of diameter can be represented by a lognormal distribution relationship, with the PDF defined by Eq. (12.31).

$$f\left(D_L;\mu,\sigma\right)=\frac{1}{\sigma\sqrt{2\pi}D_L}e^{\left(-\frac{(\ln D_L-\mu)^2}{2\sigma^2}\right)} \tag{12.31}$$

where, μ and σ are the statistical average and standard deviation characterising the distribution relationship. The numerical values of μ and σ for the lognormal distribution are 1.063 and 0.4081 for Victoria, and 1.2055 and 0.4159 for New South Wales. These values were determined with D_L expressed in the unit of centimetres. Data obtained from BoM along with the corresponding fitted lognormal distribution curves, are presented in Figure 12.7a and b, respectively.

Solving the integrals in Eq. (12.31) between the limits: $[D_1, D_2)$ provides an estimate of the cumulative probability that a hailstorm will have a maximum hail diameter within the specified range. The average rate of occurrence (ν) for each range of hailstone size can be calculated by taking the product of the cumulative probability

TABLE 12.5

Number of Hailstorms in Victoria (1960 ~ 2016) and New South Wales (1900 ~ 2019) for Different Hail Diameter Ranges

	Number of Hailstorm Events	
Hail Diameter Range (mm) ***Lower Bound Included***	**Victoria (VIC) 1960 ~ 2016**	**New South Wales (NSW) 1900 ~ 2019**
0 ~ 10	1	5
10 ~ 20	7	16
20 ~ 30	209	586
30 ~ 40	66	305
40 ~ 50	92	444
50 ~ 60	22	163
> 60	27	172

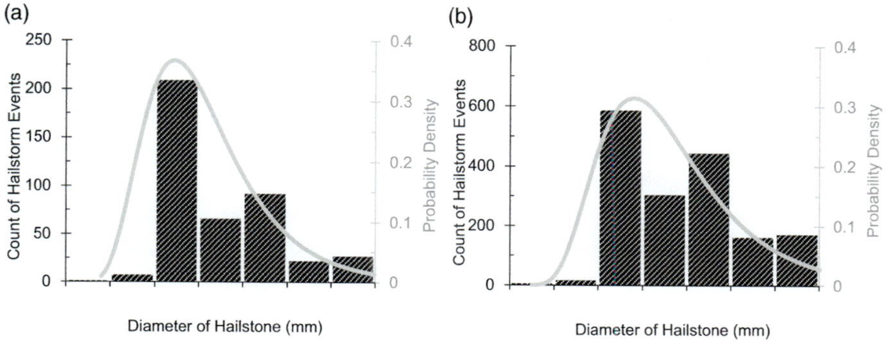

FIGURE 12.7 Distribution of hailstorm events characterised by the maximum recorded hail diameter (a) Victoria (VIC) (1960 ~ 2016) and (b) New South Wales (NSW) (1900 ~ 2019).

TABLE 12.6
Average Occurrence Rate and Return Periods of Hailstorms in Victoria and New South Wales for Different Hail Diameter Ranges

Hail Diameter Range (mm) *Lower Bound Included*	Average Occurrence Rate v (Per Year) Victoria (VIC)	New South Wales (NSW)
20 ~ 30	2.62	4.08
30 ~ 40	1.87	3.80
40 ~ 50	0.92	2.34
50 ~ 60	0.40	1.22
> 60	0.28	1.12

and the corresponding annual rate of occurrence. The latter is obtained by dividing the total number of recorded hailstorm events by the duration of the observation period expressed in years. The computed average rates of occurrence for hail diameter values exceeding 20 mm in diameter are listed in Table 12.6.

Results presented in Table 12.6 can be incorporated into the hazard function as defined by Eq. (12.30) through the use of the Poisson distribution, as expressed in Eq. (12.32).

$$h_T(t) = \frac{f_T(t)}{1 - \int_0^t f_T(t)} = \frac{ve^{-vt}}{1 - \left(1 - e^{-vt}\right)} = v \tag{12.32}$$

12.5.5 Damage Accumulation

Permanent indentation into the surface of the cladding made of aluminium alloy in a hailstorm, referred to herein as the "mark" in the marked point process model, may be represented by a positive random variable W. The value of W_n denotes the amount of indentation caused during the nth hailstorm, based on the probability distribution function $g_n(w)$. In the context of hail impact on aluminium alloy claddings, $g_n(w)$ defines the PDF of permanent indentation resulting from a given hailstorm scenario. This distribution can be derived from the PSD of hailstones (as defined by Eq. 12.12) and the deterministic relationship for quantifying permanent indentation based on known hail impact intensity (as defined by Eq. 12.25).

The main uncertainty in determining $g_n(w)$ arises from the accumulative behaviour of permanent indentation when subject to repeated impact. Experimental results obtained through repeated impact using a consistent hail impactor and fixed incident velocity indicate that the cumulative indentation at the point of impact following the nth impact, $w_{0,i=n}$, can be approximated by Eq. (12.33).

$$w_{0,i=n} = 2w_{0,i=1}\left(\frac{n^2}{1+n^2}\right) \tag{12.33}$$

where, n is the number of impacts that a panel is expected to sustain, and $w_{0,i=1}$ is the permanent indentation resulting from the very first impact, which can be calculated using Eq. (12.25). With this, the expression for $g_n(w)$ can now be fully established and is given in the form of Eq. (12.34).

$$g_n(w) = f(D) \times \frac{dD}{dw_{0,\,i=n}} \tag{12.34}$$

where, $f(D)$ is the PDF of hail size distribution, expressed as $f(D) = \lambda\exp(-\lambda D)$, which is the modified form of Eq. (12.12), and $\frac{dD}{dw_{0,i=n}}$ is the derivative of the diameter of the hail impactor with respect to the permanent indentation, which can be obtained by rearranging Eqs. (12.25) and (12.33) followed by differentiation.

At time t, the accumulated damage $S(t,n)$ and the remaining capacity $R(t,n)$, can be predicted using Eqs. (12.35) and (12.36), respectively. The extent of cumulative damage and the remaining capacity depend strongly on the passage of time (t) and the number of previously occurred hailstorm events (n).

$$S(t,n) = W_1 + W_2 + \cdots + W_n = \sum_{i=1}^{n} W_i \tag{12.35}$$

$$R(t,n) = W^* - D(t,n) = W^* - \sum_{i=1}^{n} W_i \tag{12.36}$$

Assuming that intervention occurs only after a hailstorm event, the probability of intervention (PoI) can be interpreted as the likelihood that a sufficiently large indentation is formed at time t, resulting in cumulative damage exceeding the threshold W^*. This is illustrated by the area bounded by W^*, the w-axis, and the distribution function $g_n(w)$ as shown in Figure 12.6. The time-dependent PoI can be calculated using Eq. (12.37).

$$\mathrm{PoI}(t)=\left(\int_{R(t,n)}^{\infty} g_n(w)dw\right)h_T(t)=\left(\int_{R(t,n)}^{\infty} dG_n(w)\right)h_T(t) \tag{12.37}$$

The complexity in solving Eq. (12.37) arises from the need to determine the lower limit of the integral, $R(t,n)$. While Eq. (12.36) provides a general definition for the remaining capacity $R(t,n)$, quantifying W_n requires knowledge of the preceding remaining capacity, which reflects all accumulated damages up to that point, as shown by Eq. (12.38).

$$W_n = \int_0^{R(t,n-1)} w\,dG_n(w) \tag{12.38}$$

By conditioning the number of hailstorm events (n), Eq. (12.36) can be rewritten into Eq. (12.39).

$$R(t,n)=W^* - \sum_{i=1}^{n}\left[\int_0^{R(t,i-1)} w\,dG_i(w)\right] \tag{12.39}$$

As shown by Eq. (12.39), the remaining capacity $R(t,n)$ also depends on the number of prior hailstorm events, which is treated as a random variable. If the probability of experiencing n events by time t is denoted as $P(i=n,t)$, the expression for PoI can be reformulated as a series format shown in Eq. (12.40).

$$\mathrm{PoI}(t)=\sum_{n=0}^{\infty}\left(\int_{R(t,n)}^{\infty} dG_n(w)\right)h_T(t)P(i=n,t) \tag{12.40}$$

Estimates obtained from the use of Eq. (12.40) represent the instantaneous intervention rate at a given time step. In contrast, the cumulative intervention rate (CIR), defined as the definite integral of the instantaneous rate over time interval [0, t], provides a broader assessment of cladding performance over a specified period. The formulation of the CIR is as shown in Eq. (12.41).

$$\mathrm{CIR}(t)=\int_0^t\left[\sum_{n=0}^{\infty}\left(\int_{R(t,n)}^{\infty} dG_n(w)\right)h_T(t)P(i=n,t)\right]d\tau \tag{12.41}$$

12.5.6 Worked Example – Probability of Intervention

Consider a newly installed aluminium alloy cladding panel located in the state of Victoria, Australia. The panel is fabricated from ASTM 5052 H32 alloy and shares the same material properties as discussed in Section 12.4.4. The average rate of hailstorm occurrence, characterised by the maximum hailstone diameter, can be referenced from Table 12.6. The incident velocity of hail is assumed to be 40 m/s. For new cladding panels with full initial capacity, the threshold for visible permanent indentation is set at 0.3 mm (i.e., $W^* = 0.3$ mm). Under these assumptions, the expected service life of the panel can be evaluated using the life-cycle analysis methodology as described earlier.

The analysis is initiated at time step $t = 0$ with a step size of one year. At each step, contributions from all storm events W_n, $n = 1,2,3,\ldots$ must be considered. The expected extent of damage for each hailstorm scenario is first estimated using Eqs. (12.34) and (12.38), followed by the remaining capacity calculated from Eq. (12.39). This remaining capacity $R(t,n)$ is then used as the lower limit of the integral in Eq. (12.40). The only remaining unknown in Eq. (12.40) is $P(i = n,t)$. Based on the Poisson distribution as shown in Eq. (12.32), which is used to model the hazard function and interarrival time between hailstorm events, the probability of experiencing n hailstorm events by time t can be calculated using Eq. (12.42).

$$P(i = n,t) = \frac{(\nu t)^n e^{-\nu t}}{n!} \tag{12.42}$$

The PoI estimated using Eq. (12.40) corresponds to the intervention rate at a specific time step, commonly referred to as the instantaneous intervention rate. In contrast, the CIR, as defined in Eq. (12.41), offers a more comprehensive view of the performance of the cladding over a given timeframe. This metric is obtained by integrating the instantaneous intervention rate with respect to time, and can be used to evaluate system reliability and inform decision-making over maintenance or replacement. The impact of varying hailstorm intensities on cladding performance is illustrated by the CIR curves as shown in Figure 12.8a. The combined CIR, calculated as the weighted average of probability densities derived from Eq. (12.41) across different intensity levels, is presented in Figure 12.8b.

12.5.7 Worked Example – Factors Affecting Performance

Among the various factors influencing the performance of cladding panels, three are identified as critical in the context of the methodology discussed in this chapter: (1) geographical location, (2) intervention criterion adopted, and (3) thickness of the cladding panel. Based on the data presented in Tables 12.5 and 12.6, it can be inferred that panels installed in New South Wales are more vulnerable to hail damage than those in Victoria, because of the significantly higher rate of hailstorm occurrence. This increased frequency leads to a reduced time interval of intervention. As shown in Figure 12.9a, when the same threshold is applied, the expected time for intervention is notably shorter in New South Wales than in Victoria.

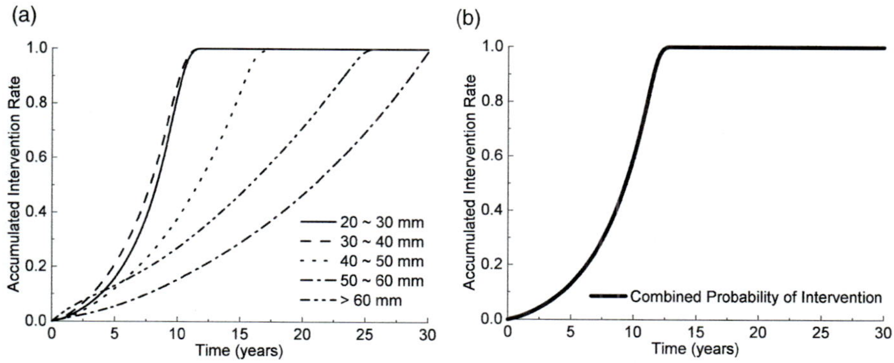

FIGURE 12.8 Cumulative intervention rate (CIR) by time t of an aluminium cladding panel located in Victoria with a thickness of 3 mm and a maximum allowable indentation of 0.3 mm (a) intervention rate curves of different hailstorm events characterised by the maximum hail diameter and (b) combined CIR.

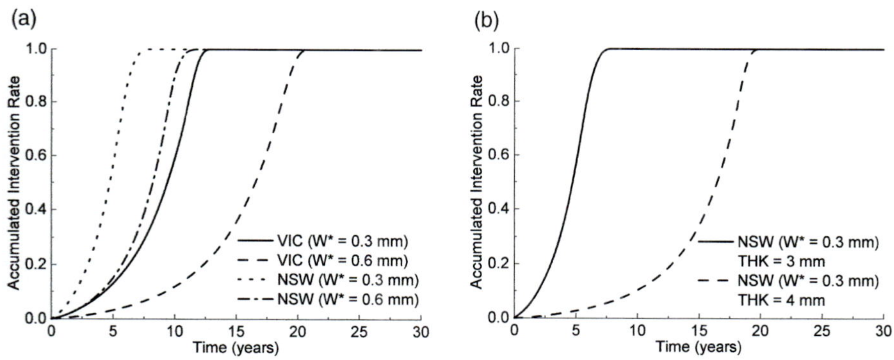

FIGURE 12.9 (a) Effect of geological region and intervention criteria on the CIR and (b) effect of panel thickness on the CIR.

Intervention criteria based on the visibility of dents can be influenced by factors such as lighting conditions and the colour of the cladding. As illustrated in Figure 12.9a, increasing the tolerance threshold from 0.3 to 0.6 mm extends the expected time to intervention by approximately 4 years in New South Wales and 8 years in Victoria.

Aluminium alloy claddings available on the market are typically manufactured with a thickness of 3 mm, balancing self-weight, construction cost, and compliance with strength requirements. In regions with elevated hail hazard exposure, such as cities in the state of New South Wales, thicker cladding panels may be warranted to enhance robustness. Although less commonly used, panels with a thickness of 4 mm offer improved durability and longer service life. As shown in Figure 12.9b, 4 mm-thick panels are expected to last nearly three times longer on average than the 3 mm counterparts when the evaluation is made under the same intervention criteria.

12.6 RESILIENCE ASSESSMENT

The concept of resilience originated in engineering mechanics for characterising material properties [13], and was later broadened and popularised for characterising the ability of a system to withstand failure and external stressors without loss of functionality. This broader interpretation has been widely adopted across the field of economics, management, and engineering [14].

In the context of the built environment, including building claddings as discussed herein, resilience is commonly understood as the capacity of a system to anticipate, respond to, recover from, and adapt to actual or potential adverse events [15]. Resilience is essential for maintaining the functionality of infrastructure systems. The ability to recover quickly following a damaging event is a highly desirable system attribute. However, recovery capacity is influenced by a combination of system properties, strategic decisions, and the specific nature of the damaging scenario, making the development of an optimal recovery strategy inherently complex. This section introduces an approach to quantify the resilience of building claddings in response to hail hazards over their service life.

12.6.1 OVERVIEW

One of the most recognised and widely adopted frameworks for resilience assessment and enhancement of civil infrastructure systems is presented in Bruneau et al. [16], with particular emphasis on natural hazards such as earthquakes. Within this framework, disaster resilience is defined as the capacity of a system to:

i. Resist extreme events, minimise operational disruptions, and prevent irreversible damage;
ii. Absorb the impact through manageable performance downtime and physical disruption, including both direct monetary losses and broader socio-economic consequences;
iii. Recover efficiently to pre-event conditions, aiming to restore adequate functionality as quickly as possible.

Quantifying this three-fold definition of resilience requires the use of a functionality metric $Q(t)$, which describes system performance through a quality index, as illustrated in Figure 12.10, depending on the system under consideration and the assessment objectives. With structural applications, common quality indices may include lateral displacement, bending moment, or other response measures. In this study, permanent indentation is selected as the performance indicator. Functionality is monitored from the time of occurrence of each event, which is denoted as T_1, T_2, ..., through to a fixed horizon time t, which serves as a reference point for comparing performance under different hazard scenarios and recovery strategies. Resilience following a disaster event is commonly defined as the area under the performance curve, as expressed in Eq. (12.43).

$$\text{Resilience} = \int_{T_1}^{t} Q(t)\,dt \qquad (12.43)$$

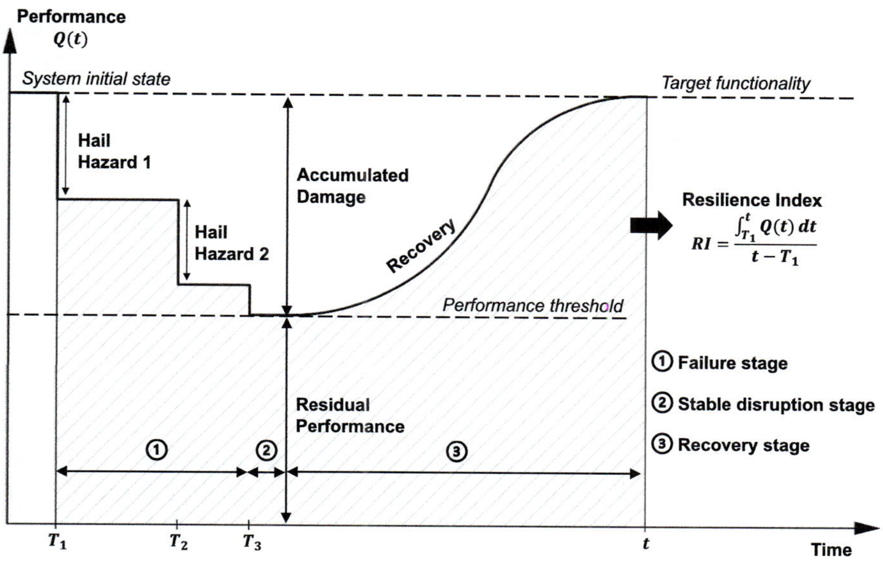

FIGURE 12.10 Schematic representation of resilience index calculation.

To enable direct comparison across different scenarios, the resilience value can be normalised with respect to the specified time horizon, resulting in a non-dimensional resilience index RI. This widely adopted quantitative measure represents the integral mean of the functionality profile from the time of hazard occurrence T_1 to the fixed horizon time t, as defined by Eq. (12.44).

$$\mathrm{RI} = \frac{\int_{T_1}^{t} Q(t)\,dt}{t - T_1} \tag{12.44}$$

The performance curve $Q(t)$ inherently involves uncertainty and is stochastic in nature. Nevertheless, it can generally be characterised by three distinct stages:

i. **Failure Stage**: also referred to as the damage accumulation stage, during which hazardous events occur randomly whilst degrading the system progressively. As a result, system performance declines from its initial state (typically normalised to unity, or 100%) until a predefined performance threshold is reached.
ii. **Stable Disruption Stage**: this stage begins once the performance of the system has fallen below the threshold at which functionality is compromised or rendered unsafe. At this point, recovery strategies are expected to be initiated. Depending on the nature of the adopted strategy, there may be a delay during which the system remains in a degraded, but stable state.
iii. **Recovery Stage**: performance is gradually or abruptly restored to the targeted level, depending on the recovery strategy implemented. This stage reflects the ability of the system to regain functionality following a disruption.

To quantify performance degradation during the failure stage, it is essential to account for the stochastic nature of hail hazards. In this framework, system performance is evaluated at discrete time steps of one year. The extent of degradation at each step is measured as the accumulated permanent indentation, normalised by the thickness of the cladding panel. The expected damage accumulation $S(t)$ is calculated as the weighted sum of W_n, as obtained from Eq. (12.38), and its associated probability, as determined from Eq. (12.42). The value of $S(t)$ is then normalised by the panel thickness, as expressed in Eq. (12.45).

$$S(t) = \frac{1}{h}\sum_{i=1}^{n}\left(\int_{0}^{R(t,i-1)} w\, dG_i(w) \times \frac{(vt)^i e^{-vt}}{i!}\right) \tag{12.45}$$

To illustrate the application of the proposed methodology, the recovery function is simplified as a linear recovery model, representing the replacement of a failed cladding panel. The typical lead time for ordering and installing a new panel is approximately 3 months or one quarter of a year. Accordingly, once the performance $Q(t)$ falls below the defined threshold, recovery is assumed to commence immediately and be completed within this period, with $Q(t)$ restored back to unity, or 100%.

12.6.2 Worked Example – Regional Case Study

Consider the same scenario as outlined in the worked examples in Sections 12.5.6 and 12.5.7. In the absence of any recovery strategy, Figure 12.11a shows that cladding panels installed in New South Wales experience significantly more rapid performance deterioration than in Victoria, which aligns with expectations given the higher hail exposure in New South Wales. Figure 12.11b presents the performance trajectories under different threshold levels for initiating recovery. Thresholds of 0.9 and 0.8 correspond to permanent indentations of 0.3 mm and 0.6 mm, respectively, consistent with the earlier examples. It is evident that in regions with more hail hazards, such as New South Wales, the frequency of recovery is higher than in regions with a lower hazard, such as Victoria.

The resilience index can be calculated using Eq. (12.44). Figure 12.12a illustrates the evolution of the resilience index over time for a cladding panel installed in New South Wales in three different scenarios: (1) no recovery implemented, (2) recovery initiated when performance drops to 0.9 corresponding to a permanent indentation of 0.3 mm, and (3) recovery initiated when performance drops to 0.8 corresponding to a permanent indentation of 0.6 mm.

In the absence of recovery, the resilience index decreases continuously over time, indicating a progressive reduction in the ability of the cladding to continue to withstand further hail hazards. When a more conservative maintenance strategy is adopted, that is, recovery at a threshold of 0.9, the panel would need to undergo maintenance or replacement, more frequently. As shown by two distinct fluctuations in the curve, the panel experiences two recovery cycles within the observed timeframe. In this case, the resilience index remains close to 0.95 reflecting a high level

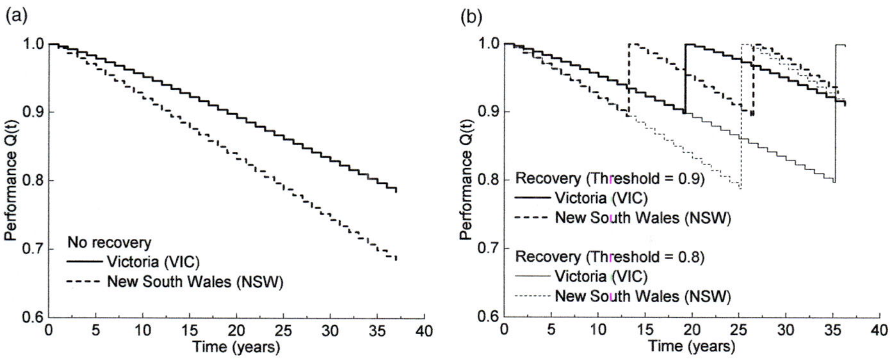

FIGURE 12.11 Realisation of performance curve $Q(t)$: (a) comparison between VIC and NSW without a recovery strategy and (b) comparison between VIC and NSW with a recovery strategy applied, using performance thresholds of 0.9 and 0.8, respectively.

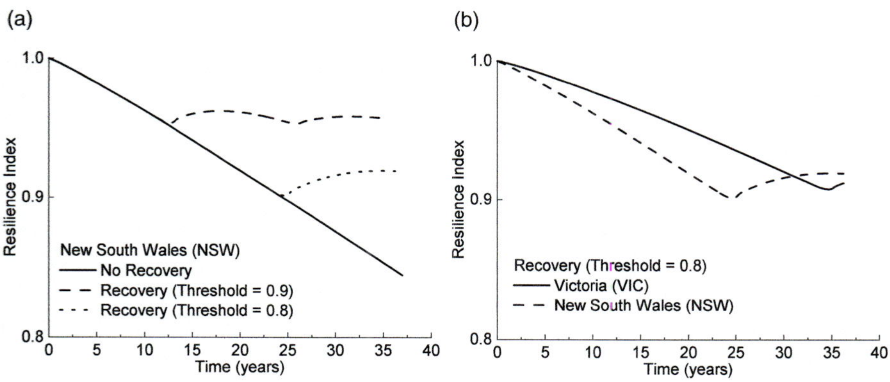

FIGURE 12.12 Resilience index results under different scenarios (a) evolution of resilience index over time for a cladding panel in NSW under different recovery strategies and (b) Comparison of resilience index between NSW and VIC based on a recovery threshold of 0.8.

of resilience to future hail hazards. However, maintaining frequent recovery may not always be practical due to cost and labour constraints. When the performance threshold is lowered from 0.9 to 0.8, fewer recovery cycles occur. The resilience index would decrease to approximately 0.9. This illustration explains the importance of identifying performance requirements for critical infrastructure and developing maintenance strategies that are both appropriate and cost-effective throughout the service life of the system.

Figure 11.12b illustrates regional differences in resilience. When recovery is initiated at a performance threshold of 0.8, hail prone regions such as New South Wales, show a more rapid decline in the resilience index requiring earlier recovery actions, approximately 10 years sooner compared to regions such as Victoria. This suggests that in areas with high hail risk, resilience-based analysis should inform both design and long-term maintenance planning to ensure satisfactory performance over time.

12.7 CONCLUSION

In this chapter, a comprehensive framework for evaluating the vulnerability, damage progression, and recovery performance of aluminium cladding panels subjected to hail hazards is presented. Starting from the modelling of contact force during hail impact, a 2DoF dynamic system was developed to simulate the interaction between the hailstone and the targeted panel. The formulation accounts for non-linear contact behaviour, fragmentation effects, and stiffness degradation, enabling a realistic representation of impact dynamics.

To characterise the hazard environment, a hail PSD model was established using meteorological data, allowing storm intensity and the rate of occurrence to be linked with local exposure conditions. This serves the purpose of informing an analytical damage model for predicting the permanent depth of indentation while incorporating the effects of strain rate and material plasticity.

The damage model has been integrated into a reliability-based framework for assessing time-dependent performance loss when subject to stochastic hailstorm exposure. Probabilistic formulations were used to estimate the likelihood of intervention based on accumulated damage and regional event frequency. In parallel, a resilience-based approach was developed to evaluate the capacity of the system to maintain, or recover, functionality following a disruption. The resilience index, derived from the performance curve over time, enables comparisons across design parameters, maintenance strategies, and geographic contexts.

To sum up, the components introduced in this chapter contribute to a structured methodology for assessing how building envelope systems respond to climatic stressors over their service life. By bridging physical modelling, statistical hazard analysis, and resilience quantification, the approach supports informed decision-making in the design and management of infrastructure systems amid increasing environmental uncertainty.

ACKNOWLEDGEMENT

The authors gratefully acknowledge the financial support provided by the Australian Research Council (ARC) through projects LP190100208, IC230100015, IH240100006, and LP240100097. The authors also thank Mr Ian Bennie (General Manager, Ian Bennie and Associates Pty Ltd) and Mr Suresh Sutrave (Managing Director, Atlite Skylights Pty Ltd) for their valuable cash and in-kind contributions as Partner Investigators.

REFERENCES

1. K. L. Johnson. *Contact Mechanics*. Cambridge University Press, Cambridge, 1987.
2. W. Goldsmith. *Impact – The Theory and Physical Behaviour of Colliding Solids*. Edward Arnold Ltd., London, England, 1960.
3. K. H. Hunt and F. R. E. Crossley. Coefficient of restitution interpreted as damping in vibroimpact. *J Appl Mech-T ASME* 1975; 42:440–445. https://doi.org/10.1115/1.3423596
4. J. Sun, N. Lam, L. Zhang, et al. Contact forces generated by hailstone impact. *Int J Impact Eng* 2015; 84:145–158. https://doi.org/10.1016/j.ijimpeng.2015.05.015

5. S. Shi, N. Lam, S. Chen, et al. An analytical approach for modelling contact forcing function of hailstone impact. *Int J Solids Struct* 2023; 269:https://doi.org/10.1016/j.ijsolstr.2023.112214
6. S. Chen, E. Gad, L. Zhang, et al. Experiments on an ice ball impacting onto a rigid target. *Int J Impact Eng* 2022; 167:e104281. https://doi.org/10.1016/j.ijimpeng.2022.104281
7. J. Grieser and M. Hill. How to express hail intensity-modeling the hailstone size distribution. *J Appl Meteorol Clim* 2019; 58:2329–2345. https://doi.org/10.1175/Jamc-D-18-0334.1
8. H. Gharababaei and A. Darvizeh. Experimental and analytical investigation of large deformation of thin circular plates subjected to localized and uniform impulsive loading. *Mech Based Des Struct Mach* 2010; 38:171–189. https://doi.org/10.1080/15397730903554633
9. N. Jones. *Structural Impact.* Second Edition. Cambridge University Press, Cambridge, 2011.
10. S. Shi, N. Lam, Y. Cui, et al. Indentation modelling of aluminium cladding panels subjected to hailstone impact. *Int J Impact Eng* 2023; 179:https://doi.org/10.1016/j.ijimpeng.2023.104638
11. T. A. Duffey. *Large Deflection Dynamic Response of Clamped Circular Plates Subjected to Explosive Loading.* Sandia Corp., Albuquerque, New Mexico, 1967.
12. S. Shi, N. Lam, Y. Cui, et al. Life-cycle performance of aluminium cladding panels in resisting hailstorms. *Struct Saf* 2024; 108:https://doi.org/10.1016/j.strusafe.2024.102439
13. E. J. Garcia and B. Vale. *Unravelling Sustainability and Resilience in the Built Environment.* Routledge, Abingdon, Oxon & New York, 2017. https://www.routledge.com/Unravelling-Sustainability-and-Resilience-in-the-Built-Environment/Garcia-Vale/p/book/9781138644045
14. D. Rehak, P. Senovsky, M. Hromada, et al. Complex approach to assessing resilience of critical infrastructure elements. 2019; 25:125–138. https://doi.org/10.1016/j.ijcip.2019.03.003
15. M. Sánchez-Silva, P. Gardoni, D. V. Val, et al. Moving toward resilience and sustainability in the built environment. 2025; 113:102449. https://doi.org/10.1016/j.strusafe.2024.102449
16. M. Bruneau, S. E. Chang, R. T. Eguchi, et al. A framework to quantitatively assess and enhance the seismic resilience of communities. *Earthq Spectra* 2003; 19:733–752. https://doi.org/10.1193/1.1623497

13 The Impact of Climate Change on Maritime Transport and Port Infrastructure Resilience

Yuhao Cao[a], *Xinjian Wang*[a,b], *and Zaili Yang*[a]
[a]Liverpool Logistics, Offshore and Marine (LOOM) Research Institute, Liverpool John Moores University, Liverpool, United Kingdom
[b]Navigation College, Dalian Maritime University, Dalian, PR China

13.1 INTRODUCTION

Maritime transport plays a central role in the global economy, facilitating over 80% of international trade and connecting nations through complex shipping networks. As the backbone of global commerce, ensuring the stability and resilience of maritime transport system is significant for cargo handling, regional connectivity, and global supply chains (Cao et al., 2025). However, due to an increasing range of extreme weather events (EWEs), for example, hurricanes and typhoons (Lawrence et al., 2020; Xu et al., 2024), flooding (Ribeiro et al., 2021), extreme wind events (Taramelli et al., 2015), heatwaves and extreme temperatures (Asariotis et al., 2024) and long-term environmental shifts, for example, sea-level rise (Abdelhafez et al., 2021), Arctic ice melt (Poo et al., 2024a) and changes in ocean currents and wave patterns (Asariotis et al., 2024), climate change has emerged as a critical threat to the maritime sector, exposing ports, shipping routes, and global supply chains to risks. Specifically, the impact of climate change on maritime transport is direct and systemic. On the one hand, direct consequences include coastal flooding, increased port downtime due to storms, and disruptions in shipping routes caused by changing oceanic conditions (Xu et al., 2024). Ports in low-lying coastal regions are particularly vulnerable to inundation and erosion, while major maritime hubs frequently experience delays, congestion, and logistical inefficiencies due to severe weather (Amer, 2024). On the other hand, climate-related disruptions can be systemic, and propagate across global trade networks, causing supply chain bottlenecks, increased shipping costs and economic instability in port-dependent industries (Wang et al., 2023). Some case studies can be found in recent research across different geographic regions. For instance, ports in the Gulf of Mexico and the U.S. East Coast are highly exposed to hurricanes and storm surges (Balakrishnan et al., 2022), while East Asian ports, particularly in

 DOI: 10.1201/9781003567332-13

China, Japan, and South Korea, frequently experience typhoon-induced disruptions (Poo et al., 2024b). European ports face growing risks from coastal flooding and extreme wave action (Ribeiro et al., 2021), and Small Island Developing States (SIDS) are disproportionately affected due to their limited infrastructure and reliance on a few transshipment hubs (Hassanali, 2017).

Given the growing risks posed by climate change, maritime transport resilience has become a critical priority for port authorities, shipping companies, and policymakers. Effective adaptation requires a deep understanding of climate risks, vulnerability assessments, and the implementation of long-term resilience strategies to ensure the continuity of maritime operations despite intensifying environmental challenges. Therefore, to provide an up to date overview of how climate change impacts maritime transport and how resilience strategies are being developed to address these challenges, this chapter delivers a comprehensive literature review, drawing on research from climate science, transport economics, disaster risk management, and maritime logistics. The data sources of this chapter incorporate the literature in the last decade from Web of Science, with the search keywords of "climate change*" AND "maritime transport*" or "marine transport*" or "port infrastructure" or "shipping industr*" or "port" AND "resilience." By selecting Science Citation Index (SCI) and Social Sciences Citation Index as the citation databases and research article as the document type, a total of 135 papers are extracted. With careful manual screening and filtering, for instance, some studies for analysing and assessing coastal ecosystem resilience, the impact of climate changes on fisheries and aquatic sciences, underwater biology, and similar are removed and finally 42 papers are obtained. The chapter synthesises research findings to highlight the key risks, assessment methodologies, and adaptation strategies relevant to climate-induced disruptions in the maritime sector, where an overview is presented in Table 13.1. Given the increasing frequency and intensity of EWEs, rising sea levels, and evolving trade network vulnerabilities, understanding how ports and shipping systems can enhance their resilience is crucial for global economic stability.

This chapter is structured to achieve the following key objectives, each of which addresses a critical aspect of climate resilience in maritime transport:

1. Examine the state-of-the-art datasets used in maritime transport resilience research. Effective resilience planning relies on robust, high-quality data that allows for accurate risk assessment and informed decision-making. Therefore, Section 13.2 of this chapter reviews multiple data sources, including: (1) Key climate and meteorological datasets, which provide essential insights into EWEs patterns, sea-level projections, and long-term climate variability; (2) Maritime trade and transport datasets, which play a role in assessing how disruptions affect global trade; (3) Policy and stakeholder data, which highlight the role of institutional planning in the climate adaptation of maritime transport.
2. Assess the methodologies employed in the resilience research of maritime transport. A wide range of quantitative and qualitative research methods are used to evaluate maritime transport resilience. Section 13.3 of this chapter categorises these methods into four main approaches: (1) Network

analysis and optimisation models assess how disruptions propagate through shipping networks and how alternative routing strategies can enhance resilience; (2) Scenario-based simulation approaches are examined for their ability to model port disruptions, vessel rerouting, and climate adaptation interventions under different risk scenarios; (3) Quantitative statistical modelling is explored to understand economic losses and infrastructure vulnerability; (4) Qualitative and stakeholder-based research is reviewed to capture insights from industry practitioners, policymakers, and maritime authorities, ensuring that resilience planning incorporates real-world perspectives and governance considerations.

3. Analyse resilience challenges through case studies from global analysis to regional exploration. Climate risks and resilience strategies vary significantly across different regions, influenced by geographical exposure, economic dependencies, and governance frameworks. Section 13.4 of this chapter provides a comparative analysis of American, European, and Asian ports, illustrating how different maritime hubs are responding to climate-induced disruptions. These case studies offer valuable insights into best practices, adaptation challenges, and region-specific policy responses.
4. Identify future research directions and policy challenges in the climate resilience of maritime transport. Based on significant progress in resilience research, Section 13.5 of this chapter further discusses emerging trends in climate adaptation, including the integration of new techniques for real-time risk monitoring and decision-making in port operations. It also highlights current gaps in resilience modelling, particularly in the economic assessment of climate disruptions and the long-term cost-benefit analysis of adaptation investments. Additionally, policy recommendations are presented to address the regulatory and institutional challenges of climate resilience planning, ensuring that both developed and developing maritime regions have the resources to enhance their adaptive capacity.

13.2 ESTABLISHED DATASETS FOR RESILIENCE ASSESSMENT

In this section, three different types of datasets are reviewed based on their nature and sources, namely climate-related and meteorological datasets, maritime and trade data, and policy and stakeholder data.

13.2.1 Climate-Related and Meteorological Datasets

Climate datasets provide essential information for assessing the impact of EWEs, sea-level rise, and environmental changes on maritime transport and port infrastructure. Typically, they contain oceanic and atmospheric details (e.g., wave height, wind (speed and direction), current, precipitation, temperature, and humidity), EWEs data (e.g., records of tropical cyclones, hurricanes, typhoons, storm surge, etc.), and different scenarios data (i.e., providing some climate models or projections to estimate and output the impacts caused by different environmental scenarios). Some of the most commonly used climate data sources include the Intergovernmental Panel on

TABLE 13.1
An Overview of Selected Literature

	Research Input			Research Method				Research Scope				
References	Climate Datasets	Maritime Transport data	Policy & Review Data	Network Analysis	Simulation	Quantitative Statistical Analysis	Qualitative Survey	Global/ General	America	Europe	Asian	Africa
Abdelhafez et al. (2021)	√	√			√	√			√			
Amer (2024)	√	√			√				√			
Asariotis et al. (2024)	√					√		√				
Azmi et al. (2023)			√			√	√				√	
Balakrishnan et al. (2022)	√	√			√	√			√			
Bathgate et al. (2022)		√				√			√			
Becker and Caldwell (2015)			√				√		√			
Becker et al. (2015)			√				√		√			
Ching-Pong Poo et al. (2024)		√		√							√	
Ding and Wu (2023)	√	√			√						√	

(*Continued*)

TABLE 13.1 (*Continued*)
An Overview of Selected Literature

	Research Input			Research Method				Research Scope				
References	Climate Datasets	Maritime Transport data	Policy & Review Data	Network Analysis	Simulation	Quantitative Statistical Analysis	Qualitative Survey	Global/ General	America	Europe	Asian	Africa
Gharehgozli et al. (2017)	√		√				√	√				
Grant et al. (2024)	√	√		√					√			
Hassanali (2017)			√				√		√			
Islam et al. (2023)		√	√		√		√		√			
Izaguirre et al. (2021)	√	√			√			√				
Kalaidjian et al. (2022)			√				√		√			
Lamers et al. (2018)			√				√			√		
Lawrence et al. (2020)			√			√			√			
León-Mateos et al. (2021)			√			√	√			√		
March et al. (2024)			√				√					√
McIntosh and Becker (2019)	√		√			√	√		√			

(*Continued*)

TABLE 13.1 *(Continued)*
An Overview of Selected Literature

References	Research Input			Research Method				Research Scope				
	Climate Datasets	Maritime Transport data	Policy & Review Data	Network Analysis	Simulation	Quantitative Statistical Analysis	Qualitative Survey	Global/ General	America	Europe	Asian	Africa
McLean and Becker (2020)			√				√		√			
Morris (2020)			√				√		√			
Polydoropoulou et al. (2025)			√			√	√			√		
Poo and Yang (2024)		√		√				√				
Poo et al. (2024a)		√		√				√				
Poo et al. (2024b)		√		√							√	
Ribeiro et al. (2021)	√				√					√		
Roukounis and Tsihrintzis (2024)			√		√	√	√			√		
Ryan-Henry and Becker (2020)			√				√		√			
Sharaan et al. (2024)			√			√	√					√

(Continued)

TABLE 13.1 (*Continued*)
An Overview of Selected Literature

	Research Input			Research Method				Research Scope				
References	Climate Datasets	Maritime Transport data	Policy & Review Data	Network Analysis	Simulation	Quantitative Statistical Analysis	Qualitative Survey	Global/ General	America	Europe	Asian	Africa
Tafur and Padgett (2024)	√	√		√	√				√			
Taramelli et al. (2015)	√					√			√			
Verschuur et al. (2020)	√	√				√			√			
Verschuur et al. (2022)		√		√	√			√				
Wan et al. (2022)		√		√	√						√	
Wan et al. (2024)		√	√			√					√	
Wang et al. (2022)					√			√				
Wang et al. (2023)	√		√				√			√		
Xia et al. (2024)		√			√						√	
Xu et al. (2024)	√	√		√		√		√				
Yin et al. (2024)		√	√			√		√				

Climate Change (IPCC) Climate Projection Datasets, the National Oceanic and Atmospheric Administration (NOAA) Climate and Extreme Weather Records, and the European Centre for Medium-Range Weather Forecasts (ECMWF) Storm Surge and Climate Projection Datasets. These datasets enable researchers to evaluate both historical trends and future risks associated with climate change impacts on global ports.

Several studies have integrated these datasets to analyse the increasing vulnerability of ports to climate-related hazards. Abdelhafez et al. (2021) employed a range of climate datasets to assess the vulnerability of seaports to hurricanes and sea-level rise. Their study incorporated historical hurricane and storm surge records from NOAA and the Federal Emergency Management Agency (FEMA) to establish a dataset of past EWEs affecting U.S. ports. Additionally, they utilised climate projections from the IPCC under the Representative Concentration Pathway (RCP) 8.5 scenario, estimating that hurricanes and storm surges would intensify due to climate change. To validate their findings, they referenced port damage assessment reports from Hurricane Katrina, using them as a benchmark to estimate potential future damages under similar storm conditions. Similarly, Asariotis et al. (2024) utilised climate hazard datasets from the IPCC, extending their analysis to include projections of sea-level rise, storm surges, and extreme heat events under both RCP 4.5 and RCP 8.5 scenarios. Their study also incorporated storm surge projection datasets from ECMWF, which simulated the increasing likelihood of extreme sea levels exceeding two metres above historical baselines by 2050 and 2100. By integrating these datasets, the study projected that over 70% of global ports would experience extreme sea levels beyond historical baselines by the end of the century, highlighting the urgency of regulatory and infrastructural adaptation measures.

Beyond global projections, some studies focus on regional assessments of port resilience. Poo et al. (2024b) conducted a study examining five major Chinese ports, evaluating the effects of cyclone-induced events on port performance, infrastructure resilience, and economic stability. Their research utilised climate and operational data from the Tropical Cyclone Data Centre of the China Meteorological Administration, which maintained historical cyclone landfall records and intensity trends from 1949 to 2023. This dataset allowed them to quantify the impact of past typhoons on port efficiency and to model potential disruptions under worsening climate conditions.

13.2.2 Maritime and Trade Data

Maritime transport datasets, such as Automatic Identification System (AIS) data, maritime trade and port performance records, provide crucial insights into shipping patterns and port efficiency, and give the chance to analyse network resilience under climate-induced disruptions. Generally, studies incorporate AIS datasets to track vessel movements before, during, and after disruptions, allowing researchers to analyse how EWEs impact maritime traffic flow. Maritime trade and port infrastructure datasets, which contain information on terminal capacity, port congestion levels, and trade flow statistics, are often used to assess the reduction in port operational capacity under different climate scenarios. For instance, Bathgate et al. (2022) examined the impact of Hurricane Harvey on Texas maritime facilities by employing a quantitative analysis of vessel movements in the Houston region. The study utilised AIS data, which recorded vessel locations, speeds, and movement patterns, spanning the period from August to November in both 2016 and 2017. This allowed for a comparative analysis between normal maritime operations and disruptions caused by the hurricane. Similarly, Wan et al. (2022) evaluated recovery strategies for disruptions in liner shipping networks from a resilience perspective, incorporating multiple empirical datasets, including AIS data, storm impact reports, and economic trade flow databases, to assess the extent of disruptions. Additionally, the study leveraged global liner service schedules from major container shipping companies to model how the shipping industry responded to climate-related disruptions.

Further to this, maritime trade dataset serves as another important resource for assessing resilience in the maritime industry in an overall perspective. For example, Poo et al. (2024a) analysed the effects of Arctic sea ice melting on global container shipping network connectivity. In the study, they utilised global shipping network data from Q4/2011 to Q3/2017 to evaluate the impact of the continued decline in Arctic sea ice and its implications for shipping routes. Their findings suggested that incorporating Arctic routes into global trade networks enhanced maritime resilience by providing alternative pathways, particularly when traditional routes faced climate-induced disruptions. Within the same research group, Poo and Yang (2024) further integrated data from various sources, including shipping route data from major liner companies, the Emergency Events Database (which documents the increasing frequency and severity of climate-related disasters), and port disruption records, with a specific focus on EWEs such as hurricanes, storms, and sea-level rise. Their study revealed that some of the busiest ports, including Singapore and Hong Kong, were projected to experience a decline in transshipment traffic due to increased climate risks. Conversely, ports with lower climate vulnerability, such as Tokyo and Lisbon, were expected to see an increase in shipping activity as trade routes shifted to mitigate risks.

13.2.3 Policy and Stakeholder Data

Policy and stakeholder data provide essential insights into understanding how maritime transport resilience is shaped by governance frameworks, regulatory policies, and stakeholder engagement. Policy documents, such as national climate strategies, Blue Economy initiatives, and international maritime regulations, provide insights into how resilience measures are embedded into long-term planning. Meanwhile, stakeholder data, including expert interviews, participatory workshops, and collaborative governance models, highlight the practical challenges and opportunities in implementing climate adaptation strategies. By analysing these sources, researchers can assess the effectiveness of current policies, identify gaps in governance, and propose more inclusive and adaptive approaches to maritime resilience.

Up to date, one significant approach to policy data collection involves analysing national and regional climate strategies, as seen in the work of March et al. (2024), which examined the integration of climate change adaptation and mitigation in the Blue Economy policies of African coastal and island nations. By reviewing official government documents, including Nationally Determined Contributions under the

Paris Agreement, their study evaluated whether climate resilience was explicitly incorporated into national maritime policies. Similarly, Hassanali (2017) focused on policy frameworks in Trinidad and Tobago, assessing the extent to which climate adaptation has been mainstreamed into productive coastal sectors, such as fisheries, tourism, and maritime transport. This study relied on policy analysis to highlight gaps in governance, particularly regarding institutional coordination and regulatory enforcement.

The use of stakeholder data, particularly through interviews and participatory engagement, is another critical method for assessing maritime transport resilience. Morris (2020) employed qualitative research data, sourcing from interviews with policymakers, port managers, and industry representatives in Norfolk, Virginia, to explore how stakeholder collaboration influenced port adaptation strategies. The study emphasised the role of multi-stakeholder governance, demonstrating that engagement among local governments, port authorities, and private-sector actors enhanced resilience planning. Similarly, McIntosh and Becker (2019) integrated expert opinions and mind-mapping exercises to develop open-data indicators for assessing seaport vulnerability to extreme weather. The research gathered insights from maritime professionals and policymakers to determine the most relevant indicators of resilience and their applicability in decision-making.

Beyond qualitative interviews and stakeholder engagement, some studies attempted to integrate policy reviews with quantitative assessments to evaluate resilience strategies. León-Mateos et al. (2021) developed a Port Resilience Index (PRI), combining infrastructure assessments with expert evaluations to measure resilience performance at the Port of A Coruña in Spain. This study gathered input from port stakeholders, including engineers, logistics operators, and policymakers, to assess resilience across governance, operational, and infrastructure dimensions.

13.3 METHODS FOR ASSESSING MARITIME TRANSPORT AND INFRASTRUCTURE RESILIENCE

This section outlines four different types of methods being reviewed, namely network analysis, scenario-based and simulation approaches, quantitative modelling and statistical analysis, and qualitative and stakeholder-based research.

13.3.1 Network Analysis

Network analysis typically involves assessing connectivity, identifying vulnerabilities, and optimising recovery strategies following disruptions in network structures. These methods leverage mathematical modelling, graph theory, and algorithmic optimisation techniques to analyse how disruptions propagate through shipping networks and how ports and logistics systems can be adapted to enhance resilience.

One of the fundamental applications of network analysis in maritime transport resilience research involves the assessment of shipping network connectivity and port centrality. Poo et al. (2024a) applied centrality-based network analysis to evaluate how Arctic shipping route influenced global container shipping network connectivity. Their study utilised betweenness centrality metrics, clustering coefficients, and efficiency measures to demonstrate that Arctic routes could improve redundancy within the network. To assess the resilience of the global liner shipping network (GLSN) to tropical cyclones, Xu et al. (2024) employed a directed co-route graph model to represent the GLSN, allowing for an assessment of alternative shipping paths and the impact of disrupted transshipment hubs on global cargo flows. In this study, the methodology involved three main components: the generation of port disruption scenarios based on historical tropical cyclone data, the computation of path redundancy within the GLSN, and the quantification of resilience at both network-wide and individual country levels. Particularly, the study applied the circle-shaped spatially localised failures model to determine the extent of port disruptions caused by tropical cyclones, using the 34-knot wind radius as the threshold for defining affected areas.

Beyond assessing network connectivity and vulnerability, optimisation models have been widely employed to enhance shipping network resilience under climate risks. Poo and Yang (2024) introduced a multi-objective optimisation model that balanced operational efficiency and climate resilience in shipping network design. By applying genetic algorithms and shortest-path optimisations, they designed adaptive routing strategies that enabled vessels to dynamically adjust their paths in response to EWEs. Moreover, network-based optimisation has been extended to intermodal transport systems to ensure resilience beyond seaports. Wang et al. (2022) introduced a two-stage optimisation model that considered the interplay between seaports and inland dry ports in risk management. Their study examined the coordination of investment strategies between maritime and inland transport hubs, incorporating network congestion modelling to identify bottlenecks and optimise cargo flow in response to climate-related disruptions. Their findings suggested that seaports and dry ports should implement joint investment strategies, as competitive pressures between ports often led to fragmented resilience planning, reducing overall supply chain efficiency.

Furthermore, the role of cascading failures in shipping networks has also been extensively studied using network-based resilience assessment models, representing an exploration of dynamic mechanism. Verschuur et al. (2022) applied systemic risk modelling to quantify how disruptions in one port propagated throughout the global shipping network. Their study employed percolation theory and fragility curve analysis to measure how individual port failures, caused by EWEs such as hurricanes, triggered multi-port disruptions. By integrating AIS vessel tracking data with climate impact datasets, they assessed how delays and rerouting patterns unfolded across multiple regions, highlighting the systemic nature of global shipping disruptions.

13.3.2 Scenario-Based and Simulation Approaches

Simulation and modelling approaches in resilience-based maritime transport and port infrastructure studies allow researchers and policymakers to reproduce real-world disruptions, evaluate different adaptation strategies, and predict system behaviour under extreme climate scenarios. These approaches provide controlled, data-driven environments where various risks and mitigation strategies can be tested without real-world consequences. So far, several studies have successfully applied scenario-based simulation approaches to assess climate resilience and disaster recovery in shipping networks. For instance, by integrating historical port congestion data and vessel rerouting records, Wan et al. (2022) simulated how different recovery strategies, such as random-based rerouting and centrality-based optimised recover, affected overall network efficiency and shipping costs. In another study, Amer (2024) developed a high-resolution Digital Elevation Model-based simulation to evaluate sea-level rise vulnerability in Southeast Texas ports. The study integrated NASA's sea-level rise projections, NOAA's historical tide data, and LiDAR survey datasets to create a geospatial inundation risk model. The hydrodynamic simulation framework incorporated multiple climate scenarios (low, medium, and high emissions) to predict the extent of port infrastructure submergence by 2100. The results provided critical insights into potential land loss and port accessibility challenges, helping to guide long-term resilience planning and adaptation investments in coastal port regions. Similarly, Ribeiro et al. (2021) used another approach called Delft3D numerical modelling to assess flood risks at Aveiro Port under projected climate change scenarios. Their study also applied hydrodynamic flood simulations to predict storm surge impacts and long-term sea-level rise effects on port infrastructure. The analysis revealed that historically classified 100-year return period flood events were likely to become 50-year return events, significantly increasing the probability of port downtime and operational disruptions. Furthermore, Islam et al. (2023) employed a scenario-based disaster simulation approach to model the impact of extreme flooding on Vancouver's maritime transport system. Their study used GIS-based flood mapping, coupled with transport demand modelling, to predict port access disruptions and freight mobility reductions under varying climate conditions. By incorporating real-time infrastructure resilience data, the study highlighted the importance of integrating digital monitoring technologies into maritime disaster response planning.

To extend the scope beyond individual ports, some studies have developed intermodal freight network simulations to assess climate resilience in broader transport systems. Specifically, Tafur and Padgett (2024) introduced a flow-based commodity-independent simulation model to evaluate the resilience of intermodal freight networks against coastal hazards. The study employed a Monte Carlo simulation approach to incorporate uncertainty from three key components, hazard models (extreme weather scenarios), structural response models (damage assessments), and recovery models (restoration trajectories of transport infrastructure). By simulating multiple disruption scenarios, the study provided a comprehensive risk assessment framework that informed logistics planning, infrastructure investment, and emergency response strategies in climate-vulnerable freight corridors.

13.3.3 Quantitative Modelling and Statistical Analysis

Quantitative modelling and statistical analysis serve as another approach to provide data-driven insights into risk assessment, disruption forecasting, and adaptation planning. These approaches leverage mathematical models, probabilistic frameworks, and statistical techniques to quantify the impacts of climate-induced disruptions, economic losses, and infrastructure vulnerabilities. One widely used approach in quantitative maritime resilience studies is probabilistic risk assessment, which estimates the likelihood and consequences of climate-related disruptions. Bayesian networks, in particular, provide a probabilistic framework for modelling interdependencies between risk factors, allowing researchers to update predictions dynamically based on new data. For example, Wan et al. (2024) developed a fuzzy evidential reasoning-based Bayesian model to evaluate the resilience of Chinese ports to typhoons. The study incorporated historical typhoon records, port infrastructure data, and emergency response measures into a Bayesian probabilistic framework. By assigning conditional probability distributions to different resilience indicators, the model estimated the likelihood of port disruptions and recovery time under varying storm intensities. Similarly, Yin et al. (2024) employed the Bayesian network model to assess seaport disruption risks, integrating maritime accident reports, AIS vessel tracking data, and expert evaluations. The study constructed the probabilistic risk graph, identifying critical failure points within port operations. Sensitivity analysis was performed to determine which factors like natural disasters, environmental disruptions, and congestion issues had the most significant impact on disruption likelihood, allowing port authorities to prioritise risk mitigation strategies based on data-driven vulnerability assessments.

Quantitative models have also been widely used to evaluate the economic consequences of climate-induced disruptions on maritime trade. Input–output modelling and econometric regression techniques enable researchers to assess direct and indirect economic losses, providing valuable insights into regional and global supply chain disruptions. Specifically, Ding and Wu (2023) integrated transport cost variations, freight rerouting strategies, and industrial production data into a multi-regional economic impact assessment. The model captured both direct damages to port infrastructure and indirect ripple effects across supply chains, revealing that ports with diversified trade connections exhibited lower economic losses due to their ability to reroute cargo efficiently. Another example of statistical economic modelling is provided by Xia et al. (2024), who examined how inter-port competition and climate uncertainty affect investment decisions in port capacity and adaptation measures. The study utilised a game-theoretic approach combined with econometric regression analysis to evaluate how different market structures influence resilience investment behaviour. The results suggested that ports facing high climate risk tended to underinvest in adaptation unless regulatory incentives or cooperative agreements were in place, highlighting the need for policy interventions to encourage long-term resilience planning.

13.3.4 Qualitative and Stakeholder-Based Research

Maritime transport and port infrastructure operate within complex socio-technical systems influenced by economic, political, and regulatory factors. While data-driven models provide objective resilience metrics, they often fail to account for institutional constraints, financial limitations, or human decision-making dynamics. At that time, qualitative research methods help bridge this gap by assessing how stakeholders perceive risks, prioritise investments, and implement adaptation measures. These approaches provide valuable insights into governance structures, stakeholder collaboration, and strategic decision-making processes, particularly in the context of climate adaptation and disaster resilience in maritime transport.

Firstly, expert-based methods are widely used in maritime transport resilience research, relying on stakeholder interviews, structured surveys, and scenario planning to assess how key actors engage with climate adaptation strategies. These approaches are particularly useful when empirical data are limited or when dealing with long-term climate risks that cannot be easily quantified. For instance, Roukounis and Tsihrintzis (2024) applied a stakeholder survey methodology, collecting responses from port authority officials, environmental agencies, and coastal engineers regarding climate adaptation strategies. Similarly, Lamers et al. (2018) adopted a practice-based qualitative approach, combining semi-structured interviews and participatory observations to assess the use of weather and sea ice information in Arctic maritime operations. The study engaged expedition cruise captains, expedition leaders, guides, port authorities, industry representatives, and policymakers in discussions about real-time decision-making under extreme environmental conditions, enabling researchers to gain first-hand insights.

Secondly, scenario-coupled analysis is another common qualitative method used to explore multiple future possibilities and identify strategic pathways for resilience. This approach allows stakeholders to evaluate potential climate impacts and alternative response strategies based on historical disruptions, policy experiences, and projected environmental changes. For example, Becker and Caldwell (2015) employed semi-structured interviews and scenario-based assessments to explore resilience planning in two U.S. seaports. The study gathered insights from port managers, municipal planners, and industry stakeholders, assessing how they perceived climate risks and integrated them into operational strategies. Qualitative coding techniques were applied to identify major challenges, including financial constraints, fragmented governance, and competing priorities between economic growth and climate adaptation. Expanding on this approach, Becker et al. (2015) further incorporated policy document reviews alongside stakeholder interviews to examine how storm scenarios influenced resilience planning at multiple ports. The study assessed port resilience plans, adaptation policies, and planning reports, revealing discrepancies between policy goals and actual implementation. It was found that uncertainties in climate risk assessments often hinder investment decisions, as stakeholders struggle to justify the long-term costs of adaptation against immediate economic pressures. Meanwhile, Sharaan et al. (2024) applied qualitative scenario analysis to assess the vulnerability of Egyptian ports to climate change. The study conducted expert interviews with port operators and climate scientists, identifying common concerns such as rising sea levels, inadequate flood control infrastructure, and limited governmental coordination. Thematic coding techniques were used to extract patterns in stakeholder perceptions, highlighting that localised climate resilience initiatives, tailored to individual port characteristics, were more effective than generic adaptation frameworks.

Furthermore, governance-based research methods examine institutional policies, international regulatory frameworks, and climate adaptation strategies to assess how governments and industries manage maritime transport resilience. These methods typically involve policy document analysis, comparative case studies, and regulatory impact assessments. For instance, Polydoropoulou et al. (2025) investigated the role of Living Labs as user-centred open innovation environments designed to enhance maritime port resilience against climate change disruptions. The study followed the Quadruple Helix Participation Model, integrating perspectives from academia, industry, government, and civil society. Through structured workshops, scenario simulations, and qualitative analysis of stakeholder discussions, the study demonstrated that multi-stakeholder collaboration fosters the development of adaptive resilience strategies tailored to specific port environments. Taking a conceptual approach, Gharehgozli et al. (2017) framed port resilience as a "wicked problem," highlighting the complex interplay of economic, environmental, and social factors in adaptation planning of maritime transport. The study acknowledged that resilience planning requires iterative, stakeholder-driven approaches rather than rigid, one-size-fits-all solutions. Meanwhile, Hassanali (2017) employed a mixed-method qualitative approach, combining policy analysis, stakeholder interviews, and document reviews to evaluate climate change adaptation in maritime and coastal sectors. The study conducted a comprehensive literature review of international best practices, assessing how climate adaptation was mainstreamed in fisheries, energy, tourism, and maritime transport policies. By analysing existing institutional structures and sector-specific strategies, the study identified key barriers to effective climate adaptation, including limited financial resources, institutional fragmentation, and a lack of technical expertise. Notably, the study assessed the role of multilateral agreements and national regulatory bodies, such as the Environmental Management Authority, in facilitating climate adaptation initiatives.

13.4 REGIONAL STUDIES IN MARITIME RESILIENCE TO CLIMATE CHANGE

13.4.1 Global Analysis

Climate change poses significant risks to global port operations, with multiple studies analysing how different climate-related hazards affect maritime infrastructure and supply chain networks across various regions. Research in this area focuses on the exposure of ports to EWEs, long-term climatic trends, and systemic vulnerabilities in global trade routes. The studies reviewed highlight high-risk regions, projected future disruptions, and systemic interdependencies that influence the resilience of the global maritime network.

As one of the most comprehensive studies, Izaguirre et al. (2021) assessed the risks posed by climate change to 2,013 coastal ports worldwide. The study examined seven major climate hazards, including waves, storm surges, wind, precipitation, temperature, tropical cyclones, and sea-level rise, to determine the degree of exposure and vulnerability of ports under a high-end warming scenario. The findings indicated that high-risk regions under current conditions were primarily located in tropical cyclone-prone areas, such as the Caribbean, the Pacific Islands, and eastern Asia. However, by 2100, climate change was going to be projected to significantly expand the number of high-risk ports, particularly in Indonesia, the Arabian Peninsula, and parts of the Mediterranean. The study also highlighted that ports in northern Europe and Saudi Arabia were expected to experience increasing operational disruptions due to extreme temperatures, overtopping, and coastal flooding. Furthermore, ports in western Africa and the Persian Gulf were identified as particularly vulnerable to worsening climatic conditions, making infrastructure adaptation and resilience planning critical priorities in these regions. Another study by Verschuur et al. (2020) further examined global port disruptions particularly caused by EWEs, integrating empirical data from the AIS data, storm surge records, and wind speed data. The study identified frequent and severe port disruptions in the Gulf of Mexico, the U.S. East Coast, and Australia due to hurricanes and tropical cyclones, while ports in China and Japan were also found to be highly vulnerable to typhoon-induced congestion and downtime. The analysis revealed that ports often experience disruptions beyond direct damage, with delays cascading through supply chains due to vessel rerouting and operational bottlenecks.

Expanding on this global-scale analysis, Verschuur et al. (2022) developed a systemic risk framework to evaluate how natural hazards impact port and supply-chain networks worldwide. The study examined multiple regional case studies, including hurricane- and typhoon-exposed ports in the United States and East Asia, as well as ports affected by inland flooding in Europe and disruptions to major maritime trade routes such as the Suez Canal. The study integrated network-based risk modelling to identify cascading vulnerabilities, demonstrating how disruptions in one region propagate through interconnected shipping networks, amplifying global supply chain risks. By considering diverse geographic contexts, the study provided a comprehensive assessment of port resilience across various climate zones and highlighted the necessity for coordinated international adaptation efforts to mitigate climate-induced disruptions.

Beyond risk exposure assessments, some studies have focused on how emerging climatic trends create new development opportunities in maritime transport while simultaneously introducing new risks. Poo et al. (2024a) analysed the impact of declining Arctic ice cover on shipping efficiency and port competitiveness. The findings indicated that the gradual opening of Arctic routes would offer significant advantages in reducing transit distances, particularly between East Asia and Europe. However, the study also highlighted potential risks, including increased exposure to extreme Arctic weather conditions, the lack of developed port infrastructure in high-latitude regions, and environmental concerns related to increased shipping activity in sensitive Arctic ecosystems.

13.4.2 Regional Findings

13.4.2.1 Case Studies in the Americas

In North America, the high frequency of EWEs (e.g., hurricanes) is continuously attracting researchers' attention. Given the developed record system of climate change data and techniques, studies focusing on the American region's maritime transport resilience are more than others, which includes disaster scenario modelling (Islam et al., 2023), vulnerability assessments (Bathgate et al., 2022; Hassanali, 2017; McIntosh and Becker, 2019) and economic loss estimation (Balakrishnan et al., 2022), particularly in areas prone to hurricanes and sea-level rise.

An instance can be found in the study by Islam et al. (2023), which developed a disaster scenario analysis to assess potential disruptions to maritime transport serving coastal communities on Vancouver Island. The study integrated consultations with marine stakeholders and a marine geologist, alongside prior studies on the Fraser River's geological and hydrodynamic conditions, to model flood risks and their implications for shipping operations. The study provided critical insights into the preparedness of maritime transport systems in Canada, which were increasingly vulnerable to storm surges, high tides, and extreme precipitation events.

Focusing on the United States, McIntosh and Becker (2019) investigated the vulnerability of seaports to climate and extreme weather impacts by developing open-data indicators for 22 major seaports in the northeastern United States. The study concentrated on the North Atlantic region, specifically ports under the jurisdiction of the U.S. Army Corps of Engineers' North Atlantic Division, which included major commercial hubs crucial to the national economy. By evaluating climate-related threats such as storm surges and sea-level rise, the study aimed to provide quantifiable resilience metrics that could inform broader climate adaptation strategies for ports in similar geographic and economic settings. Meanwhile, due to the geographical and meteorological conditions of the United States, hurricanes always pose a significant and recurrent threat to port operations. The country's eastern seaboard, Gulf Coast, and southern states were particularly vulnerable due to their exposure to Atlantic hurricanes, which formed due to the abundance of water vapour and atmospheric instability. Statistics from 1991 to 2014 indicated an annual average of more than 1,000 hurricanes of varying intensities across the United States, reinforcing the importance of studying hurricane-related port disruptions. In one such study, Balakrishnan et al. (2022) examined the impact of hurricanes on major Texas ports, including Houston, Galveston, Freeport, Corpus Christi, and Port Arthur. These ports were vital trade hubs, handling crude oil, chemicals, and manufactured goods, making them highly susceptible to hurricane-related economic losses. In the study, they identified Port Houston as the most economically vulnerable, estimating a mean economic loss of $1.96 billion for a 1-in-10-year hurricane and $4.68 billion for a 1-in-1,000-year hurricane. Additionally, industries highly dependent on port trade, particularly the chemical and pharmaceutical sectors, were found to suffer the most financial losses due to port shutdowns. With a more focused case study approach, Bathgate et al. (2022) specifically analysed the impact of Hurricane Harvey on the Houston-Galveston region, highlighting disruptions at the Port of Houston, one of the largest and most economically significant ports in the United States. Specifically, the study investigated Hurricane Harvey's influence on container terminal operations at Barbours Cut Terminal and vessel queuing behaviour at the Houston Pass anchorage. The findings revealed that port closures led to a complete halt in vessel operations at Barbours Cut Terminal, with vessel counts dropping to zero during the closure period. Following the port reopening, a backlog of ships caused a sharp increase in vessel traffic, demonstrating the extended ripple effects of EWEs on maritime transport congestion.

Turning to South America, research in this region has focused on climate vulnerability in SIDS and the impact of EWEs on maritime-dependent economies. A key study by Hassanali (2017) examined the climate risks facing Trinidad and Tobago, a highly industrialised island state where coastal and marine resources play a central role in economic sustainability. The study assessed sectoral vulnerabilities to sea-level rise, coastal erosion, and EWEs, particularly in the energy, tourism, fisheries, and maritime transport sectors. Given that the energy sector accounts for over 50% of the country's GDP, the study highlighted the paradox of a heavily fossil-fuel-dependent economy simultaneously facing some of the most severe climate change threats. Particularly, the fisheries and maritime transport sectors were found to be directly exposed to climate-related risks, threatening local livelihoods and food security. By integrating policy reviews and stakeholder engagement, the study underscored the need for targeted adaptation strategies that prioritise port infrastructure resilience, environmental conservation, and sustainable economic diversification.

13.4.2.2 Case Studies in Europe Area

European ports, particularly those in northern coastal and Arctic regions, face unique climate challenges, ranging from sea-level rise and storm surges to changing ice conditions affecting Arctic shipping routes. The following studies provide insights into how different European ports are assessing and addressing these climate-related risks.

One significant study by Ribeiro et al. (2021) investigated flooding conditions at Aveiro Port, Portugal, under projected climate change scenarios. From the study, it was indicated that flood extents would progressively expand from historical baselines to future projections, with critical port terminals and operational zones becoming highly vulnerable to inundation. Under the far-future scenario, the study predicted that extreme flood events could lead to near-total inundation of key operational areas, disrupting cargo handling and port logistics. In Greece, Polydoropoulou et al. (2025) adopted a participatory action research approach to assess climate resilience in three ports: Chios, Volos, and Heraklion. These ports were selected due to their varying exposure to climate risks, operational characteristics, and economic significance. Chios, an island port, is critical for local economic sustainability, particularly in maintaining essential goods and passenger transport. Volos, a mainland port, plays a key role in transportation for the Thessaly region, facilitating cargo and passenger flows to island destinations. Heraklion, located on Crete, is one of Greece's largest commercial and passenger ports, supporting both regional and international trade. Each of these ports faced distinct climate challenges, such as sea-level rise, EWEs, and infrastructure vulnerabilities. Furthermore, a different approach to assessing port resilience is demonstrated in the study by León-Mateos et al. (2021), which focuses on developing and validating the PRI for the external port of A Coruña in Galicia, Spain. This port was chosen due to its high exposure to EWEs and strategic role within Spain's maritime transport network. As an open-sea port, A Coruña is particularly vulnerable to sea-level rise, increasing storm intensity, and wave action, making it a key case study for resilience assessment. The study's PRI methodology evaluated resilience across multiple dimensions, including infrastructure robustness, operational adaptability, and governance effectiveness. The results indicated that A Coruña's overall resilience level was only 52%, highlighting significant vulnerabilities in both physical infrastructure and risk management frameworks. Moreover, the study mentioned that governance and operational policies required improvements, as it found that risk management strategies were still not sufficiently integrated into long-term planning efforts.

Focusing on the Arctic maritime sector, Lamers et al. (2018) examined the role of weather and sea ice services in European Arctic expedition cruise operations, with a particular focus on Svalbard. The study highlighted that weather and sea ice forecasting tools played a critical role in risk management, but information accessibility, accuracy, and real-time applicability remained major challenges. Given the increasing variability of Arctic weather conditions due to climate change, the study suggested that improving weather and sea ice monitoring infrastructure was crucial for ensuring safe and sustainable Arctic maritime operations.

13.4.2.3 Case Studies in the Asian Area

Research on climate resilience in Asian ports has focused on the economic impacts of EWEs, infrastructure vulnerabilities, and resilience assessment frameworks for major coastal ports. Given Asia's high exposure to typhoons, monsoons, and sea-level rise, studies in this region examine how climate-induced disruptions propagate through interregional trade networks and affect port operations across different maritime regions.

A significant study by Ding and Wu (2023) investigated the interregional economic impacts of extreme storm floods, specifically examining a 1,000-year return period flood scenario in China. The study modelled how transportation disruptions originating from Shanghai propagated through China's interregional trade networks, affecting economic stability in multiple provinces. The study found that in East China, including Jiangsu, Zhejiang, and Shandong, the majority of the indirect economic losses were due to its strong economic integration with Shanghai. However, the study also identified significant spillover effects on Beijing and Guangdong, despite their geographical distance from Shanghai, demonstrating the far-reaching consequences of port disruptions on national and regional economies. By incorporating transport cost variations, freight rerouting, and supply chain delays into their modelling framework, the study provided valuable insights for policymakers and port authorities seeking to enhance economic resilience against extreme climate events. Moreover, Wan et al. (2024) focused on four major coastal ports in China, strategically selected across distinct maritime regions: (1) the Bohai Bay area, known for heavy industrial exports and bulk cargo shipping; (2) the Yangtze River Delta, one of China's largest container shipping hubs, including Shanghai Port; (3) the Southeast coastal region, frequently impacted by intense typhoon activity and (4) the Pearl River Delta, a critical gateway for global trade, with extensive port infrastructure. By comparing port resilience across these diverse geographical settings, the study identified key variations in infrastructure robustness, emergency preparedness, and operational efficiency. Ports in the Pearl River Delta and Yangtze River Delta demonstrated higher resilience due to superior infrastructure and disaster response systems, whereas ports in the Bohai Bay area exhibited greater vulnerability due to exposure to EWEs and weaker emergency management protocols. The findings further underscored the importance of targeted resilience investment in high-risk coastal regions, advocating for improved governance mechanisms, enhanced forecasting capabilities, and adaptive infrastructure development. Expanding on climate-related trade disruptions, Yin et al. (2024) further highlighted that ports in the South China Sea and East China Sea faced the highest risk of operational delays and congestion, particularly when typhoons led to multi-port closures and cargo rerouting bottlenecks.

Beyond China, Xia et al. (2024) assessed how inter-port competition influences climate adaptation investments in Asian maritime hubs. Using game-theoretic modelling, the study evaluated how ports strategically allocated resources to capacity expansion, resilience upgrades, and market competitiveness under climate uncertainty. The results indicated that ports in highly competitive regions, such as Singapore, may underinvest in adaptation measures unless regulatory frameworks incentivise long-term climate resilience planning. The study emphasised the need for regional cooperation and policy coordination to avoid fragmented adaptation efforts that could weaken supply chain stability in Asia area.

13.5 FUTURE DIRECTIONS AND CHALLENGES

Overall, emerging trends in climate-resilient maritime transport indicate a growing emphasis on adaptive infrastructure, nature-based solutions, and systemic risk management. Many studies, including those by Izaguirre et al. (2021) and Ribeiro et al. (2021), highlighted the increasing vulnerability of global ports to climate change, necessitating investment in resilient port infrastructure. The trend towards nature-based solutions, such as mangrove restoration and coastal buffer zones, has been explored as an effective strategy for mitigating flooding risks while enhancing ecosystem services. Furthermore, the study by Verschuur et al. (2022) suggested a shift towards systemic risk assessment, recognising that climate-induced disruptions to ports can also create cascading failures across supply chains. The adoption of climate adaptation strategies within the Blue Economy framework, as discussed by March et al. (2024), also signalled a broader transition towards sustainability-oriented maritime transport planning. However, there remains a need for more enhanced predictive models and early-warning systems that integrate climate science with real-time operational data to improve resilience planning.

From the theoretical perspective, gaps in research and technological advancements needed for maritime transport resilience highlight several areas where current methodologies and technologies remain insufficient. Studies such as Wan et al. (2024) and Yin et al. (2024) have demonstrated the effectiveness of Bayesian and probabilistic modelling for assessing disruption risks, yet their practical implementation in port resilience planning remained limited. Similarly, while network analysis approaches, as used by Poo et al. (2024a) and Wan et al. (2022), provided valuable insights into shipping route vulnerabilities, they often failed to capture the full complexity of interdependencies between ports and hinterland logistics. The need for interdisciplinary collaboration among climate scientists, port engineers, and supply chain analysts is evident, particularly in the development of integrated risk assessment frameworks that account for both immediate and long-term climate risks. Additionally, research gaps persist in evaluating the financial feasibility of adaptation measures, as noted by Wang et al. (2022) and Xia et al. (2024), who identified inconsistencies in investment strategies across ports with varying degrees of climate exposure. A key technological challenge lies in the lack of harmonised data-sharing platforms among port authorities, shipping companies, and climate research institutions, which restricts the ability to implement coordinated resilience strategies.

From the practical standpoint, policy recommendations for enhancing maritime transport resilience should focus on regulatory reforms, enhanced international cooperation, and incentivising adaptation investments. The work of Hassanali (2017) and McIntosh and Becker (2019) highlighted the importance of incorporating climate resilience into national and regional maritime policies. Policymakers should mandate resilience assessments as part of port infrastructure development plans, ensuring that climate risk considerations are embedded into port expansion and modernisation projects. Furthermore, maritime-related national governments and International Maritime Organisations must strengthen collaboration on climate adaptation strategies, as demonstrated in the stakeholder engagement approaches discussed by Morris (2020) and Polydoropoulou et al. (2025). To address financial constraints, regulatory frameworks should encourage private sector investment in resilience-building projects through tax incentives, resilience bonds, or public–private partnerships. Additionally, trade policies should integrate climate resilience criteria, ensuring that ports with robust adaptation measures are prioritised within global supply chains. Studies by León-Mateos et al. (2021) and Gharehgozli et al. (2017) further suggested that standardised resilience indices should be adopted to benchmark and monitor climate adaptation progress across different port regions.

In future, as the development of new technologies, the integration of them, including artificial intelligence (AI), blockchain, and the Internet of Things (IoT), into resilience planning of maritime transport presents a major opportunity for the maritime sector. Many studies, such as Wan et al. (2022) and Verschuur et al. (2020), demonstrated the importance of real-time data analytics in managing disruptions, yet digital adoption in maritime resilience remains fragmented. AI-driven predictive analytics could significantly enhance early-warning systems by integrating meteorological, trade, and operational data to provide real-time risk assessments. Blockchain technology offers potential for securing supply chain data, ensuring that resilience-related policies, such as cargo rerouting during disruptions, are efficiently enforced. IoT-enabled smart ports, equipped with sensors that monitor sea-level changes, structural integrity, and cargo movements, could provide more proactive resilience measures, reducing reliance on reactive approaches. Despite these benefits, studies by Xia et al. (2024) and Yin et al. (2024) highlighted the challenges of integrating digital technologies, including concerns over cybersecurity risks, interoperability between different port systems, and resistance from traditional stakeholders within the maritime sector. Future research must address these barriers by developing industry-wide digital standards and fostering collaboration between technology developers and maritime regulators.

Collectively, these future directions emphasise the need for a multi-disciplinary and multi-stakeholder approach to maritime transport resilience. While significant progress has been made in assessing climate risks, there remains an urgent need for investment in adaptive infrastructure, regulatory reforms, and digital transformation to ensure the long-term sustainability of global maritime transport and port development.

13.6 CONCLUSION

This chapter provides a comprehensive review of the impact of climate change on maritime transport and port infrastructure resilience, including both immediate disruptions from EWEs and long-term vulnerabilities linked to environmental changes. The findings reveal that climate-induced risks to the maritime sector are highly complex and geographically uneven, affecting global trade networks through port shutdowns, supply chain bottlenecks, and increased operational costs. Methodologically, this chapter identifies four main research approaches to assessing maritime resilience: network analysis and optimisation models, scenario-based simulations, quantitative risk modelling, and qualitative stakeholder-based research. Each of these methodologies contributes to understanding how climate risks propagate through shipping networks, port operations, and global supply chains, and how resilience measures can mitigate economic losses and operational downtime. A review of state-of-the-art datasets also highlights the growing importance of climate and meteorological databases, maritime transport data, and policy-driven stakeholder assessments in developing effective adaptation strategies.

Looking forward, there is co-existence of advances and challenges in resilience assessment. In addition to those opportunities in developing economic-risk assessment models, cross-border coordination and digital technologies as discussed in Section 13.5, a greater emphasis on climate justice is needed to ensure that developing regions with high exposure to climate risks receive adequate support in resilience planning. Strengthening maritime resilience is not only essential for safeguarding global trade and economic stability, but also for ensuring sustainable and equitable adaptation to climate change in an increasingly interconnected world.

ACKNOWLEDGEMENTS

This work has received funding from the European Research Council (ERC) under the European Union's Horizon 2020 research and innovation programme (Grant Agreement No. 864724) and the China Scholarship Council under Grant: 202308060322.

REFERENCES

Abdelhafez, M.A., Ellingwood, B., Mahmoud, H., 2021. Vulnerability of seaports to hurricanes and sea level rise in a changing climate: A case study for mobile, AL. *Coastal Engineering* 167, 103884.

Amer, R., 2024. Assessing vulnerability and enhancing resilience of port systems in southeast Texas facing sea-level rise. *Frontiers in Built Environment* 10, 1369833.

Asariotis, R., Monioudi, I.N., Mohos Naray, V., Velegrakis, A.F., Vousdoukas, M.I., Mentaschi, L., Feyen, L., 2024. Climate change and seaports: Hazards, impacts and policies and legislation for adaptation. *Anthropocene Coasts* 7 (1), 14.

Azmi, M.A., Mokhtar, K., Osnin, N.A., Razali Chan, S., Albasher, G., Ali, A., Nawaz, A., Oloruntobi, O., Chuah, L.F., 2023. Enhancing coastal ecosystem resilience: Investigating the interplay between safety criteria and ferry employee's perceptions to address climate change impacts. *Environmental Research* 238, 117074.

Balakrishnan, S., Lim, T., Zhang, Z., 2022. A methodology for evaluating the economic risks of hurricane-related disruptions to port operations. *Transportation Research Part A: Policy and Practice* 162, 58–79.

Bathgate, K., Perez De La Cruz, A., Zhang, Z., 2022. Quantitative analysis of hurricane Harvey impacts on texas maritime facilities. *Transportation Research Record: Journal of the Transportation Research Board* 2676 (7), 411–422.

Becker, A., Caldwell, M.R., 2015. Stakeholder perceptions of seaport resilience strategies: A case study of Gulfport (Mississippi) and Providence (Rhode Island). *Coastal Management* 43 (1), 1–34.

Becker, A.H., Matson, P., Fischer, M., Mastrandrea, M.D., 2015. Towards seaport resilience for climate change adaptation: Stakeholder perceptions of hurricane impacts in Gulfport (MS) and Providence (RI). *Progress in Planning* 99, 1–49.

Cao, Y., Xin, X., Jarumaneeroj, P., Li, H., Feng, Y., Wang, J., Wang, X., Pyne, R., Yang, Z., 2025. Data-driven resilience analysis of the global container shipping network against two cascading failures. *Transportation Research Part E: Logistics and Transportation Review* 193, 103857.

Ching-Pong Poo, M., Yang, Z., Lau, Y.-y., 2024. Vulnerability analysis of cruise shipping in ASEAN countries facing COVID-19 pandemic. *Ocean & Coastal Management* 248, 106919.

Ding, W., Wu, J., 2023. Interregional economic impacts of an extreme storm flood scenario considering transportation interruption: A case study of Shanghai, China. *Sustainable Cities and Society* 88, 104296.

Gharehgozli, A.H., Mileski, J., Adams, A., Von Zharen, W., 2017. Evaluating a "wicked problem": A conceptual framework on seaport resiliency in the event of weather disruptions. *Technological Forecasting and Social Change* 121, 65–75.

Grant, R., Rhode-Barbarigos, L., Roy, S.S., Britton, L., Li, C., Rowe, A., Becker, A., Hope, B., Bello, M., 2024. No port stands alone: PortMiami and the resilience of its Caribbean and Mesoamerican maritime network. *Maritime Economics & Logistics* 26 (2), 342–365.

Hassanali, K., 2017. Challenges in mainstreaming climate change into productive coastal sectors in a Small Island State – The case of Trinidad and Tobago. *Ocean & Coastal Management* 142, 136–142.

Islam, S., Goerlandt, F., Sakalayen, Q.M.H., Shi, Y., Venkatesh, V.G., 2023. Developing a "Disaster Scenario" to prepare for the possibility of disruptions to maritime transportation serving coastal communities of Vancouver Island. *Marine Policy* 150, 105531.

Izaguirre, C., Losada, I.J., Camus, P., Vigh, J.L., Stenek, V., 2021. Climate change risk to global port operations. *Nature Climate Change* 11 (1), 14–20.

Kalaidjian, E., Becker, A., Pinel, S., 2022. Operationalizing resilience planning, theory, and practice: Insights from U.S. seaports. *Frontiers in Sustainability* 3, 963555.

Lamers, M., Duske, P., Van Bets, L., 2018. Understanding user needs: A practice-based approach to exploring the role of weather and sea ice services in European Arctic expedition cruising. *Polar Geography* 41 (4), 262–278.

Lawrence, J.-M., Ibne Hossain, N.U., Jaradat, R., Hamilton, M., 2020. Leveraging a Bayesian network approach to model and analyze supplier vulnerability to severe weather risk: A case study of the U.S. pharmaceutical supply chain following Hurricane Maria. *International Journal of Disaster Risk Reduction* 49, 101607.

León-Mateos, F., Sartal, A., López-Manuel, L., Quintás, M.A., 2021. Adapting our sea ports to the challenges of climate change: Development and validation of a Port Resilience Index. *Marine Policy* 130, 104573.

March, A., Woolley, M., Failler, P., 2024. Integration of climate change mitigation and adaptation in Blue Economy planning in Africa. *Mitigation and Adaptation Strategies for Global Change* 29 (5), 38.

McIntosh, R.D., Becker, A., 2019. Expert evaluation of open-data indicators of seaport vulnerability to climate and extreme weather impacts for U.S. North Atlantic ports. *Ocean & Coastal Management* 180, 104911.

McLean, E.L., Becker, A., 2020. Decision makers' barriers to climate and extreme weather adaptation: A study of North Atlantic high- and medium-use seaports. *Sustainability Science* 15 (3), 835–847.

Morris, L.L., 2020. Stakeholder collaboration as a pathway to climate adaptation at coastal ports. *Maritime Policy & Management* 47 (7), 953–967.

Polydoropoulou, A., Bouhouras, E., Papaioannou, G., Karakikes, I., 2025. Living labs for the resilience of ports against climate change disruptions. *Ocean & Coastal Management* 261, 107528.

Poo, M.C.-P., Yang, Z., 2024. Optimising the resilience of shipping networks to climate vulnerability. *Maritime Policy & Management* 51 (1), 15–34.

Poo, M.C.-P., Yang, Z., Lau, Y.-y., Jarumaneeroj, P., 2024a. Assessing the impact of Arctic shipping routes on the global container shipping network's connectivity. *Polar Geography* 47 (3), 219–239.

Poo, M.C.-P., Zhang, W., Kamalian, L., Wang, T., Lau, Y.-y., Xu, T.Z., 2024b. Resilience of Chinese ports to tropical cyclones: Operational efficiency and strategic importance. *Climate* 12 (12), 214.

Ribeiro, A.S., Lopes, C.L., Sousa, M.C., Gomez-Gesteira, M., Dias, J.M., 2021. Flooding conditions at Aveiro Port (Portugal) within the framework of projected climate change. *Journal of Marine Science and Engineering* 9 (6), 595.

Roukounis, C.N., Tsihrintzis, V.A., 2024. Climate change adaptation strategies for coastal resilience: A stakeholder surveys. *Water* 16 (11), 1519.

Ryan-Henry, J., Becker, A., 2020. Port stakeholder perceptions of Sandy impacts: A case study of Red Hook, New York. *Maritime Policy & Management* 47 (7), 885–902.

Sharaan, M., Ibrahim, M.G., Moubarak, H., ElKut, A.E., Romya, A.A., Hamouda, M., Soliman, A., Iskander, M., 2024. A qualitative analysis of climate impacts on Egyptian ports. *Sustainability* 16 (3), 1015.

Tafur, A., Padgett, J.E., 2024. A flow-based commodity-independent port capacity model for resilience assessment of intermodal freight networks subjected to coastal hazards. *Reliability Engineering & System Safety* 250, 110280.

Taramelli, A., Valentini, E., Sterlacchini, S., 2015. A GIS-based approach for hurricane hazard and vulnerability assessment in the Cayman Islands. *Ocean & Coastal Management* 108, 116–130.

Verschuur, J., Koks, E.E., Hall, J.W., 2020. Port disruptions due to natural disasters: Insights into port and logistics resilience. *Transportation Research Part D: Transport and Environment* 85, 102393.

Verschuur, J., Pant, R., Koks, E., Hall, J., 2022. A systemic risk framework to improve the resilience of port and supply-chain networks to natural hazards. *Maritime Economics & Logistics* 24 (3), 489–506.

Wan, C., Tao, J., Yang, Z., Zhang, D., 2022. Evaluating recovery strategies for the disruptions in liner shipping networks: A resilience approach. *The International Journal of Logistics Management* 33 (2), 389–409.

Wan, C., Yuan, J., Cao, D., Wang, T., Ng, A.K., 2024. A fuzzy evidential reasoning-based model for evaluating resilience of ports to typhoons. *Transportation Research Part D: Transport and Environment* 133, 104228.

Wang, B., Chin, K.S., Su, Q., 2022. Risk management and market structures in seaport–dry port systems. *Maritime Economics & Logistics* 24 (1), 114–137.

Wang, T., Poo, M.C.-P., Ng, A.K.Y., Yang, Z., 2023. Adapting to the impacts posed by climate change: Applying the Climate Change Risk Indicator (CCRI) framework in a multi-modal transport system. *Sustainability* 15 (10), 8190.

Xia, W., Mishra, J., Adulyasak, Y., 2024. Seaport adaptation and capacity investments under inter-port competition and climate-change uncertainty. *Transportation Research Part D: Transport and Environment* 130, 104183.

Xu, M., Zhu, Y., Liu, K., K. Y. Ng, A., 2024. Assessing resilience of global liner shipping network to tropical cyclones. *Transportation Research Part D: Transport and Environment* 131, 104189.

Yin, J., Khan, R.U., Wang, X., Asad, M., 2024. A data-centered multi-factor seaport disruption risk assessment using Bayesian networks. *Ocean Engineering* 308, 118338.

14 Machine Learning-Driven Resilience Assessment: *Integrating Dynamic Bayesian Networks and Machine Learning for Predicting Critical Infrastructure Resilience against Disasters*

Bawantha Rathnayaka[a], Dilan Robert[a], Varuna Adikariwattage[b], Chandana Siriwardana[c], and Sujeeva Setunge[a]

[a]Civil Engineering Department, School of Engineering, RMIT University, Melbourne, VIC, Australia
[b]Civil Engineering Department, University of Moratuwa, Moratuwa, Sri Lanka
[c]School of Built Environment, Massey University, Palmerston North, New Zealand

14.1 INTRODUCTION

Recent technological advancements have increased the complexities of Critical Infrastructures (CIs), which are vital for societal well-being (Rathnayaka et al., 2022, 2023). These infrastructures have become more interdependent, improving efficiency but also introducing vulnerabilities during disruptions, such as cascading failures and delays in restoration (Rathnayaka et al., 2022, 2024). Disaster events underscore the necessity for effective resilience assessments to manage uncertainties (Randil et al., 2022; Shehara et al., 2022). Resilience, defined as the absorptive, adaptive, and recovery capacities of a system in the face of disruptions, has gained prominence over traditional risk-based approaches, which struggle with the dynamic and unpredictable nature of CI disruptions (Hosseini et al., 2016; UNDRR, 2019). Various methodologies, including agent-based simulations, system dynamics, network theory, and Bayesian Network, have been employed to assess CI resilience (Hosseini et al., 2016; Ouyang, 2014; Rathnayaka et al., 2022, 2024).

DOI: 10.1201/9781003567332-14

The rapid evolution of Machine Learning (ML) techniques further expands the toolkit for resilience assessment by offering cutting-edge analytical capabilities that can transform disaster management practices. These capabilities enable the development of more robust, precise, and flexible frameworks than those previously attainable (Al Shafian & Hu, 2024; Kameswari, 2023; Linardos et al., 2022). ML provides data-driven insights that were once difficult to obtain using comprehensive datasets that include empirical data, real-time sensor data, and simulation outputs (Christoph Molnar, 2019; Linardos et al., 2022; Roscher et al., 2020; Zennaro et al., 2021). Advanced computational algorithms enable the detection of complex patterns, relationships, and hidden interdependencies that conventional analytical methods might overlook (Arinta & Andi, 2019; Derras & Makhoul, 2022; Kameswari, 2023; Xie et al., 2020).

The capacity to deal with non-linear relationships among the variables is one of the major advantages of ML. CIs frequently involve complex cross-sector interactions that are further shaped by external disruptions, making it essential to analyse intertwined processes under scenarios such as cascading failures, interdependent disruptions, and multi-hazard impacts (Derras & Makhoul, 2022; Kyrkou et al., 2023; Wang et al., 2022; Xie et al., 2020; Zennaro et al., 2021). Unlike traditional approaches, ML-based resilience assessment methodologies are inherently adaptive, integrating new datasets to reflect changing conditions in CIs over time. This continuous learning ability enables real-time updates and maintains the relevance of predictions and evaluations, even amidst evolving environmental, operational, or infrastructural conditions (Kameswari, 2023; Linardos et al., 2022; Soleimani & Hajializadeh, 2022; Wang et al., 2022; Xie et al., 2020). This level of flexibility is especially vital in managing disasters, as it enables prompt and precise insights that support effective decision-making and efficient allocation of resources (Arinta & Andi, 2019; Kyrkou et al., 2023; Sun et al., 2020; Wang et al., 2022; Zennaro et al., 2021).

In our previous study, we proposed a Dynamic Bayesian Networks (DBN) based methodology for assessing interdependent CI systems' resilience across multiple sectors, including transportation, electricity, and water (Rathnayaka et al., 2025). That work emphasised interdependencies and their influence on system-wide resilience dynamics. In contrast, the present chapter builds upon this prior framework by focusing on the integration of DBNs with ML techniques to enhance predictive accuracy at the single infrastructure level. This work differs by developing and testing ML models trained on DBN-generated datasets to assess resilience under varying disruption scenarios, offering a more data-driven and scalable approach to resilience prediction. By capturing the dynamic properties of CIs and employing data-driven modelling, this DBN-ML based approach delivers an enhanced resilience quantification. The framework's multidimensional nature allows it to capture shifts in system states under disaster scenarios. Ultimately, this research aims to provide a comprehensive resilience prediction tool to improve resilience in CIs.

14.2 LITERATURE REVIEW

ML methods, including supervised, unsupervised, and reinforcement learning, offer distinctive benefits across various aspects of disaster management and disaster risk reduction (Al Shafian & Hu, 2024; Arinta & Andi, 2019; Kyrkou et al., 2023;

Linardos et al., 2022; Sun et al., 2020). For example, supervised learning techniques can leverage historical data to forecast disruptions and evaluate their potential consequences (Merz et al., 2020; Venkadesh et al., 2024), while unsupervised methods uncover latent patterns or groupings within CI performance indicators (Al Shafian & Hu, 2024; Linardos et al., 2022). Reinforcement learning optimises decision-making by evaluating simulated scenarios and estimating the potential effects of various resilience improvement interventions. Moreover, ensemble learning approaches, which integrate multiple algorithms, can significantly improve both the precision and robustness of predictive outcomes (Kameswari, 2023; Zennaro et al., 2021).

Beyond prediction tasks, ML has been applied in wide-ranging applications in disaster management, such as powering early warning systems, guiding resource allocation, and informing recovery strategies (Kameswari, 2023; Kyrkou et al., 2023; Linardos et al., 2022; Wang et al., 2022; Zennaro et al., 2021). These capabilities include harnessing instantaneous data from IoTs and social media for early warning alerts, directing the deployment of emergency resources (e.g., personnel, equipment, and funding), and aiding post-disruption efforts through automated damage assessments and the prioritisation of restoration activities (Arinta & Andi, 2019; Kameswari, 2023; Kyrkou et al., 2023; Linardos et al., 2022; Zennaro et al., 2021).

Notwithstanding these advancements, integrating ML into resilience assessments and disaster response presents several challenges. Securing high-quality, extensive datasets, meeting significant computational demands, and dealing with the lack of transparency inherent in more complex models are among the key obstacles in ML-driven resilience prediction (Arinta & Andi, 2019; Kyrkou et al., 2023; Linardos et al., 2022). Addressing these issues necessitates leveraging domain specific expertise, strong data management frameworks, and advancing explainable AI to ensure that ML outputs remain trustworthy and practically interpretable. With the ongoing advancement of ML, its incorporation into resilience assessment models is poised to play a pivotal role in enabling infrastructure systems to become more flexible, resilient, and environmentally sustainable.

14.3 RESEARCH METHODOLOGY

This study has been carried out in two phases. These phases include data generation through a DBN model and ML model development. In the first phase, the training and testing data required for developing the ML model were generated using the DBN model derived from a methodology modified from a prior study by the authors (Rathnayaka et al., 2025). This model characterises the uncertainty and interconnected behaviours inherent to the system under study. The methodology used for developing the DBN model is depicted in Figure 14.1.

Step one in developing the DBN model involves defining the system under study, with an emphasis on outlining the CI sectors. Additionally, this stage includes gathering data on potential disruptions affecting the CIs. Different functionality states (S_1, S_2, S_3, and S_4 as shown in Figure 14.2) are established for the CI sector in response to various disruptive events in the second step. These states form the foundation for constructing Markov chains for each critical component within the CI. Anticipative, absorptive, adaptive, and restorative capacities of the system determine the transition

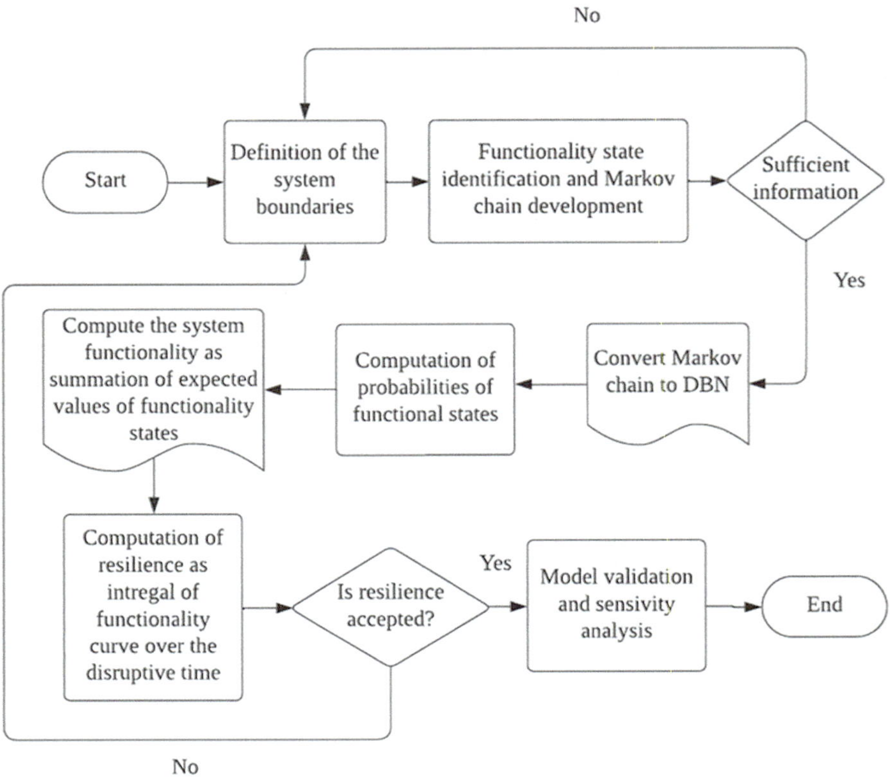

FIGURE 14.1 Modified methodology for developing a DBN model for data generation (adopted from Rathnayaka et al., 2025).

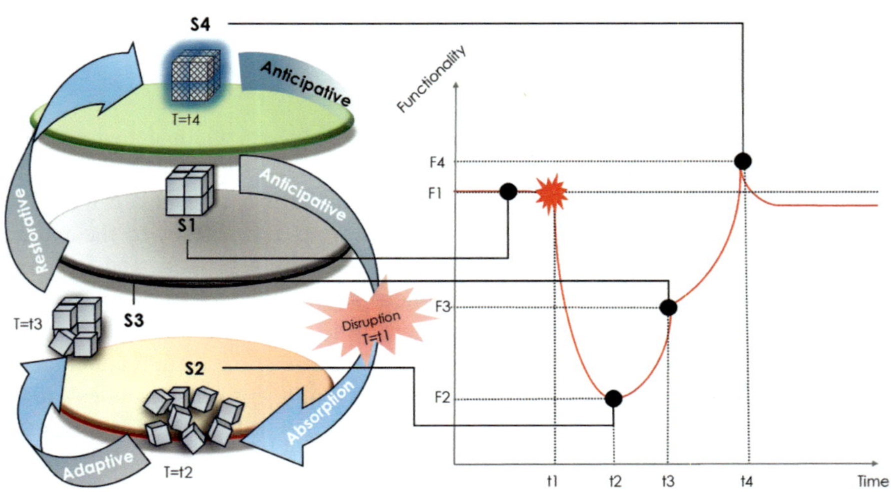

FIGURE 14.2 Illustration of the evaluation of the system's state subject to external disturbance (adopted from Rathnayaka et al., 2025).

probabilities among the states. The four key capacities are affected by both extrinsic influences and intrinsic system properties. The primary factors contributing to these capacities are determined from the authors' prior study (Rathnayaka et al., 2024).

The DBN model's structure is presented in Figure 14.3. Furthermore, Figure 14.3 presents variables for the dataset.

After, DBN models are constructed by converting the Markov chains. Then, the probabilities of the four functional states of each CI sector are estimated. Following this, the functionality of the CI is assessed by calculating the expected values of the functional states. Then, resilience can be measured by functionality loss (*R*), as represented in Eq. (14.1).

$$R = \int_{t_1}^{t_4} [100 - f(t)]\, dt \tag{14.1}$$

where

$f(t)$: Functionality curve
t_1 = time of disruption
t_4 = time the infrastructure fully recovers

In the final step, the model's robustness and accuracy were assessed through sensitivity analysis and validation. Further insight into the proposed methodology and its implementation can be found in Rathnayaka et al. (2025).

A validated DBN model, developed based on the methodology outlined above, was employed to generate the dataset for this study. The model represents a typical pump house situated within a water distribution system in the Colombo municipal area of Sri Lanka. Figure 14.4 presents a generalised schematic diagram of a typical water distribution system. Within the water distribution system, pump houses play a critical role by connecting upstream treatment facilities to the water treatment plant and from there to downstream distribution zones. These facilities typically comprise key components such as electric pumps, backup power systems, intake valves, control panels, and pressure regulation units. In the DBN framework developed for this study, the pump house was modelled as a single node that represents the collective operational functionality of these components. This abstraction enables the model to capture the cumulative impact of component interactions on the resilience and functionality of the pump house.

Figure 14.5 presents the flowchart for the data generation process used in this study. The process is initiated by inputting the DBN structure depicted in Figure 14.3. The DBN model structure is initialised using GeNIe software. Then, it was used to simulate a diverse set of scenarios, assigning different probabilistic ranges to input nodes, child nodes, disruption nodes, and state patterns.

These scenarios simulate high, low, and random probabilities, as well as both static and dynamic disruption patterns. In each simulation scenario, the model assigns probabilistic values to both input variables and disruption nodes using a uniform random distribution, while incorporating temporal patterns that reflect predefined behavioural dynamics. Subsequently, Bayesian inference is applied to compute the

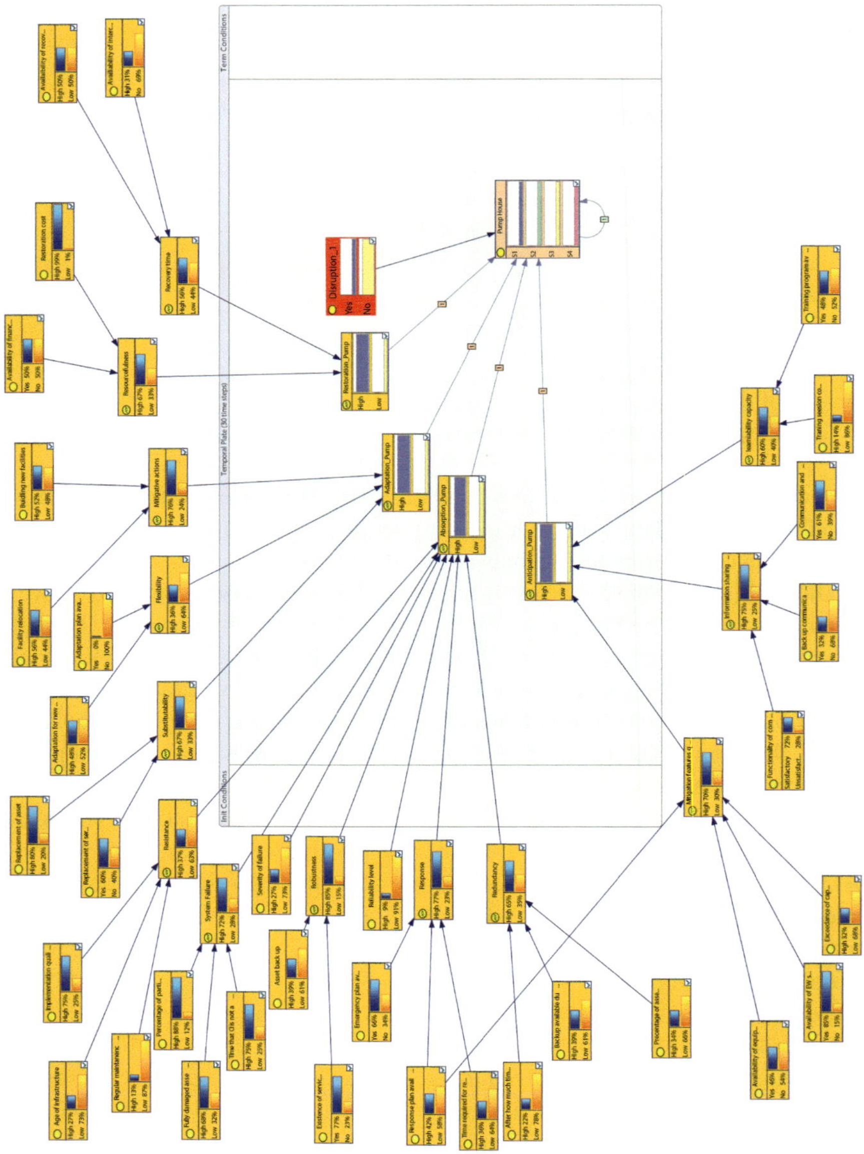

FIGURE 14.3 DBN model developed for the resilience assessment of the pump house.

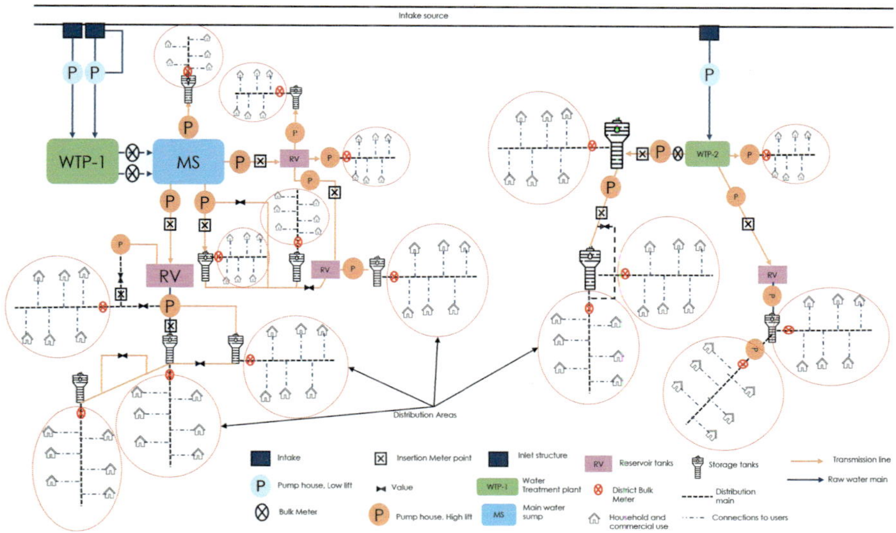

FIGURE 14.4 Schematic diagram of the typical water distribution network.

time-dependent evolution of functionality state probabilities. The output includes a comprehensive set of data comprising state probabilities, disruption statuses, and corresponding conditional probabilities, all systematically structured into a cohesive dataset. Resilience is then estimated by integrating the area under the functionality loss curve, and disruption intensity is quantified as the expected value by weighting the probability with its corresponding states. This curated dataset enables a detailed examination of how variations in input conditions influence resilience outcomes across a diverse range of scenarios.

The generated dataset was used as input for the ML model development phase, constituting the second stage of this study. Then, the dataset was split into 80/20 training and test sets to evaluate the model. During training, a 10-fold cross-validation framework was applied to strengthen the reliability of the model outcomes. Accordingly, six models were constructed using various tree-based ML algorithms, including AdaBoost, Decision Tree (DT), Extra Tree (ET), LightGBM, Random Forest (RF), and Extreme Gradient Boosting (XGBoost).

Following this, hyperparameter tuning was conducted to optimise the predictive performance of each ML model. This study adopted the "grid search" technique, systematically exploring multiple hyperparameter configurations within predefined ranges to find the optimal combination.

The search process, executed using GridSearchCV from the scikit-learn library, integrated 10-fold cross-validation to ensure a comprehensive and unbiased evaluation. Thus, the search process mitigates the overfitting to a single test set, thereby enhancing the models' generalisability. The performances of ML models were assessed using four key statistical metrics: Mean Absolute Error (MAE), Root Mean Squared Error (RMSE), Coefficient of Determination (R^2), and Pearson Correlation Coefficient (R). Each metric serves a specific purpose: R^2 and R evaluate the

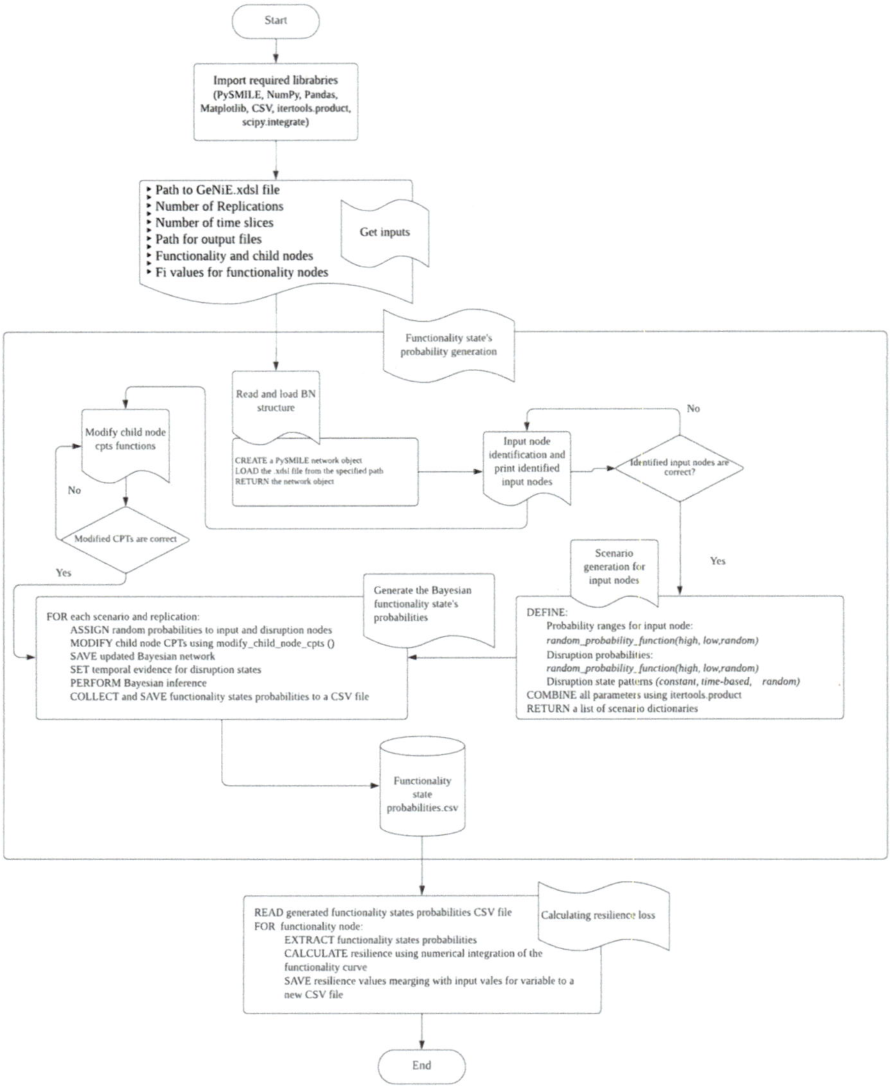

FIGURE 14.5 Data generation flow chart.

goodness of fit and linear relationship strength between predicted and actual values, where values approaching one show a strong correlation. Meanwhile, MAE measures the average magnitude of residuals, while RMSE quantifies the dispersion of prediction errors. Both training and testing datasets were analysed using R^2, while all four metrics were employed to comprehensively assess model accuracy and consistency. Additionally, visual representations were used to examine data distribution and model performance. Furthermore, an uncertainty analysis was performed to validate the robustness of the ML regression models, reinforcing their reliability and predictive capability.

14.4 RESULTS

14.4.1 Overview of the Data Generated

The dataset produced from the DBN model simulation was analysed to investigate the relationship between "Average resilience" and the associated resilience indicators. Figure 14.6 presents a correlation heatmap quantifying these relationships, revealing that the correlations between "Average Resilience" and individual resilience indicators are very low, with values close to zero. This suggests that traditional linear correlation analysis may not effectively capture the complex interactions between variables within CI system. The weak correlations indicate that the relationship between average resilience and resilience indicators is likely non-linear, underscoring the necessity of ML techniques to uncover hidden patterns and dependencies. ML models, such as tree-based regressors or deep learning approaches, can provide a more comprehensive understanding of these non-linear interactions, offering improved predictive capabilities for resilience assessment in this CI.

14.4.2 Performance Evaluation of ML Models

This section presents the hyperparameter optimisation results of the ML models. Hyperparameter tuning was conducted across six tree algorithms to optimise the predictive performance of ML models in assessing resilience loss. Table 14.1 presents the selected hyperparameter values for each model, highlighting key tuning parameters such as tree depth, the number of estimators, minimum sample constraints, and bootstrap settings. The results indicate that deeper trees (depth = 20) were required for AdaBoost, DT, and ET to capture complex non-linear dependencies. In contrast, LightGBM and XGBoost achieved optimal performance with comparatively shallower depths (12 and 14, respectively). Additionally, higher numbers of estimators (1,000) were used for AdaBoost, ET, and RF, reinforcing the necessity of ensemble averaging to improve model stability.

Figure 14.7 illustrates the variation in R^2 values for training and testing datasets as hyperparameters are tuned. LightGBM (Figure 14.7a) demonstrates consistent generalisation, with both training and testing R^2 values converging above 0.995 for tree depths greater than 6.

Similarly, XGBoost (Figure 14.7b) exhibits a high training score for increasing tree depth, while the testing remains relatively stable around 0.97, suggesting strong learning capacity with slight overfitting at deeper depths. The performance of AdaBoost (Figure 14.7c) shows an increasing trend in training and testing R^2 as tree depth increases. However, at deeper depths ($\geq$ 10), the training R^2 reaches nearly 1.0 while the testing stabilises, indicating potential overfitting. RF (Figure 14.7d) maintains stable testing across different numbers of estimators, with minimal deviation, indicating strong robustness against overfitting. The observed trends emphasise the significance of proper hyperparameter tuning in mitigating overfitting and ensuring model reliability. Furthermore, the weak correlations observed in the correlation analysis suggest that resilience indicators exhibit strong non-linearity with average resilience, further justifying the application of advanced ML learning techniques to uncover hidden dependencies.

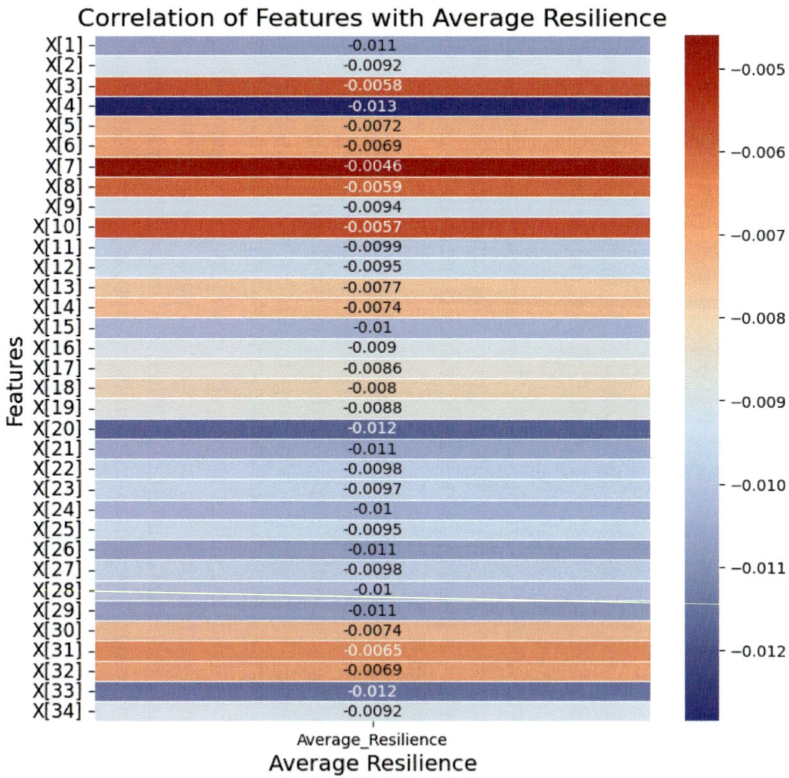

ID	Feature name
X[1]	Facility_relocation_probability
X[2]	Availiability_of_interconnected_service_probability
X[3]	Availaibility_of_recovery_plan_probability
X[4]	Restoration_cost_probability
X[5]	Availability_of_financial_support_probability
X[6]	Exceedance_of_capacity_demand_probability
X[7]	Availability_of_equipment_and_procedure_of_hazard_mitigation_probability
X[8]	Availability_of_EW_systems_probability
X[9]	Response_plan_availability_probability
X[10]	Communication_and_information_sharing_systme_avaibalility_probability
X[11]	Functionality_of_communication_system_probability
X[12]	Back_up_communication_system_availability_probability
X[13]	Training_program_availablility_probability
X[14]	Traning_seesion_conducted_per_month_probability
X[15]	Implementation_quality_of_safety_standards_probability
X[16]	Age_of_infrastructure_probability
X[17]	Regular_maintanence_level_probability
X[18]	Fully_damaged_assets_percentage_probability
X[19]	Percentage_of_partially_damaged_asset_probability
X[20]	Time_that_CI_is_not_able_serve_it_function_probability
X[21]	Severity_of_failure_probability
X[22]	Reliability_level_probability
X[23]	Asset_back_up_probability
X[24]	Existence_of_service_replacement_probability
X[25]	Backup_available_durration_probability
X[26]	After_how_much_time_backup_is_available__probability
X[27]	Precentage_of_assest_that_have_backup_probability
X[28]	Time_required_for_response_probability
X[29]	Emergency_plan_availability_probability
X[30]	Adaptation_plan_availability_probability
X[31]	Adaptation_for_new_condition_on_time_probability
X[32]	Replacement_of_asset_probability
X[33]	Replacement_of_service_probability
X[34]	Buidling_new_facilities_probability

FIGURE 14.6 Correlations of input features and average resilience.

TABLE 14.1
Summary of the Hyperparameter Optimisation

Hyperparameter	Adaboost	DT	ET	LightGBM	RF	XGBoost
Tree depth	20	20	20	12	14	14
n_estimators	1000	-	1000	-	1000	50
min_samples_leaf	-	1	2	-	-	-
min_samples_split	-	10	2	-	-	-
bootstrap	-	-	FALSE	-	-	-

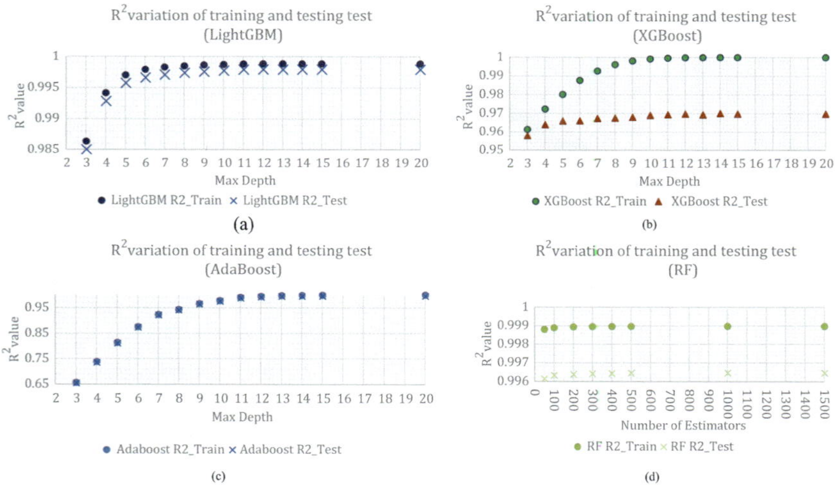

FIGURE 14.7 R^2 variation of the train and test sets with hyperparameters.

14.4.3 Model Prediction

This section provides a comprehensive analysis of the predictive performance of six ML algorithms. The predictive performance of ML models in estimating resilience loss was evaluated using actual vs. predicted resilience plots (See Figure 14.8). These plots assess how well each model aligns with actual resilience values, incorporating a perfect prediction line and 10% and 20% error bands to quantify the accuracy and reliability of predictions. The results indicate that ML models, particularly LightGBM (R=0.9990), ET (R=0.9981), and RF (R=0.9980), demonstrated the highest predictive accuracy, with most predictions falling within the 10% error band. These models exhibit strong generalisation capabilities, effectively capturing the non-linear dependencies between resilience indicators and overall system resilience.

AdaBoost (R=0.9977) also demonstrated high predictive accuracy but slightly greater variance than ET and LightGBM. The increased dispersion suggests that while AdaBoost effectively models resilience trends, it may be more sensitive to outliers. In contrast, DT (R=0.9940) exhibited noticeable deviations, with a wider

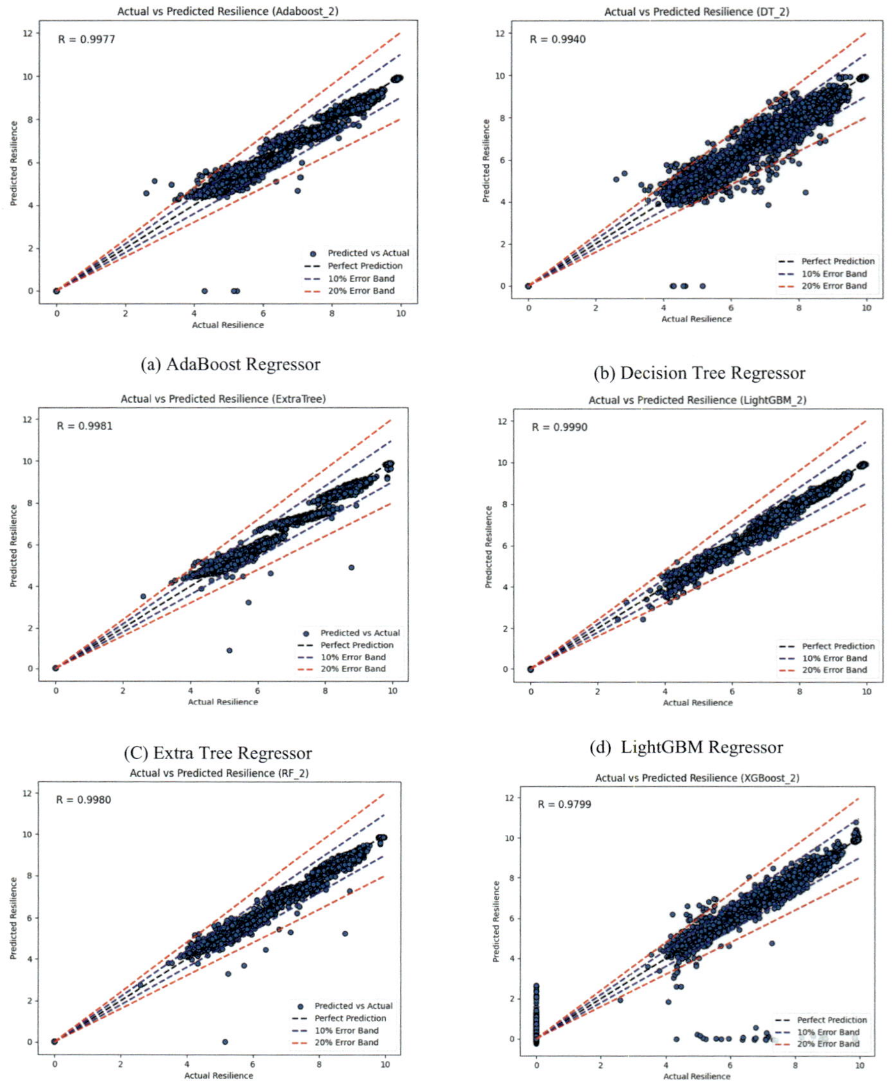

(a) AdaBoost Regressor

(b) Decision Tree Regressor

(C) Extra Tree Regressor

(d) LightGBM Regressor

(e) Random Forest Regressor

(f) XGBoost Regressor

FIGURE 14.8 ML model prediction of resilience (a) AdaBoost Regressor, (b) DT Regressor, (c) ET Regressor, (d) LightGBM Regressor, (e) RF Regressor, and (f) XGBoost Regressor.

spread of predictions and a significant number of points exceeding the 20% error margin, indicating limited generalisation due to its inherent high variance.

XGBoost (R = 0.9799) achieved the lowest correlation among the tested models, showing higher variability and a more significant number of outliers. The results highlight tree-based ML techniques' superior performance, particularly LightGBM, ET, and RF, in accurately modelling resilience loss. These models effectively minimise prediction errors while maintaining robust generalisation capabilities, making them well-suited for resilience assessment in CI systems.

14.4.4 ML Model Performance

The previously presented Figure 14.8 is unable to capture the overall uncertainty associated with the predictive models comprehensively. To address this limitation, an uncertainty analysis was conducted for the optimised models to evaluate their robustness and reliability. This section provides an in-depth assessment of the various ML models' performance metrics used in this study. Key evaluation indicators (i.e., RMSE, R^2, R, and MAE) were used to quantify model accuracy and generalisation. Figure 14.9 offers valuable insights into the predictive confidence and stability of the developed models and associated uncertainty.

The results indicate that ML models outperform traditional DTs, with LightGBM achieving the lowest MSE (0.015) and RMSE (0.122), along with the highest R^2 (0.998) and Pearson correlation (R = 0.999) values. This highlights LightGBM's superior predictive accuracy and robustness in modelling non-linear resilience dependencies. Similarly, ET and RF demonstrate strong performance, with RMSE values of 0.170 and 0.173, respectively, and high correlation coefficients (R = 0.998), reinforcing their effectiveness in resilience prediction.

In contrast, DT and XGBoost exhibit relatively lower performance, with DT having the highest MAE (0.162) and XGBoost showing the highest RMSE (0.547) and MSE (0.299), along with the lowest R-squared (R^2 = 0.960). The reduced accuracy of XGBoost suggests a higher sensitivity to data distribution and potential overfitting, necessitating further hyperparameter tuning.

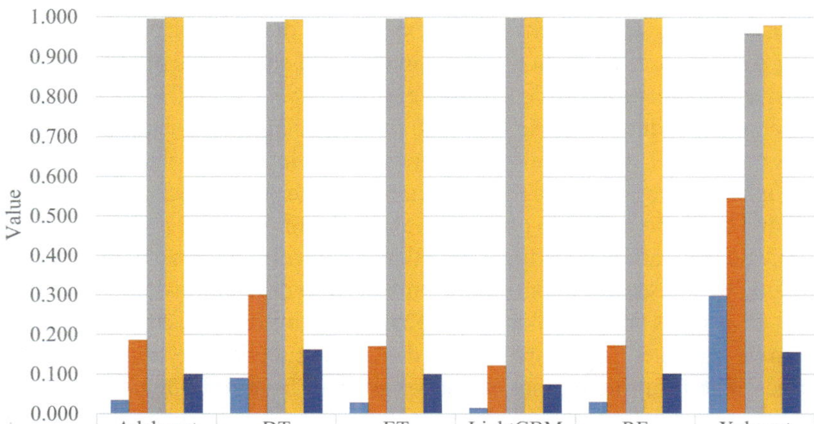

	Adaboost	DT	ET	LightGBM	RF	Xgboost
MSE	0.034	0.091	0.029	0.015	0.030	0.299
RMSE	0.186	0.301	0.170	0.122	0.173	0.547
R2	0.995	0.988	0.996	0.998	0.996	0.960
R	0.998	0.994	0.998	0.999	0.998	0.980
MAE	0.100	0.162	0.100	0.074	0.102	0.157

FIGURE 14.9 Uncertainty analysis of the ML models.

Meanwhile, AdaBoost achieves competitive performance, with an RMSE of 0.186 and an R-value of 0.998, demonstrating a balance between accuracy and computational efficiency. Overall, the findings emphasise the superiority of ensemble-based models (LightGBM, ET, RF, and AdaBoost) in predicting resilience loss. However, LightGBM is emerging as the most reliable model due to its minimal error and near-perfect correlation with the actual resilience value.

14.5 DISCUSSION

The integration of DBNs and ML in this study presents a novel and robust approach for resilience assessment in CI systems. By leveraging probabilistic modelling and advanced predictive analytics, the proposed framework effectively captures dynamic system behaviour and enhances the accuracy of resilience loss predictions. However, several areas warrant further investigation to improve the applicability and reliability of the model. This discussion focuses on four key aspects: extending ML models to capture interdependencies, applying explainable ML for interpretability, practical applications in resilience planning and resource optimisation, and model validation.

14.5.1 Extending ML Models to Capture Interdependencies in CI Systems

CI systems, such as electricity, water, and transportation, are highly interconnected, meaning that disruptions in one sector can trigger cascading failures across others. The ML models developed in this study provide accurate resilience predictions at the sector level but do not explicitly account for interdependencies. Future enhancements should incorporate advanced ML techniques, which are well-suited for modelling complex dependencies among interconnected infrastructures. Additionally, multi-task learning frameworks can be explored to develop ML models capable of making simultaneous predictions across multiple CI sectors, facilitating more comprehensive and proactive resilience strategies.

A technical roadmap (see Figure 14.10) for this extension involves first developing an integrated DBN that models the interdependencies among multiple CI sectors. This would involve identifying inter-sector nodes (e.g., electricity supply to water pumps and transport access to power stations), quantifying their probabilistic dependencies, and simulating cascading disruptions. A novel methodology introduced in a prior study by the same authors can be employed to develop this model for interdependent CI systems (Rathnayaka et al., 2025). From this expanded DBN, multi-sector data can be generated to capture the interdependent functionality states and capacity dynamics. The resulting dataset can serve as input for multi-output ML models or graph-based neural networks capable of learning joint dependencies. In this context, Graph Neural Networks and Spatio-Temporal Graph Convolutional Networks are particularly promising, as they can represent both the network topology of interdependent systems and their temporal evolution under disruptions. Feature engineering techniques can also be applied to merge sector-specific indicators into higher-level representations, enabling the ML model to infer hidden interdependencies. This approach would enable more realistic and actionable predictions of resilience across interdependent CI systems under diverse disaster scenarios.

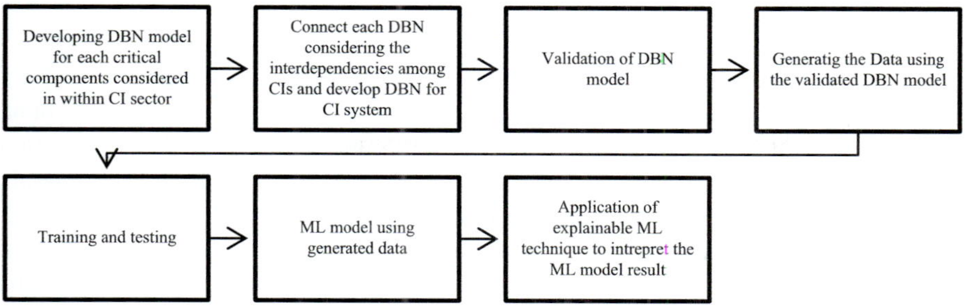

FIGURE 14.10 Technical roadmap for interdependency modelling using integrating ML–DBN.

14.5.2 Model Interpretability

Another major challenge in ML-driven resilience assessment is the interpretability of model predictions. While ML models can achieve high predictive accuracy, their complex structures often make them black-box systems, limiting their practical usability for decision-makers. To enhance transparency and trust in ML-based resilience assessments, explainable ML techniques should be integrated into the analytical workflow. Shapley Additive Explanations (SHAP), Local Interpretable Model-Agnostic Explanations, and feature importance analysis can provide insights into how different input factors contribute to resilience loss predictions. For example, using SHAP values, infrastructure planners can quantify the impact of individual resilience indicators (e.g., redundancy, recovery speed, or response efficiency) on predicted resilience outcomes. Furthermore, causal inference approaches can be leveraged to move beyond correlation-based interpretations and establish cause–and–effect relationships between CIs' resilience attributes and resilience loss, ensuring that ML-driven recommendations are grounded in actionable insights rather than statistical associations.

14.5.3 Application in Resilience Planning and Resource Optimisation

The practical applications of the developed ML–DBN framework extend beyond predicting resilience loss to enhancing disaster preparedness, response, and recovery strategies. The model can support proactive resilience planning by analysing resilience trends under various hazard scenarios and operational conditions, and by identifying infrastructure components that are most vulnerable to disruptions. For instance, CI operators can use the model to prioritise investments in adaptive measures, such as increasing redundancy, strengthening recovery mechanisms, or implementing distributed resource allocation strategies. During disasters, real-time ML predictions can optimise emergency resource distribution, ensuring personnel, equipment, and financial resources are directed to highly impacted areas. Additionally, reinforcement learning-based optimisation could be incorporated into future models to improve decision-making under evolving disaster conditions. This allows for adaptive, dynamic resource allocation in response to changing risk landscapes.

14.5.4 Validation of the Developed Model

Ensuring the validity and robustness of ML-driven resilience assessments is critical for their widespread adoption in infrastructure planning and disaster management. The proposed framework for developing the DBN model used for data generation in the study has been validated using a case study in Sri Lanka (Rathnayaka et al., 2025), demonstrating high predictive accuracy and consistency. However, broader validation efforts are required to establish the model's generalisability across diverse CI sectors, geographical regions, and hazard types. Future research should focus on multi-hazard validation, testing the framework under different disaster scenarios to assess its adaptability. Additionally, real-world implementation in collaboration with infrastructure stakeholders would provide valuable insights into the model's operational effectiveness and areas for improvement. Incorporating Bayesian uncertainty quantification into the model would enhance its reliability by providing confidence intervals for resilience predictions, ensuring robust decision support even in data-limited environments.

In conclusion, the integration of DBNs and ML techniques in this study offers a significant advancement in data-driven resilience assessment. While the developed framework demonstrates strong predictive performance, future research should focus on extending interdependency modelling, incorporating explainability features, applying the model in real-world resilience planning, and strengthening validation strategies. Addressing these aspects will enhance the applicability of the proposed methodology, ensuring that it evolves into a comprehensive, scalable, and operational resilience assessment tool capable of supporting data-driven decision-making for policymakers, emergency planners, and infrastructure managers.

14.6 CONCLUSION

A novel resilience assessment method using DBNs and ML techniques, accounting for resilience capacities, has been presented in this study to enhance predictive accuracy for CI systems. The proposed approach effectively captures CI disruptions' dynamic, probabilistic nature, addressing limitations in traditional risk-based assessment methodologies. By leveraging probabilistic modelling and data-driven learning, the framework provides a scalable, adaptive tool for quantifying resilience loss across various hazard scenarios and operational conditions.

The results demonstrate that tree-based ML models, particularly LightGBM ($R=0.9990$), ET ($R=0.9981$), and RF ($R=0.9980$), outperform conventional approaches in predicting resilience, exhibiting high accuracy and robustness. The integration of ML techniques not only enables real-time resilience assessment but also supports decision-making for infrastructure managers and policymakers. By analysing resilience performance across different functionality states and capacity indicators, the model offers actionable insights to enhance disaster preparedness, optimise resource allocation, and improve post-disruption recovery efforts.

Despite its effectiveness, the study highlights areas for further research and model refinement. Future work should incorporate interdependencies among CI sectors to enable a more comprehensive assessment of cascading failures and cross-sector resilience dynamics. Additionally, the adoption of explainable ML techniques will be essential to enhance the interpretability and transparency of ML-driven resilience assessments. Thus, ensuring decision-makers can confidently apply the model's predictions in practice. Furthermore, extensive validation is necessary to establish the generalisability of the proposed methodology. Future studies should conduct multi-hazard validation, integrate various disaster types, and expand real-world applications through collaborative testing with infrastructure sectors. The inclusion of uncertainty quantification will further enhance model reliability, enabling robust, data-driven decision-making under conditions of uncertainty.

In conclusion, the integration of DBNs and ML-based modelling constitutes a substantial methodological advance in resilience assessment for CI. This study offers a resilient and scalable framework for quantifying infrastructure resilience by integrating probabilistic modelling with data-driven methodologies, thereby supporting key applications in disaster risk management, emergency preparedness, and strategic infrastructure development. As ML techniques continue to evolve, their integration into resilience assessment is anticipated to improve the adaptability, sustainability, and robustness of infrastructure systems, thereby strengthening disaster preparedness and supporting risk-informed decision-making amid increasing global challenges.

ACKNOWLEDGEMENT

The authors acknowledge the financial support from the School of Engineering, RMIT University and the Faculty of Engineering, University of Moratuwa. This research work was conducted in collaboration with the Centre for Future Construction (CFC), School of Engineering, RMIT University, Melbourne, VIC3001.

REFERENCES

Al Shafian, S., & Hu, D. (2024). Integrating Machine Learning and Remote Sensing in Disaster Management: A Decadal Review of Post-Disaster Building Damage Assessment. *Buildings*, *14*(8), 2344. https://doi.org/10.3390/BUILDINGS14082344

Arinta, R. R., & Andi, E. W. R. (2019). Natural Disaster Application on Big Data and Machine Learning: A Review. *2019 4th International Conference on Information Technology, Information Systems and Electrical Engineering, ICITISEE 2019*, 249–254. https://doi.org/10.1109/ICITISEE48480.2019.9003984

Christoph Molnar. (2019). *Interpretable Machine Learning: A Guide for Making Black Box Models Interpretable*. LeanPub.

Derras, B., & Makhoul, N. (2022). An Overview of the Infrastructure Seismic Resilience Assessment Using Artificial Intelligence and Machine-Learning Algorithms. *Proceedings of the International Conference on Natural Hazards and Infrastructure*, Athens, Greece, July.

Hosseini, S., Barker, K., & Ramirez-Marquez, J. E. (2016). A Review of Definitions and Measures of System Resilience. *Reliability Engineering & System Safety*, *145*, 47–61. https://doi.org/10.1016/J.RESS.2015.08.006

Kameswari, Y. L. (2023). A Review on Application of Machine Learning Techniques in Climate Change and Disaster Management. *Technoarete Transactions on Climate Change and Disaster Management Research*, *2*(1), 1–10. https://www.mdpi.com/2673-8392/1/4/84

Kyrkou, C., Kolios, P., Theocharides, T., & Polycarpou, M. (2023). Machine Learning for Emergency Management: A Survey and Future Outlook. *Proceedings of the IEEE, 111*(1), 19–41. https://doi.org/10.1109/JPROC.2022.3223186

Linardos, V., Drakaki, M., Tzionas, P., & Karnavas, Y. L. (2022). Machine Learning in Disaster Management: Recent Developments in Methods and Applications. *Machine Learning and Knowledge Extraction*, *4*(2), 446–473. https://doi.org/10.3390/MAKE4020020

Merz, B., Kuhlicke, C., Kunz, M., Pittore, M., Babeyko, A., Bresch, D. N., Domeisen, D. I. V., Feser, F., Koszalka, I., Kreibich, H., Pantillon, F., Parolai, S., Pinto, J. G., Punge, H. J., Rivalta, E., Schröter, K., Strehlow, K., Weisse, R., & Wurpts, A. (2020). Impact Forecasting to Support Emergency Management of Natural Hazards. *Reviews of Geophysics*, *58*(4), e2020RG000704. https://doi.org/10.1029/2020RG000704

Ouyang, M. (2014). Review on Modeling and Simulation of Interdependent Critical Infrastructure Systems. *Reliability Engineering & System Safety*, *121*, 43–60. https://doi.org/10.1016/J.RESS.2013.06.040

Randil, C., Siriwardana, C., & Sandaruwan Rathnayaka, B. (2022). A Statistical Method for Pre-estimating Impacts from a Disaster: A Case Study of Floods in Kaduwela, Sri Lanka. *International Journal of Disaster Risk Reduction*, *76*, 103010. https://doi.org/10.1016/j.ijdrr.2022.103010

Rathnayaka, B., Robert, D., Adikariwattage, V., Siriwardana, C., Kuligowski, E., Setunge, S., & Amaratunga, D. (2025). Novel Methodology for Resilience Assessment of Critical Infrastructure Considering the Interdependencies: A Case Study in Water, Transportation and Electricity Sector. *International Journal of Disaster Risk Reduction*, 105271. https://doi.org/10.1016/J.IJDRR.2025.105271

Rathnayaka, B., Robert, D., Adikariwattage, V., Siriwardana, C., Meegahapola, L., Setunge, S., & Amaratunga, D. (2024). A Unified Framework for Evaluating the Resilience of Critical Infrastructure: Delphi Survey Approach. *International Journal of Disaster Risk Reduction*, 104598. https://doi.org/10.1016/J.IJDRR.2024.104598

Rathnayaka, B., Siriwardana, C., Amaratunga, D., Haigh, R., & Robert, D. (2023). Climate Change Impacts on Built Environment: A Systematic Review. *Lecture Notes in Civil Engineering*, *266*, 443–459. https://link.springer.com/chapter/10.1007/978-981-19-2886-4_31

Rathnayaka, B., Siriwardana, C., Robert, D., Amaratunga, D., & Setunge, S. (2022). Improving the Resilience of Critical Infrastructures: Evidence-Based Insights from a Systematic Literature Review. *International Journal of Disaster Risk Reduction*, *78*, 103123. https://doi.org/10.1016/J.IJDRR.2022.103123

Roscher, R., Bohn, B., Duarte, M. F., & Garcke, J. (2020). Explainable Machine Learning for Scientific Insights and Discoveries. *IEEE Access*, *8*, 42200–42216. https://doi.org/10.1109/ACCESS.2020.2976199

Shehara, P. L. A. I., Siriwardana, C. S. A., Amaratunga, D., & Haigh, R. (2022). *Development of a Framework to Examine the Transportation Infrastructure Resilience; Sri Lankan Context* (R. Dissanayake, P. Mendis, K. Weerasekera, S. De Silva, & S. Fernando (eds.); pp. 235–258). Springer, Singapore. https://doi.org/10.1007/978-981-16-4412-2_18

Soleimani, F., & Hajializadeh, D. (2022). Bridge Seismic Hazard Resilience Assessment with Ensemble Machine Learning. *Structures*, *38*(February), 719–732. https://doi.org/10.1016/j.istruc.2022.02.013

Sun, W., Bocchini, P., & Davison, B. D. (2020). Applications of Artificial intelligence for Disaster Management. *Natural Hazards*, *103*(3), 2631–2689. https://doi.org/10.1007/S11069-020-04124-3

UNDRR. (2019). Global Assessment Report on Disaster Risk Reduction. https://gar.undrr.org/sites/default/files/reports/2019-05/full_gar_report.pdf

Venkadesh, P., Divya, S. V., Marymariyal, P., & Keerthana, S. (2024). Predicting Natural Disasters with AI and Machine Learning. (D. Satishkumar & M. Sivaraja (eds.), *Utilizing AI and Machine Learning for Natural Disaster Management*; pp. 39–64). IGI Global Scientific Publishing. https://doi.org/10.4018/979-8-3693-3362-4.CH003

Wang, X., Mazumder, R. K., Salarieh, B., Salman, A. M., Shafieezadeh, A., & Li, Y. (2022). Machine Learning for Risk and Resilience Assessment in Structural Engineering: Progress and Future Trends. *Journal of Structural Engineering*, *148*(8), 03122003. https://doi.org/10.1061/(ASCE)ST.1943-541X.0003392

Xie, J., Alvarez-Fernandez, I., & Sun, W. (2020). A Review of Machine Learning Applications in Power System Resilience. *IEEE Power and Energy Society General Meeting*, 2020-August, 1–5. https://doi.org/10.1109/PESGM41954.2020.9282137

Zennaro, F., Furlan, E., Simeoni, C., Torresan, S., Aslan, S., Critto, A., & Marcomini, A. (2021). Exploring Machine Learning Potential for Climate Change Risk Assessment. *Earth-Science Reviews*, *220*, 103752. https://doi.org/10.1016/J.EARSCIREV.2021.103752

15 Optimizing Hurricane Evacuation Decisions under Climate Change

Yaodan Cui[a], *and Kairui Feng*[a,b]
[a]The State Key Laboratory of Autonomous Intelligent Unmanned Systems, Tongji University, Shanghai, China
[b]Computer Science, Shanghai Innovation Institute, Shanghai, China

15.1 INTRODUCTION

Tropical cyclones (TCs) are among the most devastating natural disasters, and climate change is fundamentally transforming their threat profile through multiple mechanisms: enhanced intensity, altered tracks, more frequent rapid intensification, and increased rainfall rates (Knutson et al., 2010; Emanuel, 2013). While debate continues regarding TC frequency changes, robust evidence shows that warming oceans are producing storms with higher maximum wind speeds, more destructive storm surges, and longer-lasting impacts after landfall (Mendelsohn et al., 2012). These physical changes create cascading challenges for evacuation management systems designed around historical storm patterns that no longer bound future risks.

The climate impact on evacuation extends beyond stronger storms to encompass a more insidious challenge: degraded predictability. As TCs exhibit behaviors increasingly outside historical ranges—intensifying from tropical storms to major hurricanes in 24 hours, maintaining category 4 strength over land, or executing unprecedented track shifts—forecasting models trained on past data lose reliability (Zhang and Emanuel, 2019). This dual burden of enhanced hazards and reduced forecast skill creates what we term the "climate evacuation paradox": precisely when evacuation decisions become most critical due to stronger storms, the information needed to make those decisions becomes least reliable (Regnier, 2008, 2009).

Hurricane Irma (2017) serves as an ideal baseline for examining climate impacts on evacuation management. This storm triggered Florida's largest evacuation, displacing 6.5 million people and involving 4 million vehicles across the state's constrained highway network (Feng and Lin, 2021, 2022). Critically, Irma occurred under present-day climate conditions where forecast models maintained reasonable skill and evacuation decisions could rely on established uncertainty bounds. In recent work (Cui et al., 2025), we demonstrated that reinforcement learning (RL)

 DOI: 10.1201/9781003567332-15

could optimize Irma evacuation orders, achieving an 8% improvement over historical government decisions by better coordinating evacuation timing and routing under forecast uncertainty.

This study asks a fundamental question: what happens to evacuation management when we project storms like Irma into climate-altered futures? By superimposing climate change signals from CMIP6 projections onto Irma's atmospheric environment, we create physically consistent scenarios of how similar storms might behave under 2050s and 2080s conditions. This approach reveals not just stronger storms—with wind speeds 15%–20% higher and surge 25%–30% greater—but, critically, how forecast reliability degrades when models encounter unprecedented storm behaviors.

We leverage recent breakthroughs in AI-driven weather forecasting, particularly Pangu-Weather (Bi et al., 2023), which can generate thousands of storm track ensembles in hours rather than the days required by traditional numerical models. This computational efficiency enables us to fully characterize the expanded uncertainty space created by climate change. While these models showed promise for current conditions in (Cui et al., 2025), their behavior under climate-perturbed scenarios reveals troubling degradation: track forecast uncertainty nearly doubles and intensity errors increase by 75%, fundamentally altering the decision landscape for evacuation managers.

Our analysis employs the RL framework developed in (Cui et al., 2025), modified to account for climate-enhanced uncertainty through expanded state representations and adjusted reward functions that penalize decisions under high forecast variance. We generate 20,000 climate-perturbed ensemble members across multiple warming scenarios (Shared Socioeconomic Pathway, SSP2-4.5 and SSP5-8.5), training the RL system to find optimal evacuation strategies that balance competing risks: evacuating too early imposes enormous social and economic costs, while waiting for forecast certainty risks catastrophic outcomes if storms rapidly intensify.

The results reveal a sobering reality about climate adaptation limits. While RL's relative advantage over fixed evacuation policies increases with climate severity—growing from 7% under current conditions to 17% under 2080s SSP5-8.5—absolute evacuation outcomes deteriorate dramatically. Even with optimal decision-making, evacuation performance under future climate conditions (objective value 0.239) remains 34% worse than suboptimal fixed policies achieve today (0.178). This finding challenges the assumption that technological advances in forecasting and decision support can offset climate risks (Davidson et al., 2020; Sadri et al., 2018).

This work contributes to understanding climate change impacts on disaster management by quantifying how forecast degradation compounds physical hazard increases, demonstrating the limits of optimization-based adaptation, and revealing the tragic tradeoffs that climate change imposes on evacuation decisions. As communities confront rising TC risks, our findings suggest that relying on improved evacuation management alone is insufficient—comprehensive risk reduction through infrastructure hardening, land-use reform, and potentially managed retreat may be unavoidable (Racherla and Chinthala, 2022; Nateghi and Mukherjee, 2021). Building on (Cui et al., 2025), we show that climate change doesn't simply make the evacuation problem harder; it fundamentally transforms it into a challenge where even optimal decisions yield unacceptable outcomes.

15.2 LITERATURE REVIEW

15.2.1 Climate Change Impacts on Hurricane Evacuation

Climate change introduces fundamental uncertainties into hurricane evacuation planning that existing literature has only begun to address. While early evacuation studies focused on historical storm patterns (Urbanik, 2000; Lindell and Prater, 2007; Wolshon, 2006), recent research increasingly recognizes that climate-altered storm behavior invalidates many assumptions underlying traditional evacuation models. Studies have documented how warming oceans lead to rapid intensification events (Emanuel, 2013; Zhang and Emanuel, 2019), altered storm tracks (Knutson et al., 2010), and increased precipitation rates (Mendelsohn et al., 2012), all of which complicate evacuation timing and routing decisions.

The degradation of forecast reliability under climate change represents a critical gap in evacuation planning literature. As storms exhibit increasingly unprecedented behaviors, the performance of forecasting models trained on historical data deteriorates significantly (Regnier, 2008, 2009). This forecast uncertainty compounds evacuation challenges, as authorities must make decisions with wider confidence intervals and higher false alarm rates. Our recent work (Cui et al., 2025) demonstrated effective evacuation optimization under current climate conditions, achieving an 8% improvement in system performance. However, when projected into future climate scenarios, the optimization problem fundamentally changes due to increased forecast uncertainty and extreme storm behaviors.

15.2.2 Evolution of Evacuation Modeling under Uncertainty

The progression from deterministic to probabilistic evacuation models reflects growing recognition of uncertainty in evacuation planning. Traditional macroscopic models (Chiu et al., 2007; Pel et al., 2012) and microscopic models (Chen and Zhan, 2008; Lämmel et al., 2010) were designed for well-understood storm behaviors with relatively predictable forecast errors. Hybrid approaches (Ying et al., 2016; Ahmed et al., 2019) attempted to balance computational efficiency with model fidelity, but still assumed that historical patterns would persist.

Climate change challenges these modeling paradigms by introducing nonstationary storm behaviors and forecast errors that exceed historical bounds. The evacuation model developed by Feng and Lin (2022) and refined in (Cui et al., 2025) captures both aggregate traffic flow and individual decision-making, providing a foundation for climate-adapted modeling. However, extending these models to climate-altered scenarios requires explicitly accounting for increased behavioral uncertainty, as evacuees' responses to unprecedented storm threats may differ substantially from historical patterns (Huang et al., 2016; Tanim et al., 2022; Meyer et al., 2018).

15.2.3 Behavioral Responses to Climate-Enhanced Threats

Understanding evacuation behavior under climate change requires reconsidering established behavioral models. Traditional frameworks, including rational choice models (Kyne and Donner, 2018) and bounded rationality models (Sun et al., 2017), assume that individuals can assess risks based on past experiences. However, climate change creates situations where historical experience becomes misleading (Feng et al., 2020; Karaye et al., 2022).

Recent studies emphasize how social networks and risk communication influence evacuation decisions (Yang et al., 2019, 2016; Mongold et al., 2021), but these dynamics may shift under climate change as communities experience unprecedented events. The sequential logistic regression approach in our previous work (Cui et al., 2025) captured dynamic decision-making under current conditions, but climate-altered scenarios introduce additional behavioral uncertainties that require more adaptive modeling frameworks.

15.2.4 From Traditional Optimization to Climate-Adaptive Learning

The evolution from traditional optimization to machine learning approaches reflects the increasing complexity of evacuation planning under climate uncertainty. Earlier optimization techniques for shelter location (Sherali et al., 1991), evacuation routing (Cova and Johnson, 2003), and vulnerable population management (Dulebenets et al., 2019) assumed relatively stable storm patterns and forecast accuracy. Recent advances using conflict-based heuristics (Pillac et al., 2016) and multi-objective optimization (Kabir et al., 2023) have improved computational efficiency, but remain limited when dealing with climate-induced uncertainties.

The game-theoretic framework of Yi et al. (2017) represented a significant advance by modeling government-evacuee interactions under uncertainty. However, as noted in Aldahlawi et al. (2024), these approaches relied on limited scenario trees (typically tens of branches) due to computational constraints of traditional weather models (Bauer et al., 2015). Under climate change, the assumption that experts can adequately capture future storm behaviors through manually curated scenario trees becomes increasingly tenuous.

15.2.5 AI-Driven Approaches for Climate Adaptation

Recent breakthroughs in AI-based weather forecasting (Reichstein et al., 2019; Bi et al., 2023; Lam et al., 2023; Price et al., 2024) enable the generation of thousands of storm scenarios in near real-time, fundamentally changing the optimization landscape. These capabilities, demonstrated in our recent work (Cui et al., 2025), allow for comprehensive exploration of uncertainty spaces. However, when applied to climate-perturbed scenarios, even advanced models like Pangu-Weather show degraded performance, particularly for extreme events outside historical ranges.

RL has emerged as particularly suited for climate-adaptive evacuation planning due to its ability to learn from unprecedented scenarios (Sadeghlou et al., 2020; Chen and Tang, 2020; Sharma et al., 2020). Unlike traditional optimization methods that assume stable probability distributions, RL can adapt to shifting climate patterns and degraded forecast reliability. Our previous RL framework (Cui et al., 2025) demonstrated this potential under current conditions, but climate change necessitates extensions that explicitly account for nonstationary storm behaviors and increased forecast uncertainty.

This study builds upon Cui et al. (2025) by incorporating climate projections directly into the RL training process, creating evacuation strategies robust to both current and future storm behaviors. By training on climate-adjusted scenarios with degraded forecast accuracy, our approach prepares evacuation systems for a future where historical precedents no longer provide reliable guidance.

15.3 METHODOLOGY

This study extends the RL framework from Cui et al. (2025) by incorporating climate change projections to evaluate evacuation decisions under future storm conditions. We maintain the validated evacuation and RL models while adding critical climate-adjustment components.

15.3.1 Climate-Adjusted Hurricane Simulation Framework

15.3.1.1 Climate Projection Integration

To project Hurricane Irma into future climate scenarios, we superimpose projected atmospheric and oceanic changes onto the historical storm:

$$\mathbf{X}_{\text{future}}(t) = \mathbf{X}_{2017}(t) + \Delta\mathbf{X}_{\text{climate}}(t) \tag{15.1}$$

where $\mathbf{X}$ includes sea surface temperature (SST), atmospheric temperature profiles, specific humidity, and geopotential height fields. The climate change signal $\Delta\mathbf{X}_{\text{climate}}$ is derived from CMIP6 ensemble means under SSP2-4.5 and SSP5-8.5 scenarios for the 2050s and 2080s periods.

For SST adjustments specifically:

$$\text{SST}_{\text{future}} = \text{SST}_{2017} + \Delta\text{SST}_{\text{regional}} \cdot (1 + \alpha \cdot \text{dist}_{\text{coast}}) \tag{15.2}$$

where $\Delta\text{SST}_{\text{regional}}$ ranges from 1.5°C to 3.0°C depending on the scenario, and α accounts for coastal amplification effects.

15.3.1.2 Ensemble Generation with Climate Perturbations

Following Feng et al. (2024), we generate ensemble forecasts using Pangu-Weather with climate-adjusted initial conditions. The perturbation scheme is modified to account for increased variability under climate change:

$$\mathbf{X}^{(i)}_{\text{perturbed}} = \mathbf{X}_{\text{future}} + \epsilon^{(i)} \cdot (1 + \xi \cdot \Delta T_{\text{global}}) \tag{15.3}$$

where $\epsilon^{(i)} \sim \mathcal{N}(0, \Sigma)$ represents the baseline perturbation, and ξ scales the perturbation magnitude with global temperature change.

15.3.2 Evacuation Traffic Modeling

We employ the validated evacuation model from Feng and Lin (2022), maintaining its core components as they represent human behavioral responses that remain consistent regardless of storm characteristics.

15.3.2.1 Evacuation Demand Model

Household evacuation decisions follow a sequential logit model. The probability of evacuating at time t given no prior evacuation is:

$$P_{t,i}^{e} = \frac{\exp\left(V_t^e\right)}{\exp\left(V_t^e\right) + \exp\left(V_t^s\right)} \tag{15.4}$$

$$V_t^e = \beta_0 + \beta_1 \cdot \text{Order}_{t,i} + \beta_2 \cdot \text{Cat}_t + \beta_3 \cdot \text{Dist}_{t,i} + \beta_4 \cdot \text{Surge}_i \tag{15.5}$$

The cumulative evacuation probability becomes:

$$P_{t,i} = P_{t,i}^{e} \prod_{\tau=1}^{t-1} P_{\tau,i}^{s} \tag{15.6}$$

Under climate scenarios, while the behavioral parameters (β) remain constant, the storm characteristics (Cat, Dist) reflect the climate-enhanced intensity and altered tracks.

15.3.2.2 Traffic Flow Dynamics

The network flow follows the dynamic model from Cui et al. (2025):

$$d\vec{X}_{t,\text{in}} = \mathbf{M}_t \cdot \left(\vec{P}_t \otimes \vec{N}_{t,\text{node}}\right) dt \tag{15.7}$$

$$\overrightarrow{TT}(t) = t_0 \left(1 + \alpha \left(\frac{x_{t,k}}{C_0 \cdot \text{LN}}\right)^{\beta}\right) \tag{15.8}$$

$$d\vec{X}_{t,\text{out}} = \left(1 - \frac{d}{dt}\overrightarrow{TT}(t)\right) \otimes \vec{X}_{t-\overrightarrow{TT},\text{in}} dt \tag{15.9}$$

where $\mathbf{M}_t$ is the Markov transition matrix for route choice, and traffic times follow the Bureau of Public Roads (BPR) function.

15.3.3 Reinforcement Learning Under Climate Uncertainty

15.3.3.1 Modified State Space

To capture climate-enhanced uncertainty, we expand the state representation:

$$s_t = \left\{\text{TC}_{\text{lon},t}, \text{TC}_{\text{lat},t}, \text{TC}_{\text{cat},t}, \sigma_{\text{track},t}, \sigma_{\text{int},t}, F_1, \ldots, F_{15}, w_1, \ldots, w_5, \mathcal{C}, t\right\} \tag{15.10}$$

where $\sigma_{\text{track},t}$ and $\sigma_{\text{int},t}$ represent ensemble spread metrics that increase under climate change, and $\mathcal{C} \in \{\text{current}, \text{2050s}, \text{2080s}\}$ indicates the climate scenario.

15.3.3.2 Climate-Adaptive Objective Function

The optimization objective incorporates climate-dependent risk weights:

$$F_{\text{climate}}\left(\vec{w}, \overrightarrow{\text{TC}}\right) = \sum_{i} h_i^C \cdot f_i\left(\vec{w}, \overrightarrow{\text{TC}}\right) \quad (15.11)$$

where the weights h_i^C are adjusted based on the climate scenario: - Travel time weight: $h_1^C = 0.2 \cdot (1 - 0.1 \cdot \Delta T)$ - Distance from home: $h_2^C = 0.4$ - Travel risk: $h_3^C = 0.1 \cdot (1 + 0.3 \cdot \Delta T)$ - Sheltering risk: $h_4^C = 0.3 \cdot (1 + 0.2 \cdot \Delta T)$

where ΔT is the global temperature anomaly for the scenario.

15.3.3.3 LSTM Architecture for Climate Scenarios

The Long Short-Term Memory (LSTM) network structure remains as in Cui et al. (2025) with 64 and 32 units in two layers, but the input dimension expands to include uncertainty metrics:

$$\mathbf{h}_t = \text{LSTM}\left(\mathbf{x}_t, \mathbf{h}_{t-1}; \theta\right) \quad (15.12)$$

where $\mathbf{x}_t \in \mathbb{R}^{25}$ now includes the additional uncertainty features.

15.3.3.4 Q-Learning with Climate-Dependent Rewards

The Q-learning update incorporates climate-adjusted rewards:

$$Q(s_t, a_t) \leftarrow (1 - \alpha) Q(s_t, a_t) + \alpha \left[r_t^C + \gamma \max_{a'} Q(s_{t+1}, a') \right] \quad (15.13)$$

where the reward function penalizes high-uncertainty decisions:

$$r_t^C = -F_{\text{climate}}\left(\vec{w}, \overrightarrow{\text{TC}}\right) - \lambda \cdot \sigma_t \cdot \mathbb{I}[\text{evacuation ordered}] \quad (15.14)$$

15.3.4 Training Data Generation

We generate three comprehensive datasets:

1. **Baseline (Current Climate)**: 10,000 Irma-like TC scenarios using 2017 atmospheric conditions
2. **Mid-Century Projection**: 10,000 scenarios each for SSP2-4.5 and SSP5-8.5 (2040–2060)
3. **Late-Century Projection**: 10,000 scenarios each for both SSP scenarios (2080–2100)

For each climate scenario set, we: (1). Generate perturbed ensembles using Pangu-Weather with climate-adjusted fields, (2). Compute optimal evacuation orders using deterministic optimization, (3). Train the LSTM on partial observation sequences, and (4). Fine-tune using Q-learning with climate-dependent rewards.

15.3.5 Performance Metrics under Climate Change

We evaluate performance using both traditional metrics and climate-specific indicators:

$$\Delta\text{Risk}_{\mathcal{C}} = \frac{F_{\text{climate}}^{\text{RL}} - F_{\text{current}}^{\text{RL}}}{F_{\text{current}}^{\text{RL}}} \times 100\% \tag{15.15}$$

$$\text{Robustness}_{\mathcal{C}} = 1 - \frac{\text{Var}\left[F_{\text{climate}}^{\text{RL}}\right]}{\text{Var}\left[F_{\text{climate}}^{\text{fixed}}\right]} \tag{15.16}$$

These metrics quantify how evacuation risk increases under climate change and how much RL improves robustness compared to fixed strategies.

15.4 RESULTS

15.4.1 Baseline Performance under Current Climate

We first establish the baseline performance of our RL framework using Hurricane Irma under current climate conditions, as detailed in Cui et al. (2025). This provides the foundation for understanding how climate change impacts evacuation outcomes.

Figure 15.1 illustrates the objective distribution across 10,000 simulated TC scenarios under three different policies. The evacuation objective combines four weighted components: total travel time (0.2), total distance away from home (0.4), total travel risk (0.1), and total risk of sheltering at home (0.3). Under the original governmental orders (Figure 15.1a), the mean objective is 0.1781 with a standard deviation of 0.0358. The theoretically optimal scenario with perfect TC information (Figure 15.1b) achieves 0.1606 with higher variance (0.0502), while the RL policy

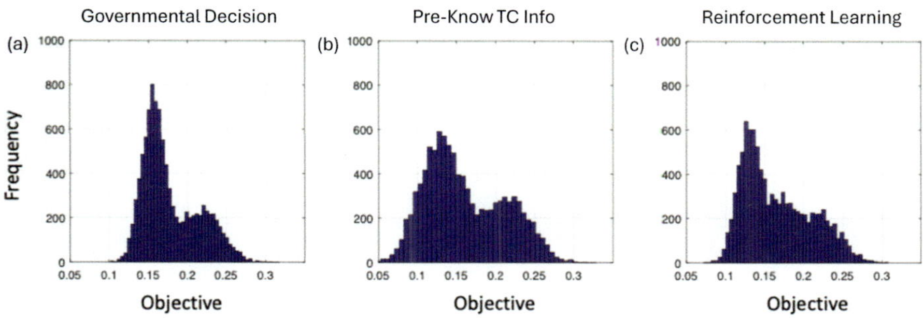

FIGURE 15.1 The objective distribution of evacuation under different policies: (a) Original governmental orders, (b) Perfect information scenario, and (c) RL policy.

(Figure 15.1c) achieves 0.1654 with a standard deviation of 0.0436, representing an 8% improvement over historical decisions.

Table 15.1 breaks down the performance across individual metrics, showing that RL achieves substantial improvements in travel time (20% reduction) and distance from home (14% reduction) while maintaining comparable risk levels.

15.4.2 The Paradox of Enhanced RL Performance under Climate Change

When Hurricane Irma is projected into future climate scenarios, we observe a striking paradox: while RL's relative advantage over fixed policies increases substantially, the absolute evacuation outcomes deteriorate significantly even with optimal decision-making. Table 15.2 reveals this dual nature of climate impacts on evacuation management.

TABLE 15.1
Comparison of Travel and Risk Metrics under Current Climate

Method	Total Travel Time	Total Distance from Home	Total Travel Risk	Total Risk of Sheltering-at-Home
Original	0.1543	0.2072	0.1537	0.2164
	(0.1154, 0.2321)	(0.1739, 0.2738)	(0.1154, 0.2321)	(0.1782, 0.2928)
TC informed	0.1011	0.1756	0.1429	0.2177
	(0.0462, 0.2109)	(0.1287, 0.2694)	(0.0462, 0.2109)	(0.1683, 0.3165)
RL	0.1235	0.1773	0.1522	0.2083
	(0.0750, 0.2205)	(0.1371, 0.2577)	(0.0750, 0.2205)	(0.1614, 0.3021)

TABLE 15.2
Evacuation Performance under Current and Future Climate Scenarios

Climate Scenario	Policy	Overall Objective	Travel Time	Distance from Home	Travel Risk	Shelter Risk
Current	Original	0.1781	0.1543	0.2072	0.1537	0.2164
	RL	0.1654	0.1235	0.1773	0.1522	0.2083
	Improvement	*7.1%*	*20.0%*	*14.4%*	*1.0%*	*3.7%*
2050s SSP2-4.5	Original	0.2134	0.1876	0.2318	0.1923	0.2651
	RL	0.1908	0.1462	0.1967	0.1834	0.2398
	Improvement	*10.6%*	*22.1%*	*15.1%*	*4.6%*	*9.5%*
2050s SSP5-8.5	Original	0.2365	0.2127	0.2472	0.2208	0.2945
	RL	0.2067	0.1671	0.2134	0.2065	0.2601
	Improvement	*12.6%*	*21.4%*	*13.7%*	*6.5%*	*11.7%*
2080s SSP5-8.5	Original	0.2871	0.2654	0.2849	0.2765	0.3428
	RL	0.2389	0.2008	0.2367	0.2518	0.2954
	Improvement	*16.8%*	*24.3%*	*16.9%*	*8.9%*	*13.8%*

The most striking finding is the systematic increase in RL's relative improvement as climate severity intensifies. Under current climate, RL provides a 7.1% improvement over fixed policies. This advantage grows to 10.6% under 2050s SSP2-4.5, 12.6% under SSP5-8.5, and reaches 16.8% by the 2080s SSP5-8.5 scenario. This pattern emerges because fixed policies become increasingly inadequate as uncertainty grows—they cannot adapt to the wider range of possible storm tracks and rapid intensity changes that characterize future climate conditions. The RL framework, by contrast, learns to hedge against this expanded uncertainty space through more flexible and distributed evacuation strategies.

However, this enhanced relative performance masks a troubling reality: absolute evacuation outcomes deteriorate dramatically regardless of decision-making quality. The RL-optimized evacuation under 2080s SSP5-8.5 achieves an objective of 0.2389, which is 44% worse than RL's performance under current climate (0.1654) and, remarkably, 34% worse than even the suboptimal fixed government policy under current climate (0.1781). This degradation occurs despite RL making objectively better decisions relative to the alternatives available under future climate conditions.

To understand this paradox, we must examine how climate change fundamentally alters the decision space. Under current climate, forecast uncertainties allow relatively concentrated evacuation windows—the 72-hour track forecast has a cross-track error of approximately 145 km, enabling targeted evacuations. By 2080s SSP5-8.5, this uncertainty balloons to 280 km, forcing evacuation orders for populations that may ultimately face no threat. Similarly, intensity forecast standard deviations increase from 12 to 21 m/s, making it impossible to distinguish between moderate storms requiring limited evacuation and major hurricanes demanding mass exodus until dangerously close to landfall.

The component metrics reveal how different aspects of evacuation suffer under climate change, even with RL optimization. Travel time shows the most severe absolute degradation, increasing from 0.1235 under current climate to 0.2008 under 2080s SSP5-8.5—a 63% increase despite RL's best efforts. This occurs because the framework must initiate evacuations earlier and for larger areas to maintain safety margins under enhanced uncertainty. The framework essentially trades efficiency for robustness, accepting guaranteed congestion to avoid the possibility of catastrophic outcomes if forecasts prove wrong.

Shelter risk metrics tell a particularly concerning story. Even with RL optimization, shelter risk increases from 0.2083 to 0.2954 (42% increase) by the 2080s, driven by storms that are 15%–20% more intense with 25%–30% higher storm surge. The RL framework responds by evacuating more people from marginal risk areas, but this creates cascading effects—more evacuees mean more congestion, longer travel times, and paradoxically, higher travel risks for those who must evacuate from genuine high-risk zones.

The increasing value of RL under climate change reflects its ability to navigate these impossible tradeoffs more skillfully than fixed policies. Where fixed policies blindly follow predetermined rules regardless of evolving conditions, RL dynamically adjusts evacuation timing and spatial distribution based on forecast updates. For instance, when ensemble forecasts show high uncertainty in track direction, RL initiates partial evacuations across multiple potential impact zones rather than waiting for clarity that may never come. This adaptive capacity becomes more valuable as uncertainty increases, explaining why RL's relative improvement grows from 7.1% to 16.8% across climate scenarios.

Yet this enhanced relative performance should not obscure the fundamental challenge: climate change creates a no-win scenario for evacuation management. The RL framework under future climate must choose between bad options—early mass evacuations that guarantee enormous social and economic costs, or delayed targeted evacuations that risk catastrophic losses if storms rapidly intensify or tracks suddenly shift. The framework becomes increasingly skilled at making these tragic choices, but the choices themselves grow progressively worse. This represents a sobering reality for coastal communities: even with perfect decision-making technology, climate change will impose unavoidable increases in evacuation burden and risk.

15.5 CONCLUSION AND DISCUSSION

15.5.1 Main Findings

This study demonstrates that climate change fundamentally alters the hurricane evacuation challenge, creating conditions where even optimal decision-making cannot prevent substantial degradation in evacuation outcomes. By projecting Hurricane Irma into future climate scenarios and applying RL to optimize evacuation orders, we reveal a critical paradox: while RL's relative advantage over fixed policies increases with climate severity (from 7.1% improvement under current climate to 16.8% under 2080s SSP5-8.5), the absolute evacuation risk worsens dramatically. The RL-optimized evacuation under 2080s conditions achieves an objective of 0.2389—44% worse than current climate performance and 34% worse than even suboptimal fixed policies under present conditions.

The degradation stems from two interconnected factors. First, forecast uncertainty nearly doubles under climate change, with 72-hour track errors increasing from 145 km to 280 km and intensity forecast standard deviations growing from 12 to 21 m/s. Second, storm characteristics become more extreme, with 15%–20% stronger peak winds, 25%–30% higher storm surge, and 65% more frequent rapid intensification events. These changes force evacuation managers into impossible tradeoffs—either evacuate larger populations earlier based on uncertain forecasts, accepting guaranteed social and economic costs, or delay decisions and risk catastrophic outcomes if storms rapidly intensify or tracks suddenly shift.

15.5.2 Implications for Climate Adaptation

Our findings have profound implications for coastal community adaptation strategies. The conventional wisdom that improved forecasting and decision-making technologies will offset climate risks proves insufficient. Even with AI-enhanced weather prediction generating thousands of ensemble members and sophisticated RL optimization, evacuation outcomes under future climate remain worse than current conditions with basic planning. This suggests that relying solely on improved evacuation management to address climate risks is fundamentally inadequate.

Communities must therefore pursue multiple adaptation pathways simultaneously. Infrastructure hardening becomes essential to reduce the population requiring evacuation. Land use policies should discourage development in high-risk areas, particularly given that our analysis shows shelter risks increasing by 42% even with optimized evacuation. Early warning systems need redesign to communicate enhanced uncertainty effectively—our results show evacuation windows extending from 36–48 hours to 60–72 hours under climate change, requiring different public communication strategies.

The increasing value of adaptive decision-making under uncertainty also implies that investment in computational infrastructure and AI capabilities becomes a climate adaptation priority. While our framework requires 4.5–5 hours of computation for climate scenarios compared to 3 hours for current conditions, this remains feasible for operational use. However, smaller communities may lack such resources, potentially creating adaptation inequities.

15.5.3 Methodological Contributions and Limitations

This work makes several methodological contributions. First, we demonstrate how to integrate climate projections with AI-based weather forecasting to generate large ensemble sets reflecting future conditions. The approach of superimposing climate change signals onto historical storms provides a computationally efficient method for exploring climate impacts on specific evacuation scenarios. Second, we show how RL frameworks can be modified to account for climate-enhanced uncertainty through expanded state spaces and adjusted reward functions that penalize decisions under high uncertainty.

However, important limitations remain. Our climate adjustments, while based on CMIP6 projections, may not capture all potential changes in TC behavior. Some studies suggest climate change could alter storm tracks more fundamentally than our perturbation approach represents. Additionally, we assume human behavioral responses to evacuation orders remain constant across climate scenarios, but risk perception and compliance may evolve as storms become more threatening.

The computational requirements, while manageable, may limit applicability. Generating 20,000 ensemble members for adequate uncertainty characterization under climate change requires substantial Graphics Processing Unit (GPU) resources. Smaller jurisdictions might need simplified versions that sacrifice some performance for computational feasibility.

15.5.4 Future Research Directions

Several research directions emerge from this work. First, extending the analysis to other hurricane-prone regions would test the generalizability of our findings. Different geographic settings—such as the narrow escape routes of the Florida Keys or the complex urban networks of Houston—may show different patterns of climate impact on evacuation effectiveness.

Second, incorporating behavioral adaptation deserves attention. As communities experience more severe storms, evacuation compliance patterns may shift. Some populations might develop "evacuation fatigue" from frequent false alarms due to increased uncertainty, while others might become hypervigilant. Understanding these behavioral dynamics could substantially affect optimal evacuation strategies.

Third, exploring hybrid strategies combining RL with human expertise could improve both performance and acceptability. Our analysis shows RL decisions can be approximated by simpler rule sets, suggesting opportunities for human-interpretable policies that maintain most of RL's performance gains while providing transparency for emergency managers.

Finally, investigating the economic implications of climate-degraded evacuation performance is crucial. Our results show total travel time increasing by 63% and evacuation windows extending by 24–36 hours under climate change. These translate to massive economic losses from extended business closures and productivity impacts that should be factored into climate damage assessments.

15.5.5 Concluding Remarks

This study reveals an uncomfortable truth about climate adaptation: there are limits to what improved decision-making can achieve when the underlying physical hazards become more severe and unpredictable. While RL and AI-enhanced weather forecasting provide valuable tools for managing climate risks, they cannot fully compensate for the fundamental challenges that climate change poses to evacuation management. The path forward requires not just better evacuation decisions but comprehensive adaptation strategies that reduce the need for mass evacuations through resilient infrastructure, appropriate land use, and possibly managed retreat from the most vulnerable areas.

The increasing relative value of adaptive decision-making under climate change—with RL's advantage growing from 7% to 17%—demonstrates that investment in these technologies remains worthwhile even as absolute outcomes worsen. However, communities must be clear-eyed about what these tools can and cannot achieve. In a climate-changed world, even optimal evacuation management will impose greater costs and risks than suboptimal management does today. This reality should inform both adaptation planning and the urgency of climate mitigation efforts to limit future warming.

DECLARATION OF INTEREST

The authors declare no conflict of interest.

AUTHOR CONTRIBUTIONS

Feng: Conceptualization, Data Curation, Formal Analysis, Methodology, Validation, Visualization, Writing—Original Draft, and Revision; Cui: Data Curation, Formal Analysis, Methodology, Validation, Visualization, and Revision.

ACKNOWLEDGMENT

Kairui Feng is supported by The Explorers Program of Shanghai (Basic Research Funding; 24TS1401600) and Xiaomi Foundation and the National Natural Science Foundation of China (Grant No. 62088101). These supports are gratefully acknowledged.

REFERENCES

S. Ahmed, K. Dey, R. Fries, Evaluation of transportation network reliability during emergency evacuation, *Journal of Transportation Safety & Security* 11 (6) (2019) 586–606.

R. Y. Aldahlawi, V. Akbari, G. Lawson, A systematic review of methodologies for human behavior mod-elling and routing optimization in large-scale evacuation planning, *International Journal of Disaster Risk Reduction* 110 (2024) 104638.

P. Bauer, A. Thorpe, G. Brunet, The quiet revolution of numerical weather prediction, *Nature* 525 (7567) (2015) 47–55.

K. Bi, L. Xie, H. Zhang, X. Chen, X. Gu, Q. Tian, Accurate medium-range global weather forecasting with 3D neural networks, *Nature* 619 (7970) (2023) 533–538.

X. Chen, F. B. Zhan, Agent-based modelling and simulation of urban evacuation: relative effectiveness of simultaneous and staged evacuation strategies, *Journal of the Operational Research Society* 59 (1) (2008) 25–33.

Y. Chen, J. Tang, Emergency resource allocation using reinforcement learning, *Knowledge-Based Systems* 205 (2020) 106270.

Y.-C. Chiu, H. Zheng, J. Villalobos, B. Gautam, Modeling no-notice mass evacuation using a dynamic traffic flow optimization model, *IIE Transactions* 39 (1) (2007) 83–94.

T. J. Cova, J. P. Johnson, A network flow model for lane-based evacuation routing, *Transportation Research Part A: Policy and Practice* 37 (7) (2003) 579–604.

Y. Cui, K. Feng, W. Ma, B. He, Reinforcement learning for optimizing hurricane evacuation decisions: Hurricane Irma case study, *Transportation Research Part D: Transport and Environment* (2025) 104882.

R. A. Davidson, L. K. Nozick, T. Wachtendorf, B. Blanton, B. Colle, R. L. Kolar, K. Dresback, W. Yi, K. Yang, N. Leonardo, Integrating forecast and observed uncertainties to estimate evacuation-related decision timing during extreme weather events, *International Journal of Disaster Risk Reduction* 46 (2020) 101510.

M. A. Dulebenets, J. Pasha, O. F. Abioye, M. Kavoosi, E. E. Ozguven, R. Moses, W. R. Boot, T. Sando, Exact and heuristic solution algorithms for efficient emergency evacuation in areas with vulnerable populations, *International Journal of Disaster Risk Reduction* 39 (2019) 101114.

K. A. Emanuel, Downscaling CMIP5 climate models shows increased tropical cyclone activity over the 21st century, *Proceedings of the National Academy of Sciences* 110 (30) (2013) 12219–12224.

K. Feng, D. Xi, W. Ma, C. Wang, Y. Li, X. Chen, Potential paradigm shift in hazard risk management: AI-based weather forecast for tropical cyclone hazards (2024). https://github.com/kelvinfkr/Perturbation_AI_weather

K. Feng, N. Lin, Reconstructing and analyzing the traffic flow during evacuation in Hurricane Irma (2017), *Transportation Research Part D: Transport and Environment* 94 (2021) 102788.

K. Feng, N. Lin, Modeling and analyzing the traffic flow during evacuation in Hurricane Irma (2017), *Transportation Research Part D: Transport and Environment* 110 (2022) 103412.

K. Feng, N. Lin, S. Xian, M. V. Chester, Can we evacuate from hurricanes with electric vehicles? *Transportation Research Part D: Transport and Environment* 86 (2020) 102458.

S.-K. Huang, M. K. Lindell, C. S. Prater, Who leaves and who stays? A review and statistical meta-analysis of hurricane evacuation studies, *Environment and Behavior* 48 (8) (2016) 991–1029.

M. Kabir, J. Mobin, M. A. Nayeem, M. A. Habib, M. S. Rahman, Multi-objective optimization and heuristic based solutions for evacuation modeling, *Transportation Research Interdisciplinary Perspectives* 18 (2023) 100798.

I. M. Karaye, N. Taylor, M. Perez-Patron, C. Thompson, J. A. Horney, Factors associated with hurricane evacuation: a statistical meta-analysis of studies, 1999-2018, *Disaster Medicine and Public Health Preparedness* 16 (3) (2022) 1064–1072.

T. R. Knutson, J. L. McBride, J. Chan, K. Emanuel, G. Holland, C. Landsea, I. Held, J. P. Kossin, A. Srivastava, M. Sugi, Tropical cyclones and climate change, *Nature Geoscience* 3 (3) (2010) 157–163.

D. Kyne, W. Donner, Kyne-Donner model of authority's recommendation and hurricane evacuation decisions: A study of hypothetical hurricane event in the Rio Grande Valley, Texas, *Population Research and Policy Review* 37 (6) (2018) 897–922.

R. Lam, A. Sanchez-Gonzalez, M. Willson, P. Wirnsberger, M. Fortunato, F. Alet, S. Ravuri, T. Ewalds, Z. Eaton-Rosen, W. Hu, et al., Learning skillful medium-range global weather forecasting, *Science* 382 (6677) (2023) 1416–1421.

G. Lämmel, D. Grether, K. Nagel, The representation and implementation of time-dependent inundation in large-scale microscopic evacuation simulations, *Transportation Research Part C: Emerging Technologies* 18 (1) (2010) 84–98.

M. K. Lindell, C. S. Prater, Critical behavioral assumptions in evacuation time estimate analysis for private vehicles: Examples from hurricane research and planning, *Journal of Urban Planning and Development* 133 (1) (2007) 18–29.

R. Mendelsohn, K. Emanuel, S. Chonabayashi, L. Bakkensen, The impact of climate change on global tropical cyclone damage, *Nature Climate Change* 2 (3) (2012) 205–209.

M. A. Meyer, B. Mitchell, J. C. Purdum, K. Breen, R. L. Iles, Previous hurricane evacuation decisions and future evacuation intentions among residents of southeast Louisiana, *International Journal of Disaster Risk Reduction* 31 (2018) 1231–1244.

E. Mongold, R. A. Davidson, J. Trivedi, S. DeYoung, T. Wachtendorf, P. Anyidoho, Hurricane evacuation beliefs and behaviour of inland vs. coastal populations, *Environmental Hazards* 20 (4) (2021) 363–381.

R. Nateghi, S. Mukherjee, A multi-paradigm framework to assess the impacts of climate change on end-use energy demand, *PLoS One* 16 (10) (2021) e0258504.

A. J. Pel, M. C. Bliemer, S. P. Hoogendoorn, A review on travel behaviour modelling in dynamic traffic simulation models for evacuations, *Transportation* 39 (1) (2012) 97–123.

V. Pillac, P. Van Hentenryck, C. Even, A conflict-based path-generation heuristic for evacuation planning, *Transportation Research Part B: Methodological* 83 (2016) 136–150.

I. Price, A. Sanchez-Gonzalez, F. Alet, T. R. Andersson, A. El-Kadi, D. Masters, T. Ewalds, J. Stott, S. Mohamed, P. Battaglia, et al., Probabilistic weather forecasting with machine learning, *Nature* 637 (8044) (2024) 84–90.

P. Racherla, J. Chinthala, Artificial intelligence for climate change adaptation: opportunities and challenges, *Frontiers in Artificial Intelligence* 5 (2022) 844425.

E. Regnier, Public evacuation decisions and hurricane track uncertainty, *Management Science* 54 (1) (2008) 16–28.

E. Regnier, Evacuation decision making and hurricane track uncertainty, *Decision Analysis* 6 (3) (2009) 187–205.

M. Reichstein, G. Camps-Valls, B. Stevens, M. Jung, J. Denzler, N. Carvalhais, F. Prabhat, Deep learning and process understanding for data-driven Earth system science, *Nature* 566 (7743) (2019) 195–204.

T. Sadeghlou, S. Galip Gok, E. Gelenbe, W. Gai, Adaptive evacuation management using reinforcement learning, *ACM Transactions on Autonomous and Adaptive Systems (TAAS)* 14 (3) (2020) 1–23.

A. M. Sadri, S. V. Ukkusuri, H. Gladwin, Integrating social network analysis with analytic network process for identifying the critical evacuation paths, *Transportation Research Part A: Policy and Practice* 114 (2018) 38–60.

J. Sharma, P.-A. Andersen, O.-C. Granmo, M. Goodwin, Deep Q-learning with Q-matrix transfer learning for novel fire evacuation environment, *IEEE Transactions on Systems, Man, and Cybernetics: Systems* 51 (12) (2020) 7363–7381.

H. D. Sherali, T. B. Carter, A. G. Hobeika, A location-allocation model and algorithm for evacuation planning under hurricane/flood conditions, *Transportation Research Part B: Methodological* 25 (6) (1991) 439–452.

D. Sun, J. E. Kang, R. Batta, Y. Song, Optimization of evacuation warnings prior to a hurricane disaster, *Sustainability* 9 (12) (2017) 2152.

S. H. Tanim, B. M. Wiernik, S. Reader, Y. Hu, Predictors of hurricane evacuation decisions: A meta-analysis, *Journal of Environmental Psychology* 79 (2022) 101742.

T. Urbanik, Evacuation time estimates for nuclear power plants, *Journal of Hazardous Materials* 75 (2–3) (2000) 165–180.

B. Wolshon, Evacuation planning and engineering for Hurricane Katrina, *Bridge* 36 (1) (2006) 27–34.

H. Yang, E. F. Morgul, K. Ozbay, K. Xie, Modeling evacuation behavior under hurricane conditions, *Transportation Research Record* 2599 (1) (2016) 63–69.

K. Yang, R. A. Davidson, B. Blanton, B. Colle, K. Dresback, R. Kolar, L. Nozick, J. Trivedi, T. Wachtendorf, Hurricane evacuations in the face of uncertainty: Use of integrated models to support robust, adaptive, and repeated decision-making, *International Journal of Disaster Risk Reduction* 36 (2019) 101093.

W. Yi, L. Nozick, R. Davidson, B. Blanton, B. Colle, Optimization of the issuance of evacuation orders under evolving hurricane conditions, *Transportation Research Part B: Methodological* 95 (2017) 285–304.

W. Ying, Y. Cheng, H. Ling, Y. Shi, A hybrid model for simulating urban evacuation, *Physica A: Statistical Mechanics and its Applications* 451 (2016) 578–592.

F. Zhang, K. A. Emanuel, Predictability and prediction of tropical cyclones on daily to seasonal time scales, *Tropical Cyclone Research and Review* 8 (3) (2019) 123–146.

16 Adoption of Green Building Strategies

A Case of Enhancing Infrastructure and Superstructure Resilience in Zimbabwe

Innocent Chirisa[a], Tinashe Magande[b], Africa Muneti[c], Mike Juru[d], Halleluah Chirisa[e], Winnet Chiganga-Ngarakana[f], and Kudzai Chitiva[g]

[a]Office of the Vice Chancellor, Zimbabwe Ezekiel Guti University, Bindura, Zimbabwe; Department of Urban and Regional Planning, University of the Free State, Bloemfontein, South Africa; Department of Urban and Regional Planning, University of Johannesburg, Johannesburg, South Africa

[b]Department of Urban & Regional Planning, University of Johannesburg, Johannesburg, South Africa

[c]Kururama Investments, Cape Town, South Africa; Firm Central Property Group, Harare, Zimbabwe

[d]Green Building Council of Zimbabwe, Harare, Zimbabwe; Integrated Properties, Pvt Ltd, Harare, Zimbabwe

[e]Directorate of Research and Innovation, Zimbabwe Ezekiel Guti University, Bindura, Zimbabwe; Department of Geography, University of the Free State, Bloemfontein, South Africa

[f]Faculty of the Built Environment, Arts and Science, BA ISAGO University, Gaborone, Botswana

[g]Green Building Council of Zimbabwe, Harare, Zimbabwe

16.1 INTRODUCTION

The adoption of green building strategies is increasingly recognised as essential for enhancing resilience in the built environment, particularly in developing countries like Zimbabwe. As climate change exacerbates extreme weather events, traditional

DOI: 10.1201/9781003567332-16

construction methods become inadequate for safeguarding lives and investments (Chavunduka et al., 2024). The built environment accounts for a significant portion of global greenhouse gas (GHG) emissions, with the United Nations Environment Programme reporting that it contributes approximately 30% of annual emissions (UNEP, 2021). In Zimbabwe, where the impacts of climate change are pronounced, the need for sustainable construction practices is urgent (Moyo et al., 2024). This article explores how integrating green building concepts can mitigate environmental impacts while promoting sustainable development. It highlights critical aspects of green building, including energy efficiency, water conservation, and the use of sustainable materials. This article aims to provide a comprehensive understanding of the potential benefits and challenges associated with green building practices in Zimbabwe (DLA Piper, 2021; Constructive Voices, 2022; Chigwenya & Zhakata, 2021; Siddiqui et al., 2024)).

Despite the acknowledged benefits of green buildings, their adoption in Zimbabwe has been slow due to various challenges. A study on the greening of the construction industry indicates that while there is some awareness of green building principles, actual implementation remains low (Adebayo, 2014; Chigwenya & Zhakata, 2021). Financial constraints, lack of institutional support, and limited knowledge among stakeholders hinder progress (CTCN, 2021; Moyo et al., 2024; Tanyanyiwa, 2017). However, notable examples such as the Eastgate Centre in Harare demonstrate that innovative designs can lead to significant energy savings and improved occupant comfort (Constructive Voices, 2022). The Green Building Council of Zimbabwe (GBCZ) plays a crucial role in promoting sustainable practices by providing guidance and resources for property developers and owners. This article will investigate these challenges and opportunities by analysing existing policies and frameworks that influence green building adoption across different contexts within Zimbabwe (Green Building Council Zimbabwe, n.d.; DLA Piper, 2021; Chavunduka et al., 2024).

By engaging with stakeholders from government agencies, private developers, and environmental organisations, this research aims to identify best practices and areas for improvement. The findings will underscore the importance of designing buildings that are resilient to climate change through strategies such as utilising durable materials and incorporating passive design techniques (Kerswill, 2017; CTCN, 2021). Overall, the article seeks to contribute to the discourse on sustainable development by illustrating how green building initiatives can enhance resilience in Zimbabwe's construction industry while addressing pressing environmental challenges.

The article is organised into several key sections to provide a thorough exploration of green building strategies in Zimbabwe. It begins with a conceptual framework that outlines the fundamental principles of green building, including energy efficiency, water conservation, and sustainable material selection. This is followed by a literature review that examines existing research on green building practices globally and within Zimbabwe, showcasing relevant case studies. The methodology section details the systematic literature review and case study approach used in this research. Findings are then presented, highlighting the current state of green building adoption and associated challenges. The discussion interprets these findings in the context of climate resilience, while the conclusion and recommendations offer actionable insights for stakeholders to advance green building initiatives in Zimbabwe.

16.2 CONCEPTUAL FRAMEWORK

Throughout a structure's lifetime, resource efficiency and environmental sustainability are given top priority in construction, and this is what is meant by green building concepts. These tactics seek to improve occupant well-being while reducing adverse environmental effects (Kerswill, 2017). A green building incorporates several techniques, including the use of sustainable materials, water conservation, and energy-efficient design. For example, the use of waste recycling systems and the integration of renewable energy sources such as solar panels are essential elements of green construction initiatives (DLA Piper, 2021). Furthermore, maximising natural ventilation and light means using fewer artificial systems, which lowers energy use and improves indoor air quality (Inso, 2023). The building sector can dramatically lower its carbon footprint and create a healthier environment by using these measures.

The basic structures and operations that enable a community or business to run are referred to as infrastructure. It falls into two primary categories: soft infrastructure and hard infrastructure. Physical structures that are necessary for economic activity, such as buildings, water supply systems, bridges, and highways, are referred to as hard infrastructure (Graham & Marvin, 2001). Soft infrastructure, on the other hand, refers to the organisations and services like regulatory frameworks, healthcare systems, and educational programs that support social and economic activity (Meyer & Miller, 2001). A subset of hard infrastructure known as "critical infrastructure" is essential to the economy, public health, safety, and national security. It consists of communication networks, energy grids, and transportation networks that need to be resilient enough to survive interruptions from emergencies or natural catastrophes. Figure 16.1 shows the conceptual diagram in this study.

Resilience theory examines how systems, whether ecological, social, or economic, can withstand disturbances while maintaining their core functions. Resilience in the context of green infrastructure and architecture refers to the capacity to adjust to the effects of climate change, such as extreme weather occurrences (Folke et al., 2010). To improve long-term sustainability, this theory highlights the significance of flexibility and adaptation in design processes (Walker et al., 2004). Incorporating

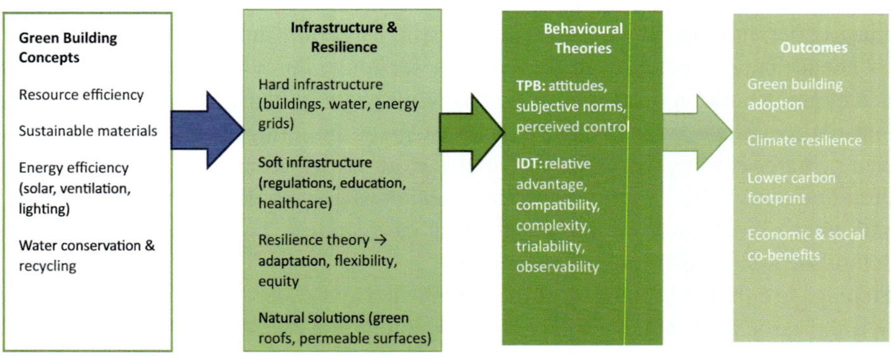

FIGURE 16.1 Conceptual framework (Source: Authors' compilations, 2025).

natural solutions, such as permeable surfaces or green roofs, can enhance a building's resilience by controlling stormwater runoff and offering supplementary environmental advantages (Barton et al., 2015). To guarantee that all stakeholders gain from sustainable practices, resilience theory also promotes a comprehensive strategy that takes social equality and community involvement into account during planning stages (Leach et al., 2010). Green building techniques that incorporate resilience principles can help communities better prepare for and recover from climate-related challenges.

The Theory of Planned Behaviour (TPB), developed by Ajzen (1991), provides a robust framework for understanding behavioural resistance and acceptance in the context of green building adoption. TPB posits that behaviour is driven by three core components: attitude towards the behaviour, subjective norms, and perceived behavioural control. In the realm of sustainable construction, individuals' attitudes towards green buildings, whether they perceive them as beneficial or burdensome, significantly influence their intention to adopt such practices (Mohd Noor &Nordin, 2023; Olubunmi et al., 2016). Subjective norms, shaped by societal expectations and peer influence, also play a critical role; when green building practices are socially endorsed, adoption rates tend to rise (Mohd Noor &Nordin, 2023; Mohd Noor et al., 2023). Perceived behavioural control reflects the ease or difficulty of adopting green practices, often influenced by access to resources, technical knowledge, and institutional support (Azis, 2021; Ong et al., 2021). TPB has been widely applied to explain why stakeholders may resist or embrace sustainable innovations, especially when faced with financial constraints or limited awareness (Ajzen, 1991). The integration of TPB into the study helps us to understand the psychological and social barriers that hinder green building adoption and identify leverage points for behavioural change through education, incentives, and policy interventions.

Innovation Diffusion Theory (IDT), introduced by Rogers (2003), offers a compelling lens for analysing how new practices such as green building technologies are adopted across social systems. IDT identifies five key attributes that influence adoption: relative advantage, compatibility, complexity, trialability, and observability. In sustainable construction, the perceived benefits of green buildings (e.g., energy efficiency, and long-term cost savings) enhance their relative advantage, encouraging uptake among developers and consumers (El Hajj & Montes, 2025; Sartipi, 2020; World Green Building Council, 2021). Compatibility with existing cultural values and construction norms also determines whether green practices are embraced or rejected (DiMaggio & Powell, 1983; Lazar & Chithra, 2022). Complexity, or the perceived difficulty of implementation, can deter adoption, especially in regions lacking technical expertise or institutional support (Qian et al., 2023; Rogers, 2003; Ibrahim et al., 2025). Trialability and observability, being able to experiment with and witness the success of green buildings, further accelerate diffusion (El Hajj & Montes, 2025; Lazar & Chithra, 2022). IDT also emphasises the role of communication channels and opinion leaders in spreading innovations. In the context of green buildings, architects, engineers, and policymakers act as change agents who influence stakeholder perceptions and behaviours (Sartipi, 2020; DiMaggio & Powell, 1983; Ibrahim et al., 2025). The application of the IDT into this study can map the pathways through which sustainable construction practices spread and identify strategies to overcome inertia and foster widespread adoption.

16.3 LITERATURE REVIEW

The literature review section will offer a thorough analysis of Zimbabwe's current situation regarding green construction practices, stressing both the obstacles to their widespread adoption and their several advantages. The benefits of using green building techniques, including increased energy efficiency, better air quality, and less environmental effect, will be discussed in this section. These factors all work together to create more resilient infrastructure. Subsequently, the review will examine the obstacles impeding the extensive adoption of these techniques in Zimbabwe, such as financial limitations, insufficient knowledge, and regulatory hurdles. The review also gives the global and regional perspectives on green building adoption.

16.3.1 Current State of Green Building Adoption in Zimbabwe

In Zimbabwe, the concept of being the "new kid on the block" in green building reflects the emerging awareness and gradual adoption of sustainable construction practices within the construction industry. Whilst the adoption of green building practices is on an upward trend globally, Zimbabwe is still in its nascent stages with a small number of buildings having achieved the green status. The country boasts various policies established to promote sustainable construction, but the adoption of and implementation is slow to non-existent. The country has been the pioneer of evident notable projects like Batanai Gardens and the Eastgate Centre, but implementation has been limited in the construction industry (Mafuku 2019). The construction industry has the sole responsibility of constructing infrastructure in the built environment like buildings, roads, and leisure centres inter alia. The construction industry is a key player in the economy through employment and infrastructure development. It contributes to approximately 2.32% of the Gross Domestic Product (ZIMSTAT, 2020). The industry has however been a threat to the environment in Zimbabwe. The use of sites like wetlands for the construction of buildings has been a cause of concern more recently. Notable areas include the Cold Comfort Zone and the Long Cheng Plaza wetlands that have been converted into shopping centres and residential areas. Such practices do not adhere to the green building principles.

16.3.2 Achievements and Institutional Efforts

The construction industry has, however, done achievements in some areas, with biomimicry being one of the notable achievements. This is a notion of adopting natural innovations and designs. Kibert (2016) refer it to as a stronger form of design that utilises the exact materials and processes used in nature. Such strategies minimise negative environmental impacts while enhancing the well-being of the occupants (Kerswill, 2017). A good example of green building constructed using this form of design in Zimbabwe is the Eastgate Centre in Harare. The building mirrored strategies that are used by termites in termite mounds for temperature control with no conventional air conditioning (Inhabitat, 2023). Other notable examples include Hurudza House and the Batanai Gardens in Harare as stated earlier on. These also exemplify energy-efficient and environmentally friendly projects.

Another achievement in the adoption and implementation of the green building principles is the existence of the GBCZ that actively pushing for sustainable building practices in Zimbabwe. The country is actively moving towards the enactment of policies and laws that enhance climate-responsive construction models in this sector. The GBCZ proposed the Climate Change Bill that emphasises the need for healthy policies that mandate sustainable construction standards. The aim of the bill is to establish a regulatory agency and strengthen institutions and funding mechanisms that will buttress the implementation of holistic climate change strategies. A paper submission that was made by the GBCZ to the government outlines a proposal for green building standards that will make them mandatory if enacted.

16.4 BENEFITS OF ADOPTING GREEN BUILDING STRATEGIES

The adoption of green building strategies offers numerous benefits that significantly enhance infrastructure resilience, particularly in the context of Zimbabwe. Green buildings are designed to optimise resource efficiency and minimise environmental impact throughout their lifecycle. This includes implementing energy-efficient systems, utilising sustainable materials, and promoting water conservation practices (Kerswill, 2017; Sagar et al., 2025; Iwuanyanwu et al., 2024). For instance, the integration of renewable energy sources such as solar panels not only reduces reliance on fossil fuels but also lowers operational costs for building occupants (DLA Piper, 2021; Umoh et al., 2024). Additionally, green buildings often incorporate features like rainwater harvesting systems and efficient plumbing fixtures that contribute to significant water savings, addressing the pressing issue of water scarcity in many regions of Zimbabwe (Constructive Voices, 2022; Didarali & Gambiza, 2019; Sikhakhane Nwokediegwu et al., 2024). By reducing energy and water consumption, green buildings can enhance the overall resilience of infrastructure against climate-related challenges.

Moreover, green building strategies contribute to improved indoor environmental quality, which is crucial for occupant health and productivity. Features such as natural ventilation and ample natural light reduce the need for artificial lighting and mechanical heating or cooling systems, leading to lower energy consumption (Inso, 2023; Ali & Akkaş, 2023; Houghton & Castillo-Salgado, 2017). Research indicates that buildings designed with these principles not only provide a healthier living environment but also improve the well-being of occupants (Agboola et al., 2023; Ifechukwu Gil-Ozoudeh et al., 2023). Furthermore, the use of non-toxic materials minimises indoor air pollutants, contributing to better air quality and reducing health risks associated with traditional construction methods (DLA Piper, 2021). As a result, green buildings not only serve as a sustainable solution to infrastructure development but also promote social well-being.

The economic benefits associated with green building strategies cannot be overlooked. While initial construction costs may be higher due to sustainable materials and technologies, long-term savings from reduced energy and water bills often outweigh these upfront investments (Agboola et al., 2023; Kholijah, 2024). Additionally, green buildings can increase property value and attract environmentally conscious tenants or buyers (Kerswill, 2017; Ifechukwu Gil-Ozoudeh et al., 2023). The implementation of green building practices fosters innovation in design and construction processes, leading to job creation in sectors related to sustainability (Constructive Voices, 2022; Didarali & Gambiza, 2019). By prioritising green building strategies, Zimbabwe can enhance its infrastructure resilience while contributing to broader economic development goals.

16.4.1 Barriers for the Adoption of Green Building Strategies

Barriers to adoption of green building technologies exist in both developed and developing nations. The primary challenge to the adoption of green building practices in Zimbabwe is lack of finance. A review by Darko and Chan (2017) showed that, compared to conventional methods of construction, funding green sustainable project requires higher initial investment making it difficult for construction companies to adopt. Chan et al. (2016) found that these green building technologies are usually hindered by higher costs. Hwang and Ng (2013) also came up with the same finding emphasising on the higher costs of green equipment and materials in Singapore. Hwang et al. (2017) also identified that higher initial costs were a cause of concern in the adoption of green parks. Zuniga-Teran et al. (2020) also found that financability is a great challenge of mainstreaming green infrastructure primarily in the built environment professions.

Lack of government policy is considered as a powerful tool in implementing sustainability initiatives primarily in urban cities (Zhang & He, 2021). Lack of such support hinders the construction of green buildings projects. Ohueri et al (2020) note that although the government is the pioneer and the powerful stakeholder of these initiatives in various countries particularly in the green retrofitting action of existing buildings, its lack of interference in policies, guidelines, and even lack of incentives discourage the action. Hanapiah et al. (2022) buttresse this point by pointing that the government has the intention but lacks the knowledge in implementing the strategies. Furthermore, regime complexes or the interplay of several functions in government agencies in green building assessment systems has caused confusion among the people (Hanapiah et al., 2022). Thus, bringing dispute on the level of effectiveness of the assessment tools (Zhang et al. 2019), and the inconsistency of the introduction of institutions.

Another key barrier to the adoption of green building strategies is the resistance to change. Chan et al. (2016) found that resistance to change was one of the major barriers to the adoption of green technologies. Cao et al (2022) contend that inadequate training in the advanced technologies related to design and construction of personnel is a significant obstacle to promoting this in China. To improve infrastructure resilience, building techniques need to be reevaluated, considering the growing frequency and intensity of climate-related catastrophes. In this context, the use of green construction strategies—which incorporate sustainable materials, energy-efficient designs, and cutting-edge technology that reduces environmental impacts—is crucial. Green buildings, according to Zuo and Zhao (2014), are made using sustainable construction materials and renewable energy sources in mind to consume as little resources as possible during their existence. These methods not only lower GHG emissions but also enhance the general robustness and efficiency of buildings, enhancing their resilience to climate-related difficulties. Features like energy-efficient HVAC systems and rainwater collecting systems, for example, reduce operating costs while improving resistance to adverse weather events (Mokhtar et al., 2020).

Furthermore, interior environmental quality, which is critical for occupant productivity and health, is greatly enhanced by green buildings. Integrating daylighting and natural ventilation lowers energy consumption by reducing dependency on artificial systems (Baker & Steemers, 2002). According to research, structures created using these ideas improve tenant well-being and offer better living spaces (Agboola et al., 2023). Furthermore, the utilisation of non-toxic materials reduces indoor air pollutants, improving air quality and lowering health hazards related to conventional construction techniques (Kerswill, 2017). Consequently, implementing green construction techniques not only tackles climate resilience but also enhances social well-being by fostering better environments for residents.

Lastly, the financial advantages of using green building techniques emphasise how crucial it is to improve the resilience of infrastructure. Sustainable materials and technology may result in higher initial building costs, but over time, these upfront expenses are frequently offset by lower energy and water expenditures (Agboola et al., 2023). Green buildings can also raise the value of a property and draw in eco-aware tenants or buyers (Zuo & Zhao, 2014). The adoption of green building methods encourages creativity in the planning and constructing phases, which results in the development of jobs in the sustainability-related industries (Mokhtar et al., 2020). Setting green building practices as a top priority can support larger objectives for economic development while strengthening the infrastructure's resilience.

16.4.2 Global and Regional Perspectives on Green Building Adoption

Windapo (2014) investigates the key drivers of green building in the Western Cape province of the Republic of South Africa and how the drivers have changed over time. The study found that the key drivers of green building in the country have not changed over time and these include the rising energy costs, the industry's Green Star rating system, competitive advantages, and legislation. Different other studies identified other key drivers of green building in different countries. In the United States of America (USA), Kibert (2016) revealed that the infiltration of the Leadership in Energy and Environmental Design (LEED) green building rating system, membership in green council leadership, state's expansion of green building programs, and the quick advancement in green building technology are the key drivers. A survey conducted by the Green Building Council of Australia indicated that 50% of the respondents indicated tenant demand, the demand by investors for proof that their assets are Green Star rated and the perception that it will be difficult to sell an asset that is not Green Star rated in future (Bond, 2010).

A study by Chebet et al. (2024) focused on the adoption of green building practices in Nairobi, Kenya. The study showed that a lot of construction workers were not aware of green building, Kenya Green Building Society in the county. Moreover, the construction workers were not trained on green building practices. Although regulations were implemented, this was happening at a moderate rate without the incorporation of green building policies into the current construction guidelines. Despite this, there is good observation of green building practices like waste management, energy, and water efficiency. The study recommended that there should be more awareness and training with government providing incentives for encouraging green building technologies. In a study conducted by Mwero and Aduda (2019), the factors that are hindering the adoption of green building strategies in Kenya include limited market options, lack of information, renewable energy reliability, uncertain returns, lack of interest, and the fear of the unknown.

According to Sichali and Banda (2017) investigated the level of awareness, attitudes, and perceptions of green building in Zambia. The study highlighted that little research is being done on green building technologies in the country. Very few buildings are under construction in the country, and this may be the source of awareness. A greater awareness was shown by those who had previously participated in sustainability projects as compared to those who had not. Others showed greater awareness through watching documentaries and television programmes. Professionals have greater awareness than artisans, spending equal time on the same job. In Nigeria, Davies et al. (2024) showed that there is a lack of an adequate institutional framework and policy development in the establishment of guidelines in line with green building technologies. This restricted how people access information on eco-friendly products. Moreover, findings showed that there is insufficient knowledge regarding sustainability concerns, limited expertise among personnel and clients, and a poor understanding of green building technologies in Nigeria. A lack of collaboration between clients and professionals in green building projects was noted. The study concluded that sustainable green technologies and practices in Nigeria are unsustainable and do not align with the major principles of sustainability.

In Spain, Okwandu et al. (2024) demonstrated that green buildings are playing a key role in sustainable development and improving the well-being of the building occupants. The study emphasised that policy and regulations are important in promoting green buildings and the government is very important in shaping the adoption of green building practices. The country uses mandatory building codes and standards to promote green buildings. Such codes are a yardstick for water efficiency, energy performance, indoor environmental quality, and for ensuring that new constructions and major renovations are done properly. The authors also emphasised the importance of incentives like financial, tax credits, grants, and subsidies for building owners and land developers who follow sustainable design features and green building certification. Pushkar (2024) explored office building certification strategies for LEED-CI v4-certified projects in the Mediterranean region. The study showed that Spain, Turkey, and Israel have the same low level of achievement in the indoor environmental quality. Iwuanyanwu et al. (2024) emphasised on the role of green building materials in advancing sustainable architecture due to the role they play in reducing environmental impacts of construction. Green building materials include energy efficiency, resource conservation, and minimal environmental harm. The study highlighted latest innovations in green building materials with advances like high performance insulation technologies, recycled and upcycled materials, biodegradable, and bio-based products.

16.5 METHODOLOGY

This chapter employed a qualitative research design, using a systematic review of literature to critically assess the adoption of green building strategies for enhancing infrastructure and superstructure resilience in Zimbabwe. The methodology involved a desk review of various sources, including peer-reviewed journal articles, policy documents, books, and grey literature such as reports and working papers. Search engines such as Google Scholar, Scopus, Web of Science, and ScienceDirect were used to gather relevant academic and policy material on green building practices, climate resilience, and infrastructure development in Zimbabwe. The study engaged a multistage review process to identify and select relevant studies. A Boolean search strategy, employing operators such as AND, OR, and NOT was used to retrieve relevant literature by combining key search terms. First, a broad search was conducted using the keywords "green building". This was followed by a more targeted search using combinations of terms like "green building strategies", "infrastructure resilience", and "construction industry". For data analysis, the study employed thematic analysis, focusing on key themes, trends, and insights that emerged from the literature. This approach allowed for a deep exploration of how green building initiatives in Zimbabwe have evolved, the challenges and opportunities associated with their adoption, and the potential role of green building in enhancing the climate resilience of infrastructure and superstructures. The combination of systematic review and thematic analysis enabled a comprehensive understanding of the current state of green building practices in Zimbabwe and their implications for sustainable development.

16.6 FINDINGS

Herald (13 April 2022) highlights that government ministries are working with the private sector to adopt green buildings in Zimbabwe in line with combating climate change effects. These include the ministries of Environment, Climate, Tourism and Hospitality Industry, National Housing and Social Amenities, and Local Government and Public Works. These are working with the GBCZ which is an NGO. With green building, precious natural resources are preserved, and the quality of infrastructure is improved. The need for green building is now key, given that Zimbabwe is prone to several climate-induced challenges affecting the country. These include extreme weather events like floods, droughts, flash floods and tropical cyclones like Cyclone Idai. Dr Juru quoted in The Herald (13 April 2022) said

> The built environment is critical sector as people spend a lot of time inside buildings, that is at home, work and even for leisure. Due to lack of mandatory building standards, the increase in urban sprawl has seen the development of infrastructure which is not environmentally responsible and sustainable

Adhering to the building standards will produce cities with resilient infrastructure that is able to adapt to different devastating effects of floods, cyclones, and other extreme weather conditions. Herald (13 April 2022) highlighted that most parts of Africa are facing challenges in fully adopting and implementing the sustainable building practices, environmentally friendly technologies, and climate change measures, and Zimbabwe is not an exception. The country needs to improve the design and environmental and economic performance of existing and new commercial, institutional, and residential buildings.

Various policies have been developed by the government ministries to adapt to the impacts of climate change. The energy ministry came up with the National Renewable Energy Policy, Energy Efficiency Policy, Net Metering Regulations, and Solar Water Heating Regulations. The environment ministry developed the National Climate Change Policy and the National Climate Change Response Strategy. The problem is however that these ministries work in silos, and in the forceable future, the problem of regime interplay may affect the effectiveness of the policies adopted. Thus, an integrated approach is needed for successful implementation of these strategies to be resilient.

Herald (21 March 2024) emphasises that incorporating sustainability principles into the real estate market is key in navigating socio-economic challenges in Zimbabwe. The source highlights that at least 40% of GHGs come from the real estate market. The article defined green building certification as assessing, rating, and certifying buildings based on environmental performance and resource efficiency. Such certification reduces the environmental impact of constructed buildings and promotes environmentally responsible construction practices. The article also mentioned various advantages that are associated with green building certification. Among these is the benefit that green building should priorities designs and technologies that enhance climate resilience, like efficient water management, passive solar design, and natural ventilation. It articulates that the country can foster resilient, equitable, and prosperous sustainable built environment through embracing and mandating green building certification.

Herald (27 August 2024), green building standards are model codes with minimum requirements for increasing environmental and health standards in buildings, sites, and structures. The Climate Technology Centre and Network (CTCN) has emphasised the need to establish new building systems and practices based on green thinking and applications.

> Effective green building standards, policy guidelines and Monitoring, Verification and Evaluation (MV&E) frameworks are critical to improving the environmental and economic performance of new and existing buildings

This aims to introduce green building standards to enhance energy efficiency in commercial, industrial, and residential buildings in the country. The first thing however to develop standards that meet the expectations of local stakeholders and are aligned with national sustainability goals.

Zimbabwe Independent (4 September 2022) emphasises that the green revolution fulfils the dictates of the UN Sustainable Development Goal (SDG) 11, with the mission to make cities that are resilient, inclusive, safe, and sustainable. A brilliant example of green building investment in Zimbabwe is the Eastgate Centre built in 1996 in the Harare Central Business District, a sustainable architecture built through its biomimicry. This was designed with ventilation systems that operate with heat handling traits of drought-resistant cactus plants. This was the first internationally acclaimed green building in Zimbabwe.

Econet wireless Zimbabwe embraced green energy, putting 216 base stations that run on solar energy saving close to 1.57 million kWh of energy in 2021 (Zimbabwe Independent, 4 September 2022). Schweppes Zimbabwe Limited installed a 1 Mega Watt rooftop solar plant at its Harare facility which is the largest solar project in sub-Saharan Africa with the exclusion of South Africa. This was done to deal with energy challenges in the country and reduce carbon emission from the use of diesel generators. Leveraging these affordable renewable options also saves financially. Old Mutual Zimbabwe also came through launching 641 kW solar plant at its Mutual Garden Business Park headquarters which is another project after they successfully launched the Eastgate Centre in 1993.

Individuals in the residential space have also joined the solar projects leading to the installation of solar geysers and solar powered homes. The solar panel has become a trademark both at home and at the workplace. Statistics reflect that building operations account for 30% of global final energy consumption and 26% of global energy-related emissions, with 8% being direct emissions in buildings and 18% indirect emissions from the production of electricity and heat used in buildings (The Standard, 24 March 2024). These can only be solved through green buildings that are designed to be energy efficient and environmentally friendly throughout their life cycle. Green building practices that involve water conservation, energy efficiency, and waste reduction, inter alia, foster economic growth and reduce damage to the environment. Energy efficiency is a solution to economic resilience through cost savings that can be redirected to several sectors. Energy efficiency has encouraged building retrofitting, infrastructure upgrade, and deployment of new technologies creating jobs. It has also seen a transition towards energy autonomy through domestic production, thus reducing reliance on imported power whilst fostering a stable energy-resilient landscape.

Green building practices play a crucial role in mitigating climate change. By reducing GHG emissions and energy consumption, sustainable architecture contributes to a more resilient and environmentally responsible built environment in Zimbabwe (African Property Magazine, 2 January 2024).

16.7 DISCUSSION

The findings reflect a collaborative effort between different stakeholders including government ministries and the private sector in Zimbabwe to adopt green building strategies. This aligns with Kerswill's (2017) assertion that sustainable construction is key in minimising negative environmental impacts and enhancing the well-being of occupants. Key ministries such as the Ministry of Environment, Climate, Tourism and Hospitality Industry, National Housing and Social Amenities, and Local Government and Public Works are working alongside the GBCZ to promote sustainable construction. This practically supports the findings from Ohueri et al. (2020) that government involvement is crucial in advancing green building initiatives. Zimbabwe's vulnerability to climate-induced challenges like floods, droughts, and cyclones emphasises the importance of adopting resilient infrastructure strategies, as discussed by Zuo and Zhao (2014), who highlighted the role of green buildings in addressing climate resilience (Table 16.1).

TABLE 16.1

Key Findings on Green Building Adoption in Zimbabwe and Their Alignment with the Literature

Finding	Evidence from Zimbabwe	Supporting Literature
Stakeholder collaboration	Ministries (Environment, Housing, and Local Government) working with Green Building Council of Zimbabwe	Government involvement is crucial in advancing green building initiatives (Ohueri et al., 2019)
Climate resilience	Vulnerability to floods, droughts, and cyclones driving resilient infrastructure adoption	Green buildings help address climate resilience (Zuo & Zhao, 2014)
Policy gaps and urban sprawl	Lack of mandatory building standards hinders sustainable adoption	Inadequate policies slow down green construction (Zhang & He, 2021)
Energy policies	Energy Ministry promotes renewable energy	Energy efficiency is essential in green buildings (DLA Piper, 2021)
Fragmented governance	Ministries operate in silos, reducing efficiency	Regime complexes hinder green initiatives (Hanapiah et al., 2022)
Certification systems	GBCZ certification encourages water management, passive design, ventilation	Certification enhances environmental performance (Kerswill, 2017; Mokhtar et al., 2020; Zuo & Zhao, 2014)
Biomimicry and innovation	Eastgate Centre as a model of biomimicry	Eastgate Centre as a model of biomimicry
Renewable energy integration	Trend towards solar and other clean energy in buildings	Green building strategies enhance sustainability (Kerswill, 2017; Mokhtar et al., 2020)
Economic resilience	Green buildings reduce carbon emissions and foster economic growth	Sustainability supports broader economic goals (Agboola et al., 2023; Constructive Voices, 2022)

Source: Authors' Compilations (2025)

The lack of mandatory building standards and the resulting urban sprawl reflect concerns raised by Zhang and He (2021), who argue that inadequate government policies hinder the adoption of sustainable construction practices. The Energy Ministry's policies promoting renewable energy support the DLA Piper's (2021) view that energy efficiency measures are essential in green buildings. However, the siloed approach among different ministries mirrors findings from Hanapiah et al. (2022), who pointed out that regime complexes can create confusion and reduce the effectiveness of green building initiatives. An integrated approach, as suggested by the findings, would facilitate collaboration among stakeholders and ensure that green building practices are adopted consistently, echoing the need for cohesive strategies (Darko & Chan, 2017).

The findings further underscore the role of green building certification in enhancing climate resilience, consistent with the analysis by Zuo and Zhao (2014) that emphasised the importance of certification in promoting climate-resilient designs.

Certification processes, such as those advocated by Kerswill (2017) and the GBCZ can promote sustainable water management, passive solar design, and natural ventilation. This will address the environmental performance of buildings as advocated by Mokhtar et al. (2020). A holistic approach through integrating the green building certification into Zimbabwe's real estate market could help the country reduce its carbon footprint, bringing this into line Zuniga-Teran et al. (2020) assertion that green infrastructure enhances sustainability in urban areas.

The need for effective green building standards and monitoring frameworks, as highlighted in the findings, resonates with Kibert (2016) view that biomimicry and environmentally responsive designs are crucial for sustainable construction. The Eastgate Centre in Harare, reviewed in the article, exemplifies the success of such green building initiatives, utilising biomimicry to enhance environmental performance, as noted by Kibert (2016) and Inhabitat (2023). Moreover, the trend towards integrating renewable energy solutions into commercial and residential buildings reflects the shift towards sustainability discussed by Kerswill (2017) and Mokhtar et al. (2020). The adoption of these strategies in Zimbabwe can improve infrastructure resilience, reduce carbon emissions, and foster economic resilience. This aligns with the broader economic development goals highlighted by Agboola et al. (2023) and Constructive Voices (2022).

16.8 CONCLUSION

The findings indicate that Zimbabwe is making significant strides in advancing the integration of green building strategies to address climate change and enhance infrastructure resilience. This move towards sustainable building is being spearheaded by cooperation between important government agencies and the corporate sector, most notably the GBCZ. More broad adoption is being impeded, nevertheless, by gaps in the execution of policies and the absence of required building requirements. Zimbabwe's efforts to address climate-related concerns will be strengthened by the creation of efficient standards and monitoring systems, and green building certification is a crucial instrument for improving climate resilience. Reducing GHG emissions and fostering energy autonomy through the integration of renewable energy sources within buildings also holds promise for creating a fairer and more sustainable built environment. This study proposes the following recommendations:

- **Adopt Mandatory Green Building Standards**: The government should develop and enforce mandatory green building codes and standards that promote sustainable construction practices across all sectors. These standards should focus on improving environmental performance and climate resilience in both new and existing buildings.
- **Strengthen Policy Coordination Among Ministries**: To overcome the current siloed approach, the government should create an integrated policy framework that fosters collaboration between ministries, ensuring that green building initiatives are consistently adopted and implemented across all sectors.

- **Expand Green Building Certification Programs**: Zimbabwe should prioritise green building certification programmes, making them a requirement for all new developments. These certifications should emphasise designs that incorporate efficient water management, passive solar energy, and natural ventilation to enhance climate resilience.
- **Develop a Comprehensive Regulatory Framework**: Establish a clear regulatory framework that not only sets standards for green building practices but also introduces financial incentives such as tax breaks and subsidies to encourage private sector investment in sustainable construction.
- **Promote Public-Private Partnerships (PPPs)**: Foster stronger collaboration among government, private sector entities, and civil society to drive innovation in green building strategies. PPPs can accelerate the adoption of renewable energy and sustainable building technologies.
- **Implement Monitoring and Evaluation Mechanisms**: Establish robust monitoring and evaluation systems to ensure compliance with green building standards. Regular assessments should be conducted to measure the environmental and economic performance of buildings, providing valuable data for improving policies and practices.

REFERENCES

Adebayo, O. (2014). Greening the Construction Industry in Zimbabwe: Challenges and Opportunities. Langaa RPCIG. https://library.gbpn.org/resource/greening-construction-industry-zimbabwe

African Property Magazine (2 January 2024). Insights Zimbabwe's Green Building Practices. https://africanpropertymagazine.com/2024/01/02/insights-zimbabwes-green-building-practices/

Agboola, O., Alotaibi, M., Dodo, A., Abuhussain, A., & Abuhussain, A. (2023). Sustainable architecture: Literature review on green buildings and sustainable development goals. *International Research Journal of Engineering and Technology*, 13(2), 1001–0104.

Ajzen, I. (1991). The theory of planned behaviour. *Organizational Behaviour and Human Decision Processes*, 50(2), 179–211.

Ali, B. M., & Akkaş, M. (2023). The green cooling factor: Eco-innovative heating, ventilation, and air conditioning solutions in building design. *Applied Sciences*, 14(1), 195. https://doi.org/10.3390/app14010195

Azis, N. (2021). Green Building Integration Building Integration in Malaysia. CID in Malaysia. CIDB Malaysia.

Baker, N., & Steemers, K. (2002). *Energy and Environment in Architecture: A Technical Guide*. Spon Press.

Barton, H., Grant, M., & Guise, R. (2015). *Shaping Neighbourhoods: A Guide for Healthier Communities*. Routledge.

Bond, S. (2010). Best of the best in green design: Drivers and barriers to sustainable development in Australia. Available at: https://researcharchive.lincoln.ac.nz/bitstreams/112d63dd-0f98-40b6-aa31-8ee1095cfeed/download

Cao, Y., Kamaruzzaman, S. N., & Aziz, N. M. (2022). Green building construction: A systematic review of BIM utilization. *Buildings*, 12(8), 1205.

Chan, A. P. C., Darko, A., Ameyaw, E. E., & Owusu-Manu, D. G. (2016). Barriers affecting the adoption of green building technologies. *Journal of Management in Engineering*, 33(3), 23–37. https://doi.org/10.1061/(ASCE)ME.1943-5479.0000507

Chebet, M., Gweyi, J., & Kitur, E. (2024). Adoption of Green Building Practices in Langata, Nairobi City County, Kenya.

Chigwenya, A., & Zhakata, T. I. (2021). Adopting green building technologies for sustainable development: Insights from Harare, Zimbabwe. *International Journal of Real Estate Studies*, 14(2), 8–17. https://doi.org/10.11113/intrest.v14n2.47

Constructive Voices. (2022). Zimbabwe Top Green Buildings: Sustainable Design Features and Contributions. https://constructive-voices.com/zimbabwe-top-green-buildings/

CTCN. (2021). Development of Green Building Standards for Zimbabwe. Climate Technology Centre & Network. https://www.ctc-n.org/technical-assistance/projects/development-green-building-standards-zimbabwe

Darko, A., & Chan, A. P. (2017). Review of barriers to green building adoption. *Sustainable Development*, 25(3), 167–179.

Davies, I. E. E., Davies, O. O. A., & Davies, R. M. (2024). Inclusive the Impact of Green Buildings on the Development of a Sustainable Environment in Nigeria. *International Journal of Housing and Human Settlement Planning*, 10(2), 40–53.

Didarali, Z., & Gambiza, J. (2019). Permaculture: Challenges and benefits in improving rural livelihoods in South Africa and Zimbabwe. *Sustainability*, 11(8), 2219. https://doi.org/10.3390/su11082219

DiMaggio, P. J., & Powell, W. W. (1983). The iron cage revisited: Institutional isomorphism and collective rationality in organizational fields. *American Sociological Review*, 48(2), 147–160.

DLA Piper. (2021). ESG Challenges in Real Estate Ahead of COP26: Green Buildings - The Call for Sustainable Retrofitting in Zimbabwe. https://www.dlapiper.com/en/insights/publications/2021/10/esg-challenges-in-real-estate-ahead-of-cop26/green-buildings-the-call-for-sustainable-retroffiting-in-zimbabwe

El Hajj, C. I., & Martínez Montes, G. (2025). Examining green building practices: The influence on building information modeling function diffusion. *Sustainability*, 17(9), 3843. https://doi.org/10.3390/su17093843

Folke, C., Carpenter, S. R., Walker, B., Scheffer, M., & Chapin III, F. S. (2010). Resilience thinking: Integrating resilience, adaptability and transformability. *Ecology and Society*, 15(4), 20.

Gil-Ozoudeh, I., Ike, C., Okwandu, A., & Iwuanyanwu, O. (2023). Sustainable urban design: The role of green buildings in shaping resilient cities. *International Journal of Applied Research in Social Sciences*, 5(10), 674–692. https://doi.org/10.51594/ijarss.v5i10.1481

Graham, S., & Marvin, S. (2001). Towards a new architecture of infrastructure. In Graham, S., & Marvin, S. (eds.), *Splintering Urbanism: Networked Infrastructures, Technological Mobilities and the Urban Condition*. Routledge.

Hanapiah, N. M., Zaki, N. M., Husain, M. A., & Mukhals, N. A. (2022). Green building in existing development: a review of current status, challenges, and implementation strategy. *Civil Engineering and Architecture*, 10, 3135–3146.

Herald (27 August 2024). Zimbabwe to Develop Green Building Standards. https://www.herald.co.zw/zimbabwe-to-develop-green-building-standards/#:~:text=ZIMBABWE%20is%20set%20to%20develop%20green%20building%20standards%20which%20are

Houghton, A., & Castillo-Salgado, C. (2017). Health co-benefits of green building design strategies and community resilience to urban flooding: A systematic review of the evidence. *International Journal of Environmental Research and Public Health*, 14(12), 1519. https://doi.org/10.3390/ijerph14121519

Moyo, T., Mudombo, M. V., Omer, M. M., & Moyo, C. (2024). Barriers to the adoption of sustainable construction materials in Zimbabwe. *International Journal of Building Pathology and Adaptation*. https://doi.org/10.1108/IJBPA-04-2024-0076

Chigwenya, A., & Zhakata, T. I. (2021). Adopting green building technologies for sustainable development: Insights from Harare, Zimbabwe. *International Journal of Real Estate Studies*, 14(2), 8–17. https://doi.org/10.11113/intrest.v14n2.47

Chavunduka, C., Mazanhi, P., & Kembo, M. (2024). Climate resilience and the New Urban Agenda in Zimbabwe: The role of the built environment disciplines and practice. In *Advances in 21st Century Human Settlements* (pp. 187–207). Springer. https://doi.org/10.1007/978-981-97-3199-2_12

Hwang, B. G., and Ng, W. J. (2013). Project management knowledge and skills for green construction: Overcoming challenges. *International Journal of Project Management*, 31(2), 272–284.

Hwang, B. G., Zhu, L., and Tan, J. S. H. (2017). Green business park project management: Barriers and solutions for sustainable development. *Journal of Cleaner Production*, 153, 209–219.

Ibrahim, A., Zayed, T., & Lafhaj, Z. (2025). Bridging barriers to lean construction adoption in megaprojects: A data-driven contribution to sustainable development using SEM. *Environment, Development and Sustainability*. https://doi.org/10.1007/s10668-025-06424-9

Inhabitat. (2023). Green Building in Zimbabwe Modeled After Termite Mounds - Inhabitat. https://inhabitat.com/building-modelled-on-termites-eastgate-centre-in-zimbabwe/

Inso. (2023). Green Building: Everything You Need to Know in South Africa. https://www.inso.co.za/blog/green-building/

Iwuanyanwu, O., Okwandu, A., Gil-Ozoudeh, I., & Ike, C. (2024). The role of green building materials in sustainable architecture: Innovations, challenges, and future trends. *International Journal of Applied Research in Social Sciences*, 6(8), 1935–1950. https://doi.org/10.51594/ijarss.v6i8.1476

Kerswill, B. (2017). Green Buildings: The Call for Sustainable Retrofitting in Zimbabwe. *DLA Piper*. https://www.dlapiper.com/en/insights/publications/2021/10/esg-challenges-in-real-estate-ahead-of-cop26/green-buildings-the-call-for-sustainable-retroffiting-in-zimbabwe

Kholijah, S. (2024). Analysis of economic and environmental benefits of green business practices in the hospitality and tourism sector. *Involvement International Journal of Business*, 1(1), 60–74. https://doi.org/10.62569/iijb.v1i1.7

Kibert, C. J. (2016). *Sustainable Construction: Green Building Design and Delivery*. John Wiley & Sons.

Lazar, N., & Chithra, K. (2022). Green building rating systems from the prospect of sustainability dimensions through the building lifecycle. *Environmental Science and Pollution Research*, 29, 51054–51064. https://doi.org/10.1007/s11356-021-16682-5

Leach, M., Raworth, K., & Saith, R. (2010). Dynamics of resilience in social-ecological systems. *The Journal of Development Studies*, 46(8), 1247–1262.

Mafuku, S. H. (2019). Greening the construction industry in Zimbabwe. In Chirisa, I. (ed.), *The Sustainability Ethic in the Management of the Physical, Infrastructural and Natural Resources of Zimbabwe* (pp. 361–381), Langaa Publisher, Bamenda, Cameroon. (https://www.langaa-rpcig.net/).

Meyer, M. D., & Miller, E. J. (2001). *Urban Transportation Planning*. McGraw-Hill.

Mohd Noor, N.H., Azd Noor, N.H., Azizan, N.S., & Abuizan, N.S., & Abu Hasan, N.S. (2023). Hasan, N.S. (2023). Universiti Tekn Universiti Teknologi MARA Repositoryologi MARA Repository.

Mohd Noor, N. H., & Nordin, N. B. (2023). Reassuring pro-environmental behaviour: a goal-framing theory perspective. *Bioresources and Environment (BioEnv)*, 1(3), 14–26.

Mokhtar, M., Ali, A., & Rahman, I. (2020). The role of green building in enhancing sustainability: A review. *Sustainability*, 12(4), 1123.

Mwero, J. N., & Aduda, V. O. (2019). End-users' perspective on green building: A study in Nairobi. *International Journal of Scientific and Research Publications (IJSRP)*, 9(4), p8880.

Nwokediegwu, Z., Ugwuanyi, E., Dada, M., Obaigbena, A., & Majemite, M. (2024). Urban water management: A review of sustainable practices in the USA. *Engineering Science & Technology Journal*, 5(2), 517–530. https://doi.org/10.51594/estj.v5i2.829

Ohueri, C. C., Enegbuma, W. I., & Habil, H. (2020). MyCREST embedded framework for enhancing the adoption of green office building development in Sarawak. *Built Environment Project and Asset Management*, 10(2), 215–230.

Okwandu, A. C., Esho, A. O. O., Iluyomade, T. D., & Olatunde, T. M. (2024). The role of policy and regulation in promoting green buildings. *World Journal of Advanced Research and Reviews*, 22(1), 139–150. https://doi.org/10.30574/wjarr.2024.22.1.1047

Olubunmi, O. A., Xia, P. B., & Skitmore, M. (2016). Green building incentives: A review. *Renewable and Sustainable Energy Reviews*, 59, 1611–1621.

Ong, Y. S., Yusof, N. A., & Osmadi, A. (2021). Challenges of green office implementation: A case study in Penang, Malaysia. *International Journal of Sustainable Construction Engineering and Technology*, *12*(1), 153–163.

Pushkar, S. (2024). Leadership in energy and environmental design commercial interior version 4 (LEED-CI v4) gold-certified office space projects: A pairwise comparative analysis between three mediterranean countries. *Buildings*, 14(3), 815. https://doi.org/10.3390/buildings14030815

Qian, Y., Liu, H., Mao, P., & Zheng, X. (2023). Evaluation of safety management of smart construction sites from the perspective of resilience. *Buildings*, 13(9), 2205.

Rogers, E. M. (2003). *Diffusion of Innovationsusion of Innovations* (5th ed.). Free Press.

Sagar, M., Arya, Y., & Singhal, P. (2025). Energy efficient green building design utlilising renewable energy and low-carbon development technologies. *Science and Technology for Energy Transition*, 80, 25. https://doi.org/10.2516/stet/2025004

Sartipi, F. (2020). Diffusion of innovation theory in the realm of environmental construction. *Journal of Construction Materials*, 1(4), 2–4. https://doi.org/10.36756/JCM.v1.4.2

Sichali, M., & Banda, L. J. (2017). Awareness, attitudes and perception of green building practices and principles in the Zambian construction industry. *International Journal of Construction Engineering and Management*, 6(5), 215–220.

Siddiqui, R. A., Adamu, Z., Ebohon, O. J., & Aslam, W. (2024). Factors affecting intention to adopt green building practices: A journey towards meeting sustainable goals. *Construction Innovation*, 25(6), 1797–1820.

Tanyanyiwa, V. I. (2017). Weather forecasting using local traditional knowledge (LTK) in the midst of climate change in Domboshawa, Zimbabwe. In Leal Filho, W., Manolas, E., Azul, A. M., Azeiteiro, U. M., & McGhie, H. (eds.). *Handbook of Climate Change Communication: Vol. 2: Practice of Climate Change Communication* (pp. 1–20). Cham: Springer International Publishing.

The Herald. (April 13, 2022). Zim speeds up green buildings adoption. Available at: https://www.heraldonline.co.zw/zim-speeds-up-green-buildings-adoption/

The Herald. (August 8, 2024). Real estate sector shows resilience, poised to grow. Available at: https://www.heraldonline.co.zw/real-estate-sector-shows-resilience-poised-to-grow/

The Standard (24 March 2024). Green Buildings, Energy Efficiency and Economic Development: The Nexus. https://www.newsday.co.zw/thestandard/renewable-energy/article/200024740/green-buildings-energy-efficiency-and-economic-development-the-nexus

Umoh, A., Ilojianya, V., Adefemi, A., Nwokediegwu, Z., Etukudoh, E., & Ibewe, K. (2024). Green architecture and energy efficiency: A review of innovative design and construction techniques. *Engineering Science & Technology Journal*, 5(1), 185–200. https://doi.org/10.51594/estj.v5i1.743

UNEP. (2021). Global Status Report on Buildings and Construction: Towards a Zero-emission, Efficient and Resilient Buildings and Construction Sector. https://www.unep.org/resources/report/global-status-report-buildings-and-construction-towards-zero-emission-efficient-and-resilient-buildings-and-construction-sector

Walker, B., Holling, C. S., Carpenter, S. R., & Kinzig, A. (2004). Resilience, adaptability and transformability in social–ecological systems. *Ecology and Society*, 9(2), 5.

Windapo, A. O. (2014). Examination of green building drivers in the South African construction industry: Economics versus ecology. *Sustainability*, 6(9), 6088–6106.

World Green Building Council. (2021). Annual Report 2021. https://worldgbc.org/wp-content/uploads/2022/08/WorldGBC-2021-Annual-Report.pdf

Zhang, G., & He, B. J. (2021). Towards green roof implementation: Drivers, motivations, barriers and recommendations. *Urban Forestry & Urban Greening*, *58*, 126992.

Zhang, Y., Wang, H., Gao, W., Wang, F., Zhou, N., Kammen, D. M., & Ying, X. (2019). A survey of the status and challenges of green building development in various countries. *Sustainability*, *11*(19), 5385.

Zimbabwe Independent (4 September 2022). Zim's Green Buildings Revolution in Motion. https://www.bing.com/search?pglt=41&q=herald%2C+Green+Building+Strategies+in+Zimbabwe&cvid=aa1a17ab9f054a00815f14f1a1e3f109&gs_lcrp=EgZjaHJvbWUyBggAEEUYOTIGCAEQABhAMgYIAhAAGEAyBggDEAAYQDIGCAQQABhAMgYIBRAAGEAyBggGEAAYQDIGCAcQABhAMgYICBAAGEDSAQkxNTY2MWowajGoAgCwAgA&FORM=ANNTA1&PC=U531&ntref=1

ZIMSTAT (Zimbabwe National Statistics Agency) (2020). Monitoring COVID-19 Impact on Households in Zimbabwe. Report No. 2. Harare, Zimbabwe: ZIMSTAT.

Zuniga-Teran, A. A., Staddon, C., de Vito, L., Gerlak, A. K., Ward, S., Schoeman, Y., ... & Booth, G. (2020). Challenges of mainstreaming green infrastructure in built environment professions. *Journal of Environmental Planning and Management*, *63*(4), 710–732.

Zuo, J., & Zhao, Z. (2014). Green building research–current status and future agenda: A review. *Renewable and Sustainable Energy Reviews*, 29, 1–10.

Index

Note: Italic page numbers refer to figures.